Advances in Pharmaceutical Technology

Advances in Pharmaceutical Technology

Dr. S. C. Dinda

Professor & Director
School of Pharmaceutical Education & Research,
Berhampur University,
Orissa- 760 007, India.

Pharma Med PRESS
An Imprint of Pharma Book Syndicate

PharmaMed Press
An imprint of Pharma Book Syndicate

An unit of **BSP Books Pvt., Ltd.**

4-4-316, Giriraj Lane,
Sultan Bazar, Hyderabad - 500 095.

Published by

PharmaMed Press

An imprint of Pharma Book Syndicate

An unit of BSP Books Pvt., Ltd.

4-4-316, Giriraj Lane, Sultan Bazar, Hyderabad - 500 095.
Phone: 040-23445605, 23445688; Fax: 91+40-23445611
E-mail: info@pharmamedpress.com
www.pharmamedpress.com/pharmamedpress.net

Printed at

Aditya Offset Process (I) Pvt. Ltd.
Hyderabad.

ISBN : 978-81-7800-256-9

Preface

The book, **Advances in Pharmaceutical technology,** is designed to present a comprehensive coverage on the advanced area of Pharmaceutical Technology utilized towards designing of Drug Delivery and Drug Targeting systems. It is a compilation of Pharmaceutical Biotechnology, Nanotechnology, Immunotechnology and such other Drug Delivery Technology, utilized for drug delivery and drug targeting through the oral as well as parenteral route. The volume is unique in that it covers the basic scientific principles involved in designing approach of drug delivery systems and focuses on its current area of biomedical applications as well as highlight on its ongoing research at present scenario, which may encourage the readers to develop their basic knowledge, update the present area of research output and will create interest towards the research involvement.

The author hopeful that the information cited in the book, mostly covered in all frontiers of pharmaceutical technology, will satisfy the requirements of scientific, teaching, student and research community. The untiring effort of the author will be useful in academic as well as scientific development.

The suggestions from readers for improving the quality of this book and subsequent volumes are encouraged and acknowledged.

-Author

Acknowledgements

The author of the book, **Advances in Pharmaceutical Technology** thanks to the professionals, scientists from academic institutions as well as from industries for sharing their published research out puts as well as literature reviews, retrieved from the websites, books and journals for disseminating the same among the scientific, teaching, student and research community for academic as well as scientific development. Thanks are also due to the pharmaceutical and chemical industries for utilizing and citing some of their product profile, its design aspect, technical know-how, biomedical applications, and ongoing research information in therapeutic management.

The author expresses his deep sense of gratitude to Prof. Jayanta Kumar Mohapatra, Vice-Chancellor, Berhampur University for his inspiration and encouragement to shape this book.

The author also expresses his heartfelt thanks to his parents, wife and son for extending their cooperation and inspiration for successful completion of writing of the book.

Thanks are also due to all individuals, whose contributions are cited in the book for the academic as well as scientific development.

-Author

Contents

CHAPTER 2

Bio-Degradable Polymers in Drug Delivery Systems

CHAPTER 3

Nanotechnology in Drug Delivery

CHAPTER 4

Aerosol Technology in Drug Delivery

Chapter 5

Biotechnology in Drug Delivery

Chapter 6

Immunotechnology in Drug Delivery

CHAPTER 7

Therapeutic Careers in Drug Delivery

CHAPTER 7A

Liposome as Therapeutic Drug Career

CHAPTER 7B

Niosomes as Therapeutic Drug Career

CHAPTER 8

Oral Controlled Release Systems

CHAPTER 9

Parenteral Controlled Release Systems

Chapter 10

Transdermal Drug Delivery Systems

1

Pharmacokinetic Approaches in Designing of Drug Delivery Systems

Introduction

Pharmacokinetics is the study of those rate process involved in the absorption, distribution, metabolism, and excretion of drugs and their relationship to the pharmacological, therapeutic, or toxic response in animals or humans. Pharmacokinetics technique attempts mathematically to define the time course of drug in the body by assaying the drug and it's metabolite in readily accessible fluids such as blood and urine. The goal is to quantify the amount of drug available to the body fluid (bioavailability) from the time of administration to the time of total clearance.

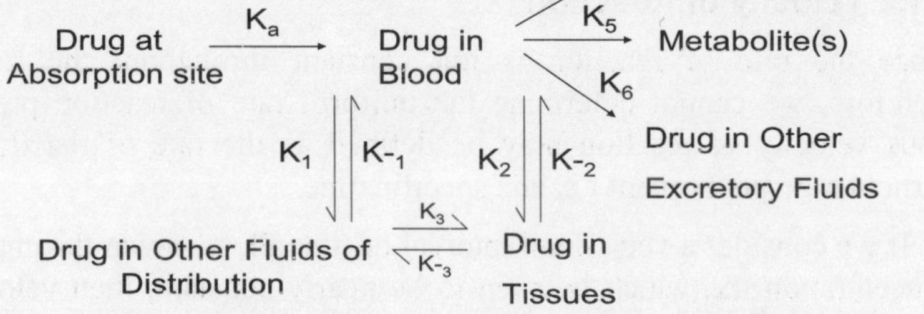

Schematic representation of drug absorption, distribution and elimination.

The study of pharmacokinetics involves experimental approaches to develop the biologic sampling techniques; analytical method development for the measurement of the drugs and metabolites; data collection and manipulation; as well as theoretical aspects to develop pharmacokinetic models to predict the drug disposition after drug administration. The estimation and interpretation of data result in predicting dosage regimen for individual or group of patients.

Pharmacokinetic concepts are utilized at all stage of drug development. Clinical application of pharmacokinetics have resulted in the improvement of drug utilization and consequently, direct benefit to patients. Clinically, two of the important applications of pharmacokinetic principles are: design of an optimal dosage regimen and clinical management of individual patient and therapeutic drug monitoring.

1.1 Rate, Rate Constant, and Order

1.1.1 Rate of a Chemical Reaction

It is defined as the quantity of a reactant consumed or the quantity of a product formed in unit time in a chemical reaction.

Mathematically:

$$\text{Rate of Reaction} = \frac{\text{Decrease in concentration of reactants}}{\text{Time in which the change takes place}}$$

In other words,

$$\text{Rate of Reaction} = \frac{\text{Increase in concentration of products}}{\text{Time in which the change takes place}}$$

1.1.2 Velocity of Reaction

Since the rate of reaction is not constant throughout the reaction, therefore, we cannot determine the uniform rate of reaction precisely. Thus velocity of reaction may be defined as the rate of reaction at a particular given moment i.e. at a specific time.

If we consider a very small interval of time **dt**, in which the change in concentration **dx**, which is taken to be nearly constant, then velocity of reaction is given by:

$$\text{Velocity Reaction} = \frac{dx}{dt}$$

Rate expression and rate constant

Consider a general reaction:

Reactants $\rightarrow$ products

According to the law of mass action, rate of reaction is directly proportional to the active mass, hence for the above reaction becomes:

Rate of reaction α [Reactants]

Rate of reaction = K [Reactants]

This expression is called rate expression and **K** is called rate constant or velocity constant.

Characteristics of rate constant

(i) It has a fixed value at a particular temperature.

(ii) Value of **K** varies with temperature.

(iii) Value of **K** remains unaltered with the change in concentration of reactants.

1.1.3 Order of Reaction

The order of reaction is defined as the sum of all the exponents of the reactants involved in the rate equation.

It should be noted down that all the molecules shown in a chemical equation do not determine the value of order of reaction but only those molecules whose concentrations are changed are included in the determination of the order of a reaction. In other words:

"The number of reacting molecules whose concentration alters as a result of chemical reaction is termed as the order of reaction."

For example:

$$2NO + O_2 \rightarrow 2NO_2$$

$dx/dt = k[NO]^2[O_2]$, the reaction is of third order as 2 + 1 = 3

For a reaction, maximum order is three and the minimum is zero.

First order reactions

The reaction in which only one molecule undergoes a chemical change is called first order reactions. Example:

$$N_2O_5 \rightarrow 2NO_2 + \frac{1}{2} O_2$$

Second order reactions

The reaction in which two molecules undergo a chemical change is called second order reactions. Example:

$$2CH_3CHO \rightarrow CH_4 + 2\,CO$$

Third order reactions

The reaction in which only three molecules undergo a chemical change is called third order reactions. Example:

$$2NO + O_2 \rightarrow 2NO_2$$

In chemical kinetics, the order of reaction with respect to a certain reactant, is defined as the power to which its concentration term in the rate equation is raised.

For example, given a chemical reaction $2A + B \rightarrow C$ with a rate equation

$$r = k\,[A]^2[B]^1$$

The reaction order with respect to A would be 2 and with respect to B would be 1; the total reaction order would be $2 + 1 = 3$. It is not necessary that the order of a reaction be a whole number i.e. zero and fractional values of order are possible, but they tend to be integers. Reaction orders can be determined only by experiment. Their knowledge allows conclusions about the reaction mechanism.

The reaction order may not necessarily be related to the stoichiometry of the reaction, unless the reaction is elementary. Complex reactions may or may not have reaction orders equal to their stoichiometric coefficients.

For example:

- The alkaline hydrolysis of ethyl acetate is:

  $$CH_3COOC_2H_5 + OH^- \rightarrow CH_3COO^- + C_2H_5OH.$$

  It has the following rate equation: $r = k\,[CH_3COOC_2H_5]\,[OH]$

- The rate equation for imidazole catalyzed hydrolysis may be:

  $$r = k\,[\text{imidazole}][CH_3COOC_2H_5]$$

  Although, there may not be presence of imidazole in the stoichiometric chemical equation.

- In the reaction of aryldiazonium ions with nucleophiles in aqueous solution i.e., $ArN_2^+ + X^- \rightarrow ArX + N_2$, the rate equation may be:

  $$r = k\,[ArN_2^+]$$

Reactions can also have an undefined reaction order with respect to a reactant, for example one can not talk about reaction order in the rate equation, when dealing with a bimolecular reaction between adsorbed molecules, the rate equation will be:

$$r = k\frac{K_1K_3C_AC_B}{(1+K_1C_A+K_2C_B)^2}$$

If the concentration of one of the reactants remains constant (because of it is a catalyst or it is in great excess with respect to the other reactants) its concentration can be included in the rate constant, obtaining a pseudo constant: if B is the reactant, whose concentration is constant then the eq. becomes:

$$r = k[A][B] = k'[A]$$

The second-order rate equation has been reduced to a pseudo-first-order rate equation. This makes the treatment to obtain an integrated rate equation much easier.

Zero-order reactions are often seen for thermal chemical decompositions where the reaction rate is independent of the concentration of the reactant (changing the concentration has no effect on the speed of the reaction).

Broken-order reactions

This order is a non-integer typical of reactions with a complex reaction mechanism. For example, the chemical decomposition of ethanol into methane and carbon monoxide proceeds with an order of 1.5 with respect to ethanol. The decomposition of phosgene to carbon monoxide and chlorine has order-1, with respect to phosgene itself and order 0.5 with respect to chlorine.

Mixed-order reaction

This order of a reaction changes in the course of a reaction as a result of changing variables such as pH. An example is the oxidation of an alcohol to a ketone by a ruthenate (RuO_4^{2-}) and a hexacyanoferrate, the latter serving as the sacrificial catalyst converting Ru(IV) back to Ru(VI): the disappearing-rate of the ferrate is zero-order with respect to the ferrate at the onset of the reaction (when its concentration is high and the ruthenium catalyst is quickly regenerated) but changes to first-order, when its concentration decreases.

Negative-order reactions are rare, for example, the conversion of ozone (order 2) to oxygen (order 1).

1.1.4 Rate Equation

The **rate law** or **rate equation** for a chemical reaction is an equation, which links the reaction rate with concentrations or pressures of reactants and constant parameters (normally rate coefficients and partial reaction orders). To determine the rate equation for a particular system one can combine the reaction rate with a mass balance for the system. For a generic reaction *m*A + *n*B → C with no intermediate steps involved in its reaction mechanism (that is, an elementary reaction), can be expressed as:

$$r = k[A]^m [B]^n$$

Where, [A] and [B] expresses the concentration of the species A and B, respectively [(usually in moles per liter (molarities)]; *m* and *n* are the respective stoichiometric coefficients of the imbalanced equation; they must be determined experimentally. *k* is the *rate coefficient* or *rate constant* of the reaction. The value of this coefficient *k* depends on conditions such as temperature, ionic strength and surface area of the adsorbent or light irradiation. For elementary reactions, the rate equation can be derived from first principles using collision theory. Again, *m* and *n* are not always derived from the balanced equation.

The rate equation of a reaction with a multi-step mechanism cannot, in general, be deduced from the stoichiometric coefficients of the overall reaction; it must be determined experimentally. The equation may involve fractional exponential coefficients, or it may depend on the concentration of an intermediate species.

The rate equation is a differential equation, and it can be integrated to obtain an **integrated rate equation**, which links concentrations of reactants or products with time.

If the concentration of one of the reactants remains constant (because it is a catalyst or it is in great excess with respect to the other reactants), its concentration can be grouped with the rate constant, obtaining a **pseudo constant**. If, B is the reactant, whose concentration is constant, then $r = k[A][B] = k'[A]$. The second order rate equation has been reduced to a **pseudo first order** rate equation. This makes the treatment to obtain an integrated rate equation much easier.

Zero-order reactions

Zero-order kinetics has a rate which is independent of the concentration of the reactant(s). Increasing the concentration of the reacting species will not speed up the rate of the reaction. Zero-order reactions are typically found when a material that is required for the reaction to proceed, such as a catalyst, which is saturated by the reactants. The rate law for a zero-order reaction can be expressed as:

$$r = k$$

Where, r is the reaction rate, and k is the reaction rate coefficient with units of concentration/time. It occurs with a condition such as: 1) the reaction occurs in a closed system; 2) there is no net build-up of intermediates; and 3) there are no other reactions occurring simultaneously, which can be shown by solving a mass balance for the system using the following equation:

$$r = -\frac{d[A]}{dt} = k$$

If, this differential equation is integrated it gives an equation, which is often called the **integrated zero-order rate law** as:

$$[A]_t = -kt + [A]_0$$

Where, $[A]_t$ represents the concentration of the chemical of interest at a particular time, and $[A]_0$ represents the initial concentration.

A reaction is zero order, if concentration data are plotted versus time and the result is a straight line. The slope of this resulting line is the negative of the zero order rate constant k.

The half-life of a reaction describes the time need for half of the reactant to be depleted (same as the half-life involved in nuclear decay, which is a first-order reaction). For a zero-order reaction the half-life is given by

$$t_{\frac{1}{2}} = \frac{[A]_o}{2k}$$

Example of a zero-order reaction is:

$$2NH_3(g) \rightarrow 3H_2(g) + N_2(g)$$

It should be noted that the order of a reaction cannot be deduced from the chemical equation of the reaction.

First-order reactions

A **first-order reaction** depends on the concentration of only one reactant (a **unimolecular reaction**). Other reactants can be present, but each will be zero-order. The rate law for an elementary reaction, which is first order with respect to a reactant 'A' is:

$$r = -\frac{d[A]}{dt} = k[A]$$

Where, k is the first order rate constant, which has units of 1/time.

The **integrated first-order rate equation** is

$$\text{In}\,[A] = -kt + \text{In}\,[A]_0$$

A plot of ln[A] Vs. time t gives a straight line with a slope of $-k$.

The half life of a first-order reaction is independent of the starting concentration and is given by

$$t_{\frac{1}{2}} = \frac{In(2)}{k}$$

Examples of some reactions, which are first-order with respect to the reactant, are:

- $H_2O_2 \rightarrow H_2O(l) + \frac{1}{2}\,O_2\,(g)$
- $SO_2Cl_2(l) \rightarrow$ SO2 (g) + Cl2 (g)
- $2N_2O_5(g) \rightarrow 4NO_2\,(g) + O_2\,(g)$

Further Properties of First-Order Reaction Kinetics

The integrated first-order rate equation, In [A] = $-kt$ + In $[A]_0$, may usually be written in the form of the exponential decay equation as:

$$A = A_{0e}^{-kt}$$

A different (but equivalent) way of considering first order kinetics is as follows:

The exponential decay equation can be rewritten as:

$$A = A_0\left(e^{-k\Delta t_p}\right)^n$$

where, Δt_p corresponds to a specific time period and n is an integer corresponding to the number of time periods. At the end of each time period, the fraction of the reactant population remaining relative to the amount present at the start of the time period, f_{RP}, will be:

$$\frac{A_n}{A_{n-1}} = f_{RP} = e^{-k\Delta t_p}$$

Such that after n time periods, the fraction of the original reactant population will be:

$$\frac{A}{A_0} \equiv \frac{A_n}{A_0} = \left(e^{-k\Delta t_p}\right)^n = \left(f_{RP}\right)^n = \left(1 - f_{BP}\right)^n$$

where, f_{BP} corresponds to the fraction of the reactant population that will break down in each time period. This equation indicates that the fraction of the total amount of reactant population, which will break down in each time period is independent of the initial amount present. The chosen time period may corresponds to:

$$\Delta t_p = \frac{\text{In}(2)}{k}$$

Where, the fraction of the population, which will break down in each time period will be exactly ½ the amount present i.e. the time period corresponds to the half-life of the first-order reaction).

The average rate of the reaction for the nth time period may be given by:

$$r_{avg,n} = -\frac{\Delta A}{\Delta t_p} = \frac{A_{n-1} - A_n}{\Delta t_p}$$

Therefore, the amount remaining at the end of each time period will be related to the average rate of that time period and the reactant population at the starting time period follows the equation:

$$A_n = A_{n-1} - r_{avg,n}\Delta t_p$$

Then the fraction of the reactant population, which will break down in each time period can be expressed as:

$$f_{BP} = 1 - \frac{A_n}{A_{n-1}}$$

The amount of reactant, which break down in each time period in relation to the average rate over a time period may be expressed by:

$$f_{BP} = \frac{r_{avg,n}\Delta t_p}{A_{n-1}}$$

Such that the amount that remains at the end of each time period will be related to the amount present at the starting time according to the equation as:

$$A_n = A_{n-1}\left(1 - \frac{r_{avg,n\Delta t_p}}{A_{n-1}}\right)$$

This equation is a precursor, allowing the calculation of the amount present after any number of time periods, without need of the rate constant, provided that the average rate for each time period is known.

Second-order reactions

A **second-order reaction** depends on the concentrations of one second-order reactant, or two first-order reactants.

For a second order reaction, its reaction rate is given by:

$$r = 2k[\mathrm{A}]^2 \quad \text{or} \quad r = 2k\,[\mathrm{A}][\mathrm{B}] \quad \text{or} \quad \mathrm{r} = 2\mathrm{k}[\mathrm{B}]^2$$

In several popular kinetic, the rate law for second-order reaction is defined as:

$$-\frac{d[A]}{dt} = k[A]^2$$

Conflating the 2 inside the constant for the first derivative form may make it the second integrated form (presented below). The option of keeping the 2 out of the constant in the derivative form is considered more correctly as it is almost always used in universally.

There by the **integrated second-order rate equations** may be written as:

$$\frac{1}{[A]} = \frac{1}{[A]_0} + kt$$

Or,

$$\frac{[A]}{[B]} = \frac{[A]_0}{[B]_0} e^{([A]_0 - [B]_0)kt}$$

$[A]_0$ and $[B]_0$ must be different to obtain that integrated equation.

The half-life equation for a second-order reaction may be:

$$t_{\frac{1}{2}} = \frac{1}{k[A]_0}$$

It may be considered that for a second-order reaction, half-lives progressively double.

Another way to present the above rate equation by taking the log on both sides as:

$$\text{In } r = \text{In } k + 2 \text{ In } [A]$$

Example of a Second-order reaction is:

$$2\ NO_2(g) \rightarrow 2NO(g) + O_2\ (g)$$

Pseudo first order

Measuring a second order reaction rate with reactants A and B can be problematic. The concentrations of the two reactants must be followed simultaneously, which is more difficult; or measurement of one of them and calculate the other as a difference, which is less precise. A common solution for that problem is the **pseudo first order approximation**

If *either* [A] or [B] remain constant as the reaction proceeds, then the reaction can be considered **pseudo first order** because of it only depends on the concentration of one reactant. For example, [B] remains constant then:

$$r = k[A][B] = k'[A]$$

Where, $k' = k[B]_0$ (k' or k_{obs} with units s^{-1}) and an expression is identical to that of the first order expression above.

One way to obtain a pseudo first order reaction is to use a large excess of one of the reactants ([B]>>[A]), so that, as the reaction progresses only a small amount of the reactant is consumed and its concentration can be considered to stay constant. By collecting *k'* for many reactions with different (but excess) concentrations of [B]; a plot of *k'* versus [B] gives *k* (the regular second order rate constant) and the slope.

Reactions with order 3 (called **ternary reactions**) are rare and unlikely to occur.

***Summary for reaction orders 0, 1, 2 and* n**

	Zero-Order	**First-Order**	**Second-Order**	***n*th-Order**
Rate Law	$-\frac{d[A]}{dt}=k$	$-\frac{d[A]}{dt}=k[A]$	$-\frac{d[A]}{dt}=k[A]^2$	$-\frac{d[A]}{dt}=k[A]^n$
Integrated Rate Law	$[A]=[A]_0-kt$	$[A]=[A]_0e^{-kt}$	$\frac{1}{[A]}=\frac{1}{[A]_0}+kt$	$\frac{1}{[A]^{n-1}}=\frac{1}{[A]_0{}^{n-1}}+(n-1)kt$ [Except first order]
Units of Rate Constant (k)	$\frac{M}{s}$	$\frac{1}{s}$	$\frac{1}{M.s}$	$\frac{1}{M^{n-1}.s}$
Linear Plot to determine k	$[A]$ vs. t	$\text{In}([A])$ vs. t	$\frac{1}{[A]}$ vs. t	$\frac{1}{[A]^{n-1}}$ vs. t [Except first order]
Half-life	$t_{1/2}=\frac{[A]_0}{2k}$	$t_{1/2}=\frac{\text{In }(2)}{k}$	$t_{1/2}=\frac{1}{k[A]_0}$	$t_{1/2}=\frac{2^{n-1}-1}{(n-1)k[A]_0{}^{n-1}}$ [Except first order]

Where, M stands for concentration in molarity (mol · L^{-1}), t for time, and k for the reaction rate constant. Half life is often expressed as $t_{1/2}=0.693/k$ for a first order reaction (as In 2=0.693).

1.2 Pharmacokinetic Models

A model in pharmacokinetics is a hypothetical structure, which can be used to characterize with reproducibility, the behavior and fate of a drug in biological system when administered through certain route of administration and in a particular dosage form.

1.2.1 One-Compartment Model

Following drug administration, the body is depicted as a kinetically homogeneous unit (see Fig 1.1). This assumes that the drug achieves instantaneous distribution throughout the body and that the drug equilibrates instantaneously between tissues. Thus the drug concentration–time profile shows a monophasic response (i.e., it is monoexponential; Figure 1.1a). It is important to note that this does not imply that the drug concentration in plasma (Cp) is equal to the drug concentration in the tissues. However, changes in the plasma concentration quantitatively reflect changes in the tissues. The relationship described in Fig 1.1(a) can be plotted on a log *Cp* Vs time graph (Fig. 1.1b) will show a linear relation, represents a one-compartment model.

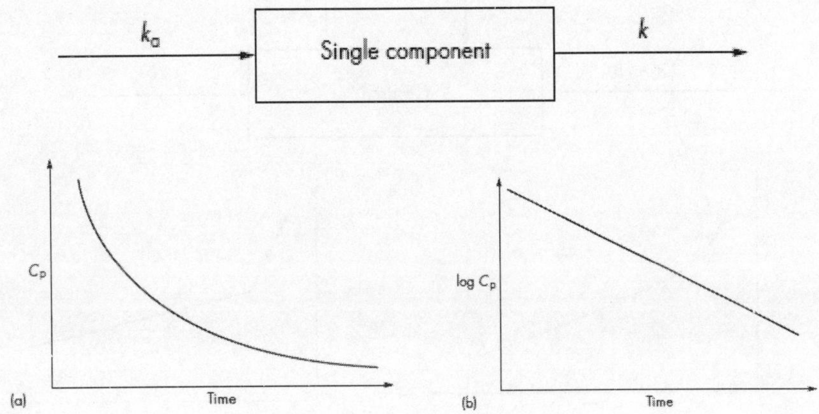

Fig. 1.1 One-compartment model; Ka = absorption rate constant (h=1), K = elimination rate constant (h=1). **(a)** Plasma concentration (Cp) versus time profile of a drug showing a one-compartment model. **(b)** Time profile of a one-compartment model showing log Cp versus time.

1.2.2 Two-Compartment Model

The two-compartment model resolves the body into a central compartment and a peripheral compartment (see Fig 1.2). Although these compartments have no physiological or anatomical meaning, it is assumed that the central compartment comprises tissues that are highly perfused such as heart, lungs, kidneys, liver and brain. The peripheral compartment comprises less well-perfused tissues such as muscle, fat and skin. A two-compartment model assumes that, following drug administration into the central compartment, the drug distributes between the central compartment and the peripheral compartment. However, the drug does not achieve instantaneous distribution, i.e. equilibration, between the two compartments. The drug concentration–time profile shows a curve (Fig.1.2 a), but the log drug concentration–time plot shows a biphasic response (Fig.1.2 b)

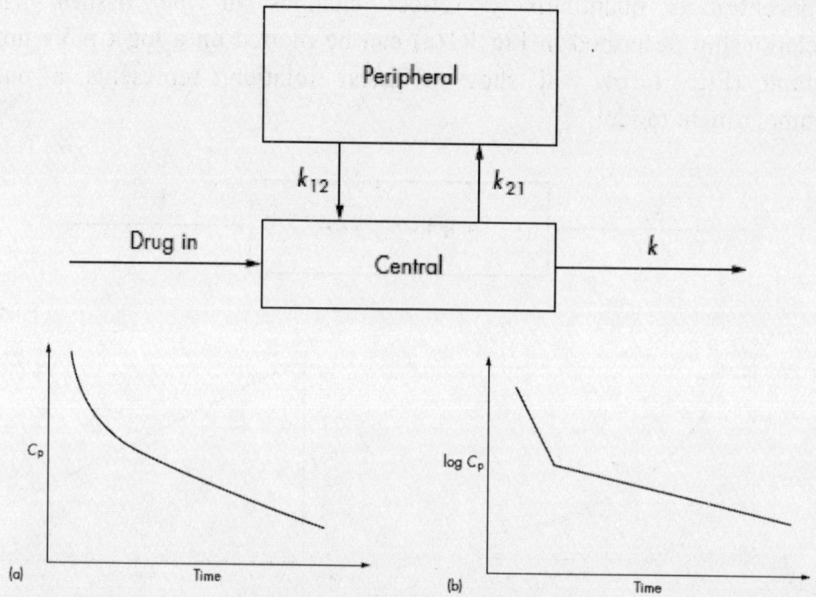

Fig. 1.2 Two-compartment model; K_{12}, K_{21} and K are first-order rate constants: K_{12} = rate of transfer from central to peripheral compartment; K_{21} = rate of transfer from peripheral to central compartment; K= rate of elimination from central compartment **(a)** Plasma concentration versus time profile of a drug showing a two compartment model. **(b)** Time profile of a two-compartment model showing log Cp versus time.

Fig 1.2(b) can be used to distinguish whether a drug shows a one- or two-compartment model. Fig 1.2(b) shows a profile in which initially there is a rapid decline in the drug concentration owing to elimination from the central compartment and distribution to the peripheral compartment. Hence during this rapid initial phase, the drug concentration will decline rapidly from the central compartment, rise to a maximum in the peripheral compartment. After a time interval (*t*), distribution equilibrium is achieved between the central and peripheral compartments, and elimination of the drug is assumed to occur from the central compartment as that of the one compartment model. All the rate processes may be described by first-order rate process.

1.2.3 Multi-compartment Model

In this model, the drug distributes into more than one compartment and the concentration–time profile shows more than one exponential (Fig 1.3a). Each exponential on the concentration–time profile describes a compartment. For example, gentamycin can be described by a three-compartment model following a single IV dose (see Fig 1.3b).

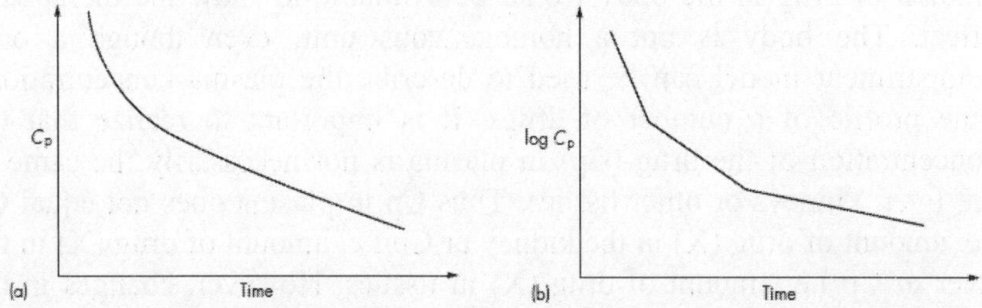

Fig. 1.3 (a) Plasma concentration versus time profile of a drug showing Multi-compartment model. **(b)** Time profile of a multi-compartment model showing log Cp versus time.

1.3 Pharmacokinetic Parameters in Dosage Regimen Fixation

For a majority of drugs subject to pharmacokinetic monitoring, the goal is to design individualized dosage regimen for patients, in order to keep the plasma concentrations of the drug within a preset minimum (Cmin) and maximum (Cmax) for multiple dosing regimens or at steady state (Css) for constant input regimens. For most drugs, the Cmin and Cmax values correspond to the minimum effective (MEC) and minimum toxic (MTC) concentrations. For example, the therapeutic range of digoxin for cardiac dysfunction is between 0.8 and 2ng/mL. This implies that the concentrations above 2ng/mL are more likely to be associated with toxicity and concentrations below 0.8ng/mL are more likely to produce little or no effect. Therefore, it is desired to fix a dosage regimen for digoxin to produce plasma concentrations within this therapeutic range.

The required data to design a dosage regimen depends on the kinetics of the drug in the patient and the reported therapeutic range of the drug. The kinetic parameters are derived from the plasma sample(s) taken from the patient (patient-specific values), adjustment of the reported average kinetic values with patient characteristics such as renal function (adjusted population values), or both the population and patient-specific data i.e. disease conditions, age, sex, body weight, etc.

1.3.1 Volume of Distribution

The volume of distribution (Vd) has no direct physiological meaning; it is not a 'real' volume and is usually referred to as the apparent volume of distribution. It is defined as the volume of plasma to which the total amount of drug in the body would be available to show the therapeutic effect. The body is not a homogeneous unit, even though a one-compartment model can be used to describe the plasma concentration–time profile of a number of drugs. It is important to realize that the concentration of the drug (Cp) in plasma is not necessarily the same in the liver, kidneys or other tissues. Thus Cp in plasma does not equal Cp i.e. amount of drug (X) in the kidney or Cp i.e. amount of drug (X) in the liver or Cp i.e. amount of drug (X) in tissues. However, changes in the drug concentration in plasma (Cp) are proportional to changes in the amount of drug (X) in the tissues.

Since, Cp (plasma) α C_p (tissues) i.e. C_p (plasma) α X (tissues)

Then C_p (plasma) = V_d x X (tissues)

Where, V_d is the constant of proportionality and is referred to as the volume of distribution, which thus relates the total amount of drug in the body at any time to the corresponding plasma concentration.

Thus $$V_d = \frac{X}{C_p}$$

And, V_d can be used to convert drug amount X *to* concentration.

Since, $$X = X_0 \exp(-kt)$$

Then $$\frac{X}{V_d} = \frac{X_0 \exp(-kt)}{V_d}$$

Thus $$C_{pt} = C_p^0 \exp(-kt)$$

This formula describes a mono-exponential decay (see Fig 1.2), where C_{pt} = plasma concentration at any time *t*.

The curve can be converted to a linear form (Fig 1.4) using natural logarithms (In):

$$\text{In } C_{pt} = \text{In } C_p^0 - kt$$

Where, the slope = – k, the elimination rate constant and the y, is the intercept = In C_p^0.

Since, $$V_d = \frac{X}{C_p}$$

Then, at zero concentration (C_p^0), the amount administered is the dose, *D*,

So that C_p^0 will be: $$C_p^0 = \frac{D}{V_d}$$

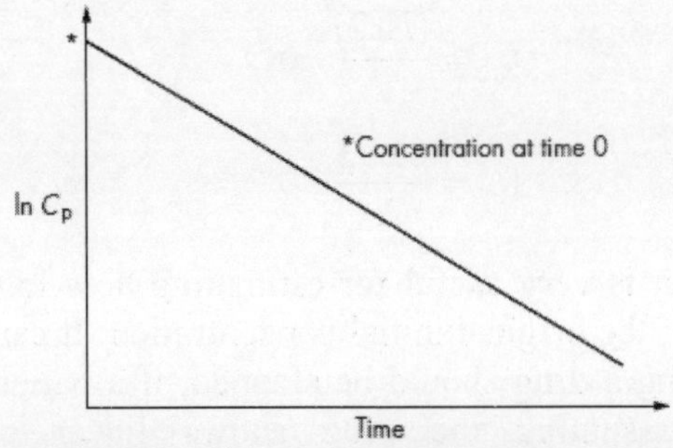

Fig. 1.4 In C_p Vs time profile.

If the drug has a large V_d, which does not equate to a real volume, e.g. total plasma volume, this suggests that the drug is highly distributed in tissues. On the other hand, if the V_d is similar to the total plasma volume, which will suggest that the total amount of drug is poorly distributed and is mainly in the plasma.

1.3.2 Elimination Rate Constant

Consider a single IV bolus injection of drug X (see Fig 1.2). As time proceeds, the amount of drug in the body is eliminated. Thus the rate of elimination can be described (assuming first-order elimination) as:

$$\frac{dX}{dt} = -kX$$

Hence $$X = X_0 \exp(-k_t)$$

Where, X = amount of drug X, X_0 = dose and k = first-order elimination rate constant.

1.3.3 Half-Life

The time required to reduce the plasma concentration to one half its initial values and is defined as the *half-life* ($t_{1/2}$). Consider

$$\text{In}\, C_{pt} = \text{In}\, C_P^0 - kt$$

Let C_P^0 decay to $C_P^0 / 2$ and solve for $t = t_{1/2}$:

$$\text{In}\left(C_P^0 / 2\right) = \text{In}\, C_P^0 - kt_{1/2}$$

Hence

$$kt_{1/2} = \text{In}\, C_P^0 - \text{In}\left(C_p^0 / 2\right)$$

And
$$t_{1/2} = \frac{(l\text{n}\, 2)}{k} \quad \text{or,}$$

$$t_{1/2} = \frac{0.693}{k}$$

This parameter is very useful for estimating how long it will take to reduce the half of the original/initial concentration. It can also be used to estimate, how long a drug should be stopped, if a patient showing toxic manifestations, assuming the drug shows linear one-compartment pharmacokinetics.

1.3.4 Clearance

Drug clearance (CL) is defined as the volume of plasma in the vascular compartment cleared off drug per unit time by the processes of metabolism and/or excretion. Rate of clearance of a drug is constant, if the drug is eliminated following first-order kinetics. Drug can be cleared by renal excretion or through metabolic processes or following both. With respect to the kidney and liver, etc., clearances are additive, which can be determined as:

$$CL_{total} = CL_{renal} + CL_{nonrenal}$$

Mathematically, clearance is the product of the first-order elimination rate constant (k) and the apparent volume of distribution (V_d).

Thus,
$$CL_{total} = k \times Vd$$

Hence, the clearance is the elimination rate constant (i.e. the fractional rate of drug loss) from the volume of distribution.

Clearance can be expressed in relation to half-life by the rate process following 1st order kinetics as:

$$t_{1/2} = \frac{0.693 \times V_d}{CL}$$

If a drug has a CL of 2L/h, which tells that the 2 liters of the V_d is cleared off drug per hour. If the C_p is 10 mg/L, then 20 mg of drug is cleared per hour.

1.4 Relationship among Pharmacokinetic Parameters

Before discussing a modification of dosage regimens, it is necessary to realize, how the three major kinetic parameters (V, Cl, and $t_{1/2}$ or K) are related to each other. This is important because of these kinetic parameters determine the shape of the plasma concentration-time profiles and therefore, affect the fluctuation between the C^{∞}_{max} and C^{∞}_{min} values. The mathematical relationship among these three parameters is demonstrated below:

$$Cl = K \bullet V$$

If, two of these parameters are known, the third can easily be estimated from the above equation. However, the use of the above equation without an understanding of the underlying physiological relationship among these three parameters may result in erroneous conclusions. This is because of V and Cl, which are independent parameters, while the elimination half life (or rate constant) is a composite parameter dependent on both V and Cl as described below.

Clearance is a measure of the efficiency of the organ(s) of elimination and is dependent on certain physiologic parameters in the organ (*e.g.*, organ blood flow and drug intrinsic clearance and free fraction in the blood). For instance, for a drug eliminated by renal excretion, clearance is dependent on how well the kidneys can excrete the drug in urine. Therefore, in the elderly with reduced renal function, the clearance of renally excreted drugs would be reduced.

The extent of distribution of drugs, however, is independent of their clearance. Distribution is dependent on certain physiologic parameters such as perfusion and permeability of tissues to drugs and the level of binding proteins in the blood and tissues. Therefore, a reduction in the renal clearance of a drug in the elderly does not necessarily mean that the volume of distribution of the drug will also be different in this population. In other words, clearance and volume of distribution are independent of each other.

On the other hand, the elimination half life is dependent on both V and Cl. An increase in Cl (elimination efficiency) results in a reduction in $t_{1/2}$ (or an increase in K). This is easy to understand that the more efficient elimination pathway may result in a faster decline in the plasma concentrations. An increase in V, however, results in prolongation of $t_{1/2}$, because of the drug is distributed more extensively into the tissues (where it is safe to be eliminated). However, because the distribution is a reversible process, as the drug gets eliminated and plasma concentrations decline, the drug in the tissue will return to plasma, resulting in a more sustained level in plasma (increased half life). Therefore, the half life is dependent on both the clearance and volume of distribution, and a more appropriate presentation of the relationship among these three parameters is:

$$t_{1/2} = \frac{0.693V}{Cl}$$

Or,

$$K = \frac{Cl}{V}$$

For example, if the volume of distribution changes, the half life (or rate constant) changes proportionally while clearance remains the same. To demonstrate this point, consider the following scenario. The volume of distribution and elimination rate constant of a drug in a patient are 35 L and 0.091 hr-1, respectively. While under treatment with this drug, the patient receives a second drug, which increases the drug V to 70 L. What is the clearance of the drug in the absence and presence of the interacting drug?

The mathematical relationship Cl = K × V may be used to estimate Cl in the absence of drug interaction:

$$Cl = 0.091 \times 35 = 3.2 \text{ L/hr}$$

Clearance is independent of V changes. Therefore, when V is increased to 70 L due to a drug interaction, Cl remains the same (3.2 L/hr). However, the above equation without an understanding of this fundamental concept may be misleading. This is because one may erroneously conclude from the equation Cl = K × V that doubling V would result in doubling Cl. This conclusion, however, is not valid as a two-fold increase in V would result in a two-fold decrease in K without any effect on Cl.

1.5 Steady State

Steady state occurs, when the amount of drug administered (in a given time period) is equal to the amount of drug eliminated in that same period. At steady state, the plasma concentrations of the drug (C_p^{ss}) at any time during any dosing interval, as well as the peak and trough, are similar. The time to reach steady-state concentrations is dependent on the half-life of the drug under consideration.

Effect of Dose on Steady state

The higher the dose, the higher the steady-state levels, but the time to achieve steady-state levels is independent of dose (see Fig 1.5). Note that the fluctuations in $C_{p\,max}$ and $C_{p\,min}$ are greatest with higher doses.

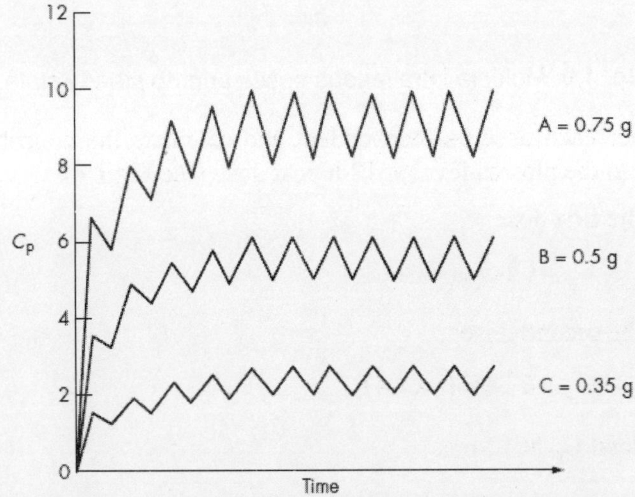

Fig. 1.5 Time profile of multiple IV doses – reaching steady state using different doses.

Time to reach steady state

For a drug with one-compartment characteristics, the time to reach steady state is independent of the dose, the number of doses administered, and the dosing interval, but it is directly proportional to the half-life.

Prior to steady state

As an example, estimate the plasma concentration at 12 h after therapy commences with a drug **A** given 500 mg three times a day.

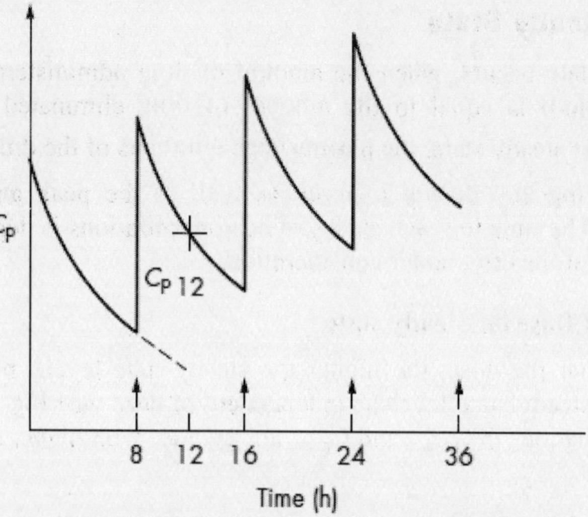

Fig. 1.6 Multiple intravenous doses prior to steady state.

Consider each dose as independent and calculate the contribution of each dose to the plasma level at 12 h post dose (see Fig 1.6).

From the first dose:

$$C_{p^1} = C_P^0 \exp(-k \times 12)$$

From the second dose:

$$C_{p^2} = C_P^0 \exp(-k \times 4)$$

Thus, total C_{pt} at 12 h is

$$C_{pt} = C_P^0 \exp(-k \times 12) + C_P^0 \exp(-k \times 4)$$

Remember that $C_P^0 = D / V_d$

This method uses the principle of superposition. The following equation can be used to simplify the process of calculating the value of C_p at any time t after the nth dose:

$$C_{Pt} = \frac{D \times [\exp(-kn\tau) \times (\exp(-kt)]}{Vd \times \exp(-k\tau)}$$

Where, n = number of doses, τ = dosing interval and t = time after the nth dose.

At steady state

To describe the plasma concentration (C_p) at any time (t) within a dosing interval (τ) at steady state (see Fig 1.7):

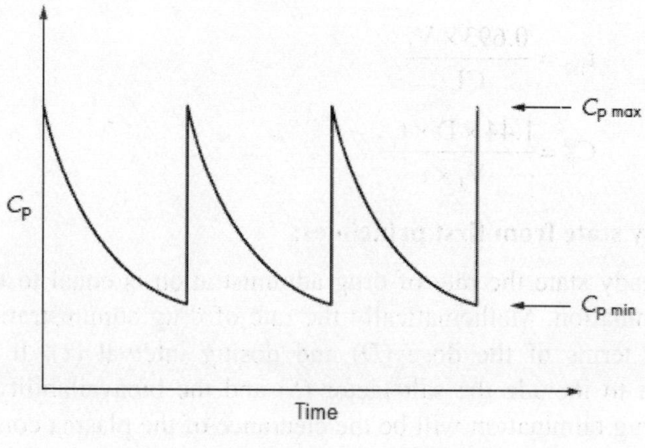

Fig. 1.7 Time profile at steady state and the maximum and minimum plasma concentration within a dosage interval.

$$C_{pt} = \frac{D \times [\exp(-kt)]}{V_d \times [1 - \exp(-kt)]}$$

Remember that $C_P^0 = D / V_d$. Alternatively, for some drugs, it is important to consider the salt factor (S). Hence, if salt factor applicable, $C_P^0 = SD / V$ and C_{pt} will be:

$$C_{pt} = \frac{S \times D \times [\exp(-kt)]}{V_d \times [1 - \exp(-k\tau)]}$$

To describe the maximum plasma concentration at steady state (i.e. t =0 and exp ($-k_t$) = 1):

$$C_{p\max} = \frac{D \times 1}{V_d \times [1 - \exp(-k\tau)]}$$

To describe the minimum plasma concentration at steady state (i.e. $t = \tau$):

$$C_{p\min} = \frac{D \times [\exp(-kt)]}{V_d \times [1 - \exp(-k\tau)]}$$

To describe the average steady-state concentration, C_P^{ss}

$$C_P^{ss} = \frac{D}{CL \times \tau} \text{ or } C_P^{ss} = \frac{S \times D}{CL \times \tau}$$

Since $$t_{1/2} = \frac{0.693 \times V_d}{CL}$$

Then $$C_P^{ss} = \frac{1.44 \times D \times t_{1/2}}{V_d \times \tau}$$

Steady state from first principles:

At steady state the rate of drug administration is equal to the rate of drug elimination. Mathematically the rate of drug administration can be stated in terms of the dose (*D*) and dosing interval (τ). It is always important to include the salt factor (*S*) and the bioavailability (*F*). The rate of drug elimination will be the clearance of the plasma concentration at steady state:

$$\text{Rate of drug administration} = \frac{S \times F \times D}{\tau}$$

$$\text{Rate of drug elimination} = CL \times C_P^{ss}$$

At steady state:

$$\frac{S \times F \times D}{\tau} = CL \times C_p^{ss}$$

Rearranging the equation:

$$C_P^{ss} = \frac{S \times F \times D}{CL \times \tau}$$

In practice, steady state is assumed to be reached in 4–5 half-lives.

1.6 Dosage Regimen Fixation

1.6.1 Intravenous Infusion

Some drugs are administered as an intravenous infusion rather than as an intravenous bolus. To describe the time course of the drug in the plasma during the infusion prior to steady state, one can use:

$$C_{pt} = \frac{R\left[1 - \exp(-kt)\right]}{CL}$$

where, $R = \frac{D}{\tau}$

or, $R = \frac{S \times D}{\tau}$

(If a salt form of the drug is given)

During the infusion, at steady state, as the rate in = rate out,

$$R = CL \times C_P^{ss}$$

$$C_P^{ss} = \frac{D}{\tau \times CL}$$

Where, $R = D/\tau$ = infusion rate (dose/h)

Loading dose

The time required to obtain steady-state plasma levels by IV infusion will be long, if the drug has a long half-life. It is, therefore, useful in such cases to administer an intravenous loading dose to attain the desired drug concentration immediately and then attempt to maintain this concentration by a continuous infusion.

To estimate the loading dose (LD), where C_p^{ss} is the final desired concentration, one can use to calculate LD by the equation:

$$LD = V_d \times C_p^{ss}$$

If the patient has already received the drug, then the loading dose should be adjusted accordingly as:

$$LD = V_d \times \left(C_p^{ss} - C_p^{initial}\right)$$

or,

$$LD = \frac{V_d \times \left(C_p^{ss} - C_p^{initial}\right)}{S}$$

if, the salt form of the drug (salt factor *S*) is used.

Now consider the plasma concentration–time profile following a loading dose and maintenance infusion (see Fig 1.8).

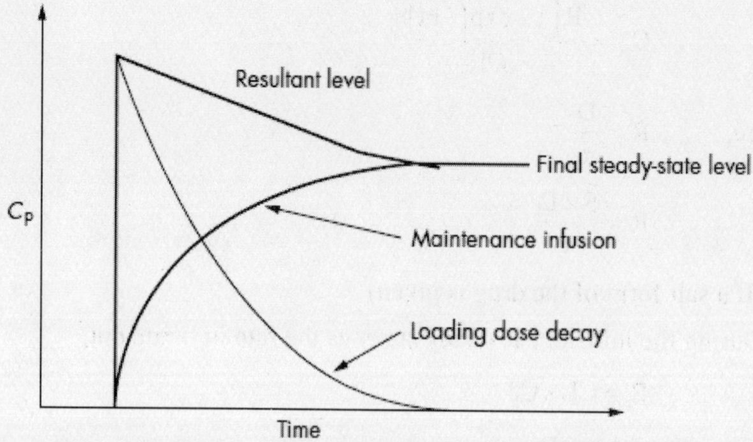

Fig. 1.8 Profile following a loading dose and maintenance infusion.

The equation to describe the time course of the plasma drug concentrations following simultaneous administration of an IV loading dose (LD) and initiation of infusion (*D*) is the sum of the two equations describing these two processes individually as:

$$C_P = \frac{LD\exp(-kt)}{V_d} + \frac{D[1-\exp(-kt)]}{\tau \times CL}$$

The final plasma concentration achieved may not be the 'true' steady state concentration, because it may require about 4 half-lives to reach the steady state, but depending on the accuracy of the loading dose. However, this regimen allows the concentration some where near steady state to be achieved more rapidly. If the salt form of the drug is used, then the C_p can be calculated by the equation:

$$C_P = \frac{S \times LD\exp(-kt)}{V_d} + \frac{S \times D[1-\exp(-kt)]}{\tau \times CL}$$

1.6.2 Single Oral Dose

The plasma concentration–time profile of a large number of drugs can be described by a one-compartment model with first-order absorption and elimination. Consider the concentration versus time profile following a single oral dose (Fig 1.9). Assuming that the first-order absorption and first-order elimination.

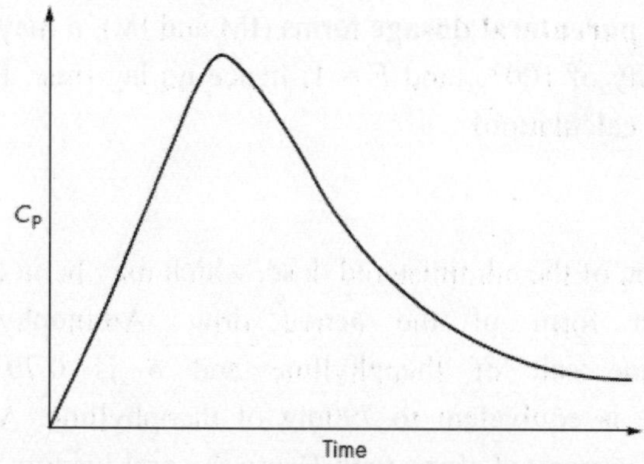

Fig. 1.9 Single oral dose profile.

Then, the rate of change of amount of drug (X) in the body is described by:

$$\frac{dX}{dt} = k_a X_a - kX$$

where, k_a = absorption rate constant; k = elimination rate constant; X = amount of drug in the body; and X_a =amount of drug at the absorption site (X_0 if all is available). Following integration:

$$X = \frac{X_0\, k_a\, [\exp(-kt) - \exp(-k_a t)}{k_a - k}$$

To convert X to C_p one can use the apparent volume of distribution (V_d). Further following oral administration, the bioavailability (F) and salt factor (S) (see below) must be considered.

If the loading dose is to be administered orally, then the bioavailability term (F) needs to be introduced. Thus:

$$LD = \frac{V_d \times C_P}{F}$$

Fractional bioavailability, F

F is the fraction of an oral dose that reaches the systemic circulation, following oral administration, which may be less than 100%. Thus, if $F = 0.5$ then 50% of the drug is absorbed.

(In case of **parenteral dosage** forms (IM and IV), it may assume to be a bioavailability of 100%, and F = 1; hence no lag time, for which it is omitted in the calculation)

Salt factor, S

S is the fraction of the administered dose, which may be in the form of an ester or salt form of the active drug. Aminophylline is the ethylenediamine salt of theophylline and *S* is 0.79. Thus 1 g aminophylline is equivalent to 790mg of theophylline. Accordingly, *S* needs to be incorporated along with *F* into the oral loading dose equation and the equation that describes the plasma concentration Cp at any time *t* following a single oral dose. Thus loading dose can be calculated as:

$$LD = \frac{V_d \times C_P}{S \times F}$$

and

$$C_{pt} = \frac{SFD}{V_d} \times \frac{k_a\left[\exp(-kt) - \exp(-k_a t)\right]}{k_a - k}$$

N.B.: The *S* factor may need to be considered during IV infusion administration.

1.6.3 Multiple Oral Dosing

Prior to steady state

Consider a patient on medication prescribed three times a day. The profile in Fig 1.10 shows the administration of three doses. If we consider a time 28 h into therapy and all three doses would have been administered.

To calculate C_p at 28 h post dose, one can use the single oral dose equation and consider the contributions of each dose as:

Contribution from dose 1; t_1 = 28 h:

$$C_{p1} = \frac{SFD}{V_d} \times \frac{k_a\left[\exp(-kt_1) - \exp(-k_a t_1)\right]}{k_a - k}$$

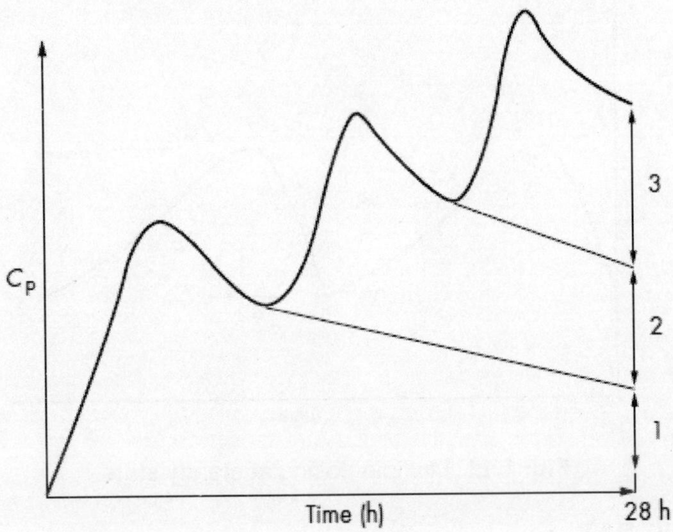

Fig. 1.10 Multiple dosing prior to steady state.

Contribution from dose 2, t_2 = 18 h:

$$C_{p2} = \frac{SFD}{V_d} \times \frac{k_a\left[\exp(-kt_2) - \exp(-k_a t_2)\right]}{k_a - k}$$

Contribution from dose 3; t_3 =8h:

$$C_{p^3} = \frac{SFD}{V_d} \times \frac{k_a\left[\exp(-kt_3) - \exp(-k_a t_3)\right]}{k_a - k}$$

Thus, $C_{p28h} = C_{p^1} + C_{p^2} + C_{p^3}$

The above method uses the principle of superposition to calculate the C_p at any time t after the nth dose. The following equation can simplify the process as:

$$C_{pt} = \frac{SFDk_a}{V_d(k_a - k)} \times \left\{\frac{\left[1-\exp(-nk\tau)\right](\exp(-kt))}{1-\exp(-k\tau)} - \frac{\left[1-\exp(-nk_a\tau)\right](\exp(-k_a t))}{1-\exp(-k_a\tau)}\right\}$$

Where, n = number of doses, τ = dosage interval and t = time after the nth dose.

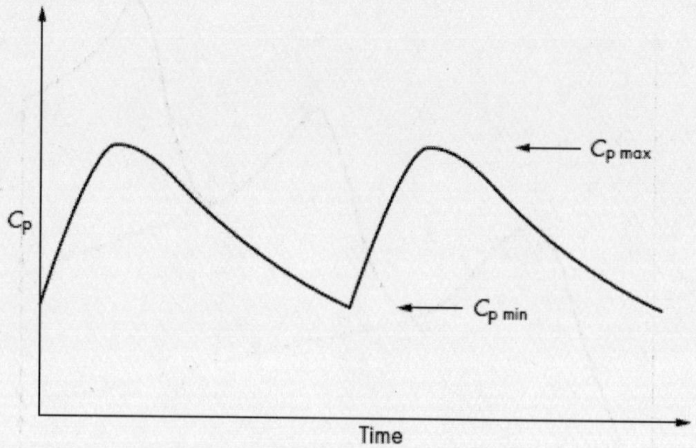

Fig. 1.11 Multiple dosing at steady state.

At steady state

At steady state the plasma concentration–time profile can be described by the equation:

$$C_{pt} = \frac{SFDk_a}{V_d(k_a - k)} \times \left\{ \frac{\exp(-kt)}{1-\exp(-k\tau)} - \frac{\exp(-k_a t)}{1-\exp(-k_a \tau)} \right\}$$

When the half-life of a drug is long, the fluctuations between the peak and trough are small, and the equation derived above under Intravenous infusion can be used to describe the average steady-state concentration:

$$C_p^{ss} = \frac{D}{T \times CL}$$

Practice Problem

Example 1: A patient D has a potentially toxic digoxin level of 4.5μg/L. Given that the half-life of digoxin in this patient is 60 h, and assuming that renal function is stable and absorption is complete, for how long should the drug be stopped to allow the level to fall to 1.5μg/L?

(a) Calculate elimination rate constant (k):

$$k = \frac{0.693}{60} = 0.0116\,h^{-1}$$

(b) Time for decay (t) from C_{p1} to C_{p2}

$$t = \frac{\ln C_{p^1} - \ln C_{p^2}}{k}$$

$$t = \frac{\ln 4.5 - ln 1.5}{0.0116}$$

$$= 94.7 \text{ h}$$

Hence, t = 4 days

Example 2: What is the loading dose required for a drug if, Target concentration is 10 mg/L?

- VD is 0.75 L/kg
- Patients weight is 75 kg

Answer Dose = Target Concentration x VD.

VD = 0.75 L/kg x 75 kg = 56.25 L.

Target Conc. = 10 mg/L.

Dose = 10 mg/L x 56.25 = 565 mg.

This would probably be rounded to 560 or even 500 mg.

What maintenance dose is required for a drug if, Target average SS concentration is 10 mg/L?

- CL of drug A is 0.015 L/kg/hr
- Patient weighs 75 kg

Answer Maintenance Dose = CL x C_{pSS}av.

Cp_{SS}av is the target average steady state drug concentration.

The units of CL are in L/hr or L/hr/kg.

Maintenance dose will be in mg/hr, so for total daily dose will need multiplying by 24

CL = 0.015 L/hr/kg x 75 = 1.125 L/hr

Dose = 1.125 L/hr x 10 mg/L = 11.25 mg/hr

So, will need 11.25 x 24 mg per day = 270 mg

1.7 Nonlinear Pharmacokinetics

Drugs such as phenytoin will show nonlinear drug handing. The process of metabolism is nonlinear and the rate of metabolism shows zero order. In practice, **Michaelis–Menten** pharmacokinetics may be applied. If a patient receives different doses of phenytoin, e.g. 200 mg/day, 250 mg/day, 300 mg/day or 400 mg/day, the steady-state plasma concentration varies exponentially with time; that is, a small change in the total daily dose of phenytoin shows a disproportionate increase in the steady state concentration (C_p^{ss}) (Fig 1.12).

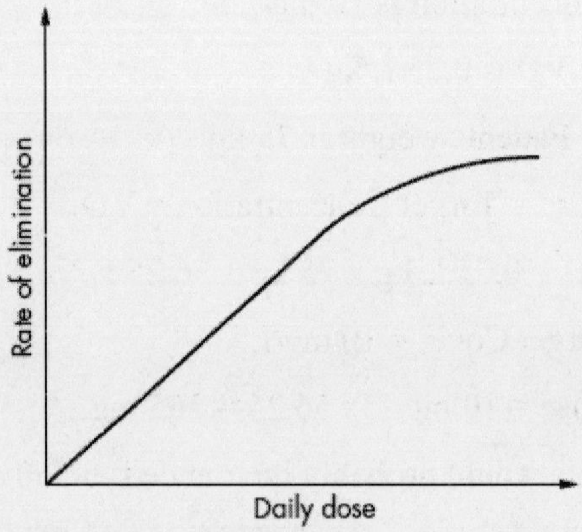

Fig. 1.12 Profile of elimination following phenytoin administration.

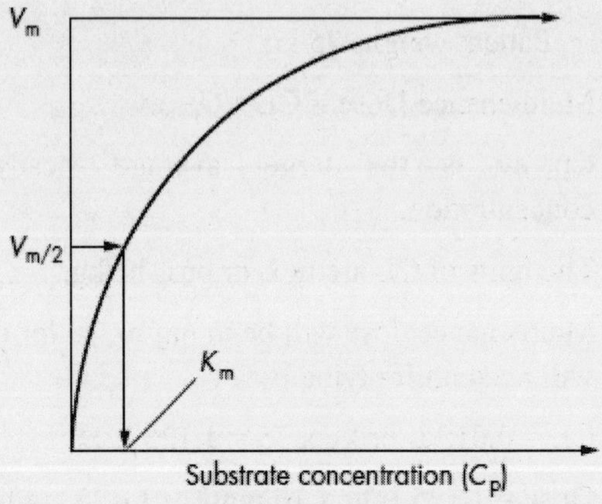

Fig. 1.13 Relationship between rate of metabolism (V) versus substrate concentration (C_p) for a drug showing nonlinear pharmacokinetics.

Fig 1.13 describes the profile of the rate of metabolism of phenytoin given at different dosages. As the dose of phenytoin increases, the rate of elimination increases until it reaches a plateau, where the rate of elimination is constant despite increases in the total daily dose of the drug. The profile can be described as follows.

Rate of elimination:

$$\frac{-dX}{dt} = \frac{V_m \times C_P^{ss}}{K_m + C_p^{ss}}$$

Hence the model that appears to fit the pattern for the metabolic elimination of phenytoin is not linear and is the one proposed by Michaelis and Menten. The velocity (V) or rate at which an enzyme can metabolize a substrate (C_p) can be described by the following equation:

$$V = -\frac{V_m \times C_p}{K_m + C_p}$$

where, V is the rate of metabolism, V_m (sometimes referred to as V_{max}) is the maximum rate of metabolism and Km is the substrate concentration (C_p) at which V will be half V_m, i.e., when half the total enzyme is complexes with the substrate (See Fig 1.13.).

At steady state, we know that the rate of administration is equal to the rate of elimination; hence, in the clinical situation, the daily dose (R or D) is substituted for velocity (V), and the steady-state phenytoin concentration (C_P^{ss}) is substituted for substrate concentration (S). Further equations can be described for steady-state concentrations. At steady state the rate of administration is equal to the rate of elimination. The rate of administration can be expressed as SFD / τ, where D/τ can be equal to R. Hence

$$RSF = \frac{V_m \times C_p^{ss}}{K_m + C_p^{ss}}$$

Where, V_m is the maximum metabolic capacity, i.e. the total amount of drug that can be eliminated at saturation. K_m is the Michaelis' constant, which by definition is the concentration at which the metabolism is operating at half the maximum capacity.

All drugs will show nonlinear handling if they are administered in high enough doses. However, only a small number of drugs show nonlinear handling at the doses used clinically. Whether a drug will show linear or nonlinear drug handling in therapeutic doses depends on the Michaelis' constant K_m of the drug. For example, consider a drug, which has a *K*m, much greater then C_p^{ss}, i.e. the plasma levels seen at normal therapeutic doses of the drug. Then the rate of elimination can be described as:

$$\frac{-dX}{dt} = \frac{V_m \times C_p^{ss}}{K_m + C_p^{ss}}$$

Since, *K*m is much more than C_p^{ss}, the equation simplifies to

$$\frac{-dX}{dt} = \frac{V_m \times C_p^{ss}}{K_m}$$

Since V_m and K_m are constants, which represents a first-order rate process. In another simulation a drug has a *K*m, which is much less than C_p^{ss}, i.e. the plasma levels seen with normal therapeutic doses of the drug, then the equation becomes:

$$\frac{-dX}{dt} = \frac{V_m \times C_p^{ss}}{K_m + C_p^{ss}}$$

Since, *K*m is much less than C_p^{ss}, the equation simplifies to

$$\frac{-dX}{dt} = V_m$$

Since, V_m is a constant; this now represents a zero-order process.

Hence, the relationship between the Michaelis constant (K_m) of the drug and the plasma levels of the drug normally achieved with therapeutic dosages, which determine, whether the drug will show linear first order or zero-order saturation pharmacokinetics.

1.8 Effects of Clearance, Volume of Distribution, and Plasma Half Life on the Magnitude of Dose, Dosage Interval, and/or Dosing Rate

Clearance: Among the primary kinetic parameters, Cl determines the magnitude of dosing rate (Dose/ τ) or infusion rate (R_0) to achieve a certain average concentration at steady state $\left(C_{ave}^{ss} \text{ or } C_{ss}\right)$.

$$C_{ave}^{ss} = \frac{F(Dose/\tau)}{Cl}$$

$$Dose/\tau = \frac{Cl.C_{ave}^{ss}}{F} \text{ or } Dose = \frac{Cl.C_{ave}^{ss}.\tau}{F}$$

(For multiple Extra-vascular and IV doses)

and, $R_0 = Cl \cdot C_{ss}$ (For constant IV infusion)

The above relationships indicate that when the clearance of a drug is altered because of a drug interaction and/or disease state, C_{ave}^{ss} or C_{ss} would change inversely relative to the change in Cl. For example, the clearance of theophylline decreases by > 50 percent because of some drug interactions (*e.g.*, enoxacin) or disease states (*e.g.*, hepatic cirrhosis), resulting in higher C_{ave}^{ss} or C_{ss} values, if the usual doses of theophylline are administered.

To illustrate the significance of a change in Cl, assume that an interacting drug added to the regimen of the patient receiving the example drug (Cl of 3.2 L/hr) results in a two- fold reduction in the drug Cl (to 1.6 L/hr). Administration of the same dose of the example drug (280 mg every 6 hr or 46 mg/hr as constant IV infusion) would be expected to result in C_{ave}^{ss} or C_{ss} value of 29 mg/L (instead of ~14.5 mg/mL) which is above the upper limit of the desired plasma concentration for this drug (20 mg/L) (Fig 1.15). In addition to an increase in C_{ave}^{ss} or C_{ss}, a two-fold decrease in Cl would result in the following changes in the kinetics of the drug (Fig 1.14):

- The drug half life in the above case would be expected to increase by a factor of 2 (15 hr instead of 7.6 hr).
- The time to reach steady state would be expected to double because of a two-fold increase in the half life.
- The fluctuation in the plasma concentrations after multiple dosing would be reduced in the presence of reduced Cl.

The modification necessary in the presence of the interacting drug to decrease in its dosing rate or infusion rate (23 mg/hr instead of 46 mg/hr), which is proportional to the decrease in Cl (two-fold). For the multiple dosing method, the modification could be in the form of a reduction in the dose (140 mg every 6 hr), a longer interval (280 mg every 12 hr) or both (180 mg every 8 hr), keeping the dosing rate the same (~23 mg/hr) for all three scenarios and half of that in the absence of the drug interaction (~46 mg/hr).

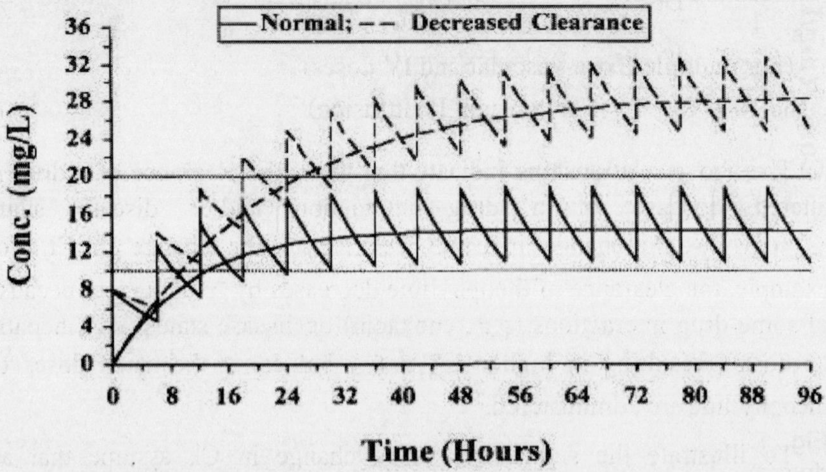

Fig. 1.14 Simulated plasma concentration-time profile of an example drug after IV multiple bolus and constant infusion dosing in the absence and presence of a drug interaction resulting in a two-fold reduction in drug clearance.

Volume of Distribution: The volume of distribution **V,** depends on the magnitude of loading dose (D_L), because it relates the drug plasma concentration to the amount of drug in the body:

$$DL = C_{ss}(\text{or } C_{max}) \cdot V$$

However, as demonstrated in the clearance section, the dosing rate (maintenance dose) of the drug is only dependent on Cl and not on V. On the other hand, the loading dose is not affected by a change in Cl as demonstrated in the above equation. To clarify these concepts, assume that a drug interaction results in a two-fold increase in the V of the example drug (from 35 L to 70 L). The C^{ss}_{ave} or C_{ss} values of the drug in the presence of this interaction would not be different from those in the absence of the drug interaction (Fig 1.15). Therefore, the dosing rate of the drug need not be altered. However, it should be noted that an increase in V results in a proportional increase in half life (from 7.6 to 15 hr), a proportional increase in the time to reach steady state and a decrease in the plasma concentration fluctuation (Fig 1.15).

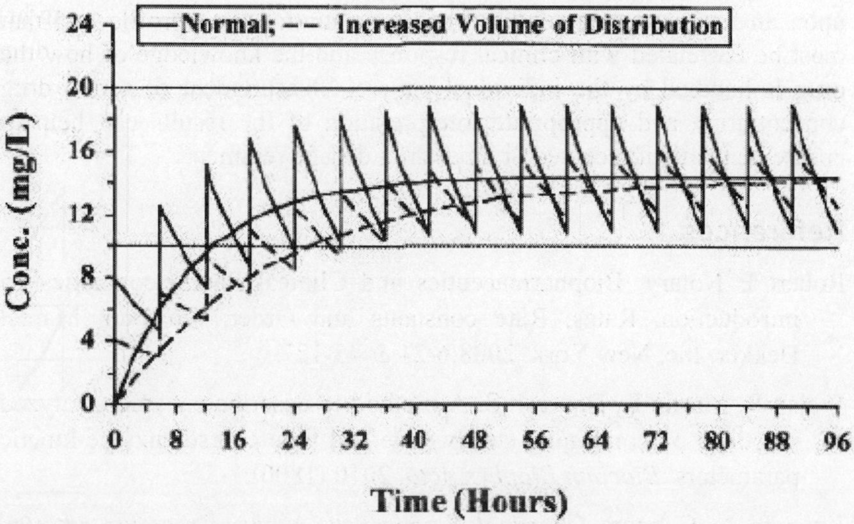

Fig. 1.15 Simulated plasma concentration-time courses of an example drug after IV multiple bolus and constant infusion dosing in the absence and presence of a drug interaction resulting in a two-fold increase in drug volume of distribution.

Although the dosing rate should be the same when V is increased, the reduced fluctuation may allow administration of larger doses at longer intervals (*e.g.*, 560 mg every 12 hr) for multiple dosing method. Additionally, because of the larger V, the loading dose should be increased. For example, for constant IV infusion, the loading dose should be increased from 500 mg (14.4 mg/L x 35 L) in the absence of drug interaction to 1000 mg (14.4 mg/L x 70 L) in the presence of drug interaction.

Half Life: As discussed above, the half life may change as a result of a change in V and/or Cl. Generally, a change in the half life affects the time to reach steady state and the shape of the plasma concentration-time profile (Figures 1.14 and 1.15).

However, as discussed under clearance and volume of distribution sections, a change in the half life does not necessarily affect C^{ss}_{ave} but it will affect the C^{ss}_{max} and C^{ss}_{min} values (the degree of fluctuation).

In addition to the pharmacokinetic characteristics of the patient, concentration time profile of the drugs depend on the route of administration and on the impute rate, which can be altered by the formulation. Individualization of drug dosage regimens, which depends upon an understanding of the drug concentration-time profile and that must be correlated with clinical response and the knowledge of how the drug is handled by the individual patient. Measurement of serum drug concentration and appropriate interpretation of the results can help to ensure that patient received optimal drug dosage regimen.

References

Robert E Notary. Biopharmaceutics and Clinical Pharmacokinetics an introduction, Rates, Rate constants and Order, 4th Edn, Marcel Dekker, Inc. New York. 2008:6-21 & 45-127.

Walsh R, Martin E, Darvesh S. A method to describe enzyme-catalyzed reactions by combining steady state and time course enzyme kinetic parameters. *Biochim Biophys Acta*. 2010 (1800):1-5

Kenneth A. Connors *Chemical Kinetics, the study of reaction rates in solution*, 1991, VCH Publishers

IUPAC Gold Book, Rate Low (empirical differential equation), 1994: 1077-1157

Leon Shargel, S. W. Pong, Andrew B. C. Yu. Applied Biopharmaceutics and Pharmacokinetics, Mathematical Fundamentals in Pharmacokinetics. 5th Edn. (International Edition) Mc Grow Hill (Asia). 2005: 21-50.

Christos Capellos and Bennon H. Bielski "Kinetic systems: mathematical description of chemical kinetics in solution" 1972, Wiley-Interscience (New York)

José A. Manso et al. "A Kinetic Approach to the Alkylating Potential of Carcinogenic Lactones" *Chem. Res. Toxicol*. 2005, 18 (7): 1161-1166.

Reza Mehvar. Pharmacokinetic-Based Design and Modification of Dosage Regimens. *American Journal of Pharmaceutical Education,* 1998 (62): 189-195.

Peck, C.C., D'Argenio, D.Z. and Rodman, J.H., "Analysis of pharmacokinetic data for individualizing drug dosage regimens," in Applied Pharmacokinetics: Principles of Therapeutic Drug Monitoring, (edits. Evans, W.E., Schentag, J.J. and Jusko, W.J.), Applied Therapeutics, Vancouver, WA , 1992:1-31.

Alison Thomson. *The Pharmaceutical Journal*, 2004 (273):188-190.

Gibaldi M, Nagashima R, Levy G; Relationship between drug concentration in plasma or serum and anount of drug in the body. *J Pharma Sci*, 1969 (58):193-197.

Hurley S. F.., McNeil J. J.. A comparison of the accuracy of a least-squares regression, a Bayesian, Chiou's and the steady-state clearance method of individualizing theophylline dosage. *Clin Pharmacokinet,* 1988 (14):311-320.

Wagner J, A safe method for rapidly achieving plasma concentration plateaus. *Clin Pharmacol Ther*, 1974 (16):691-700.

Gibaldi M, Estimation of the pharmacokinetic parameters of the two-compartment open model from postinfusion plasma concentration data. *J Pharma Sci,* 1969 (58):1133-1135.

Koup J, Greenblatt D, Jusko W, et al., Pharmacokinetics of digoxin in normal subjects after intravenous bolus and infusion dose. *J Pharmacokinetic Biopharm*, 1975 (3):181-191.

Clark, B (1986). In Clark B, Smith D A, eds. *An Introduction to Pharmacokinetics*, 2nd ed. Oxford: Blackwell Scientific.

Evans W E, Schentag J J, Jusko W J, Harrison. In Evans W E, Schentag J J, eds. *Applied Pharmacokinetics: Principles of Therapeutic Drug Monitoring*, 3rd edn., 1992.

Gibaldi M, Prescott L. *Handbook of Clinical Pharmacokinetics*. New York: ADIS Health Science Press, 1983.

Taylor W J, Diers-Caviness M H. *A Textbook of the Clinical Application of Therapeutic Drug Monitoring*. Irving, TX: Abbott Laboratories Ltd, Diagnostic Division, 2003.

Winter M E. *Basic Clinical Pharmacokinetics*, Philadelphia: Lippincott Williams and Wilkins, 4th edn., 2003.

Mehvar, R., "Computer-assisted generation and grading of pharmacokinetics assignments in a problem-solving course," *Am. J. Pharm. Educ.*, 1997 (61):436-441.

Hyatt, J.M., Mckinnon, P.S., Zimmer, G.S. and Schentag, J.J., "The importance of pharmacokinetic/pharmacodynamic surrogate markers to outcome: Focus on antibacterial agents." *Clin. Pharmacokinet.* 1995 (28):143-160.

Hyatt, J.M., Mckinnon, P.S., Zimmer, G.S. and Schentag, J.J., "The importance of pharmacokinetic/pharmacodynamic surrogate markers to outcome: Focus on antibacterial agents." *Clin. Pharmacokinet.*, 1995 (28):143-160.

Peck, C.C., D'Argenio, D.Z. and Rodman, J.H., "Analysis of pharmacokinetic data for individualizing drug dosage regimens," in *Applied Pharmacokinetics: Principles of Therapeutic Drug Monitoring*, (edits. Evans, W.E., Schentag, J.J. and Jusko, W.J.), *Applied Therapeutics*, Vancouver, WA; 1992, 3(1):3-31.

Holford. N.H.G. and Sheiner, L.B., "Kinetics of pharmacologic response," *Pharmacol. Ther.*, 1982 (10):143-166.

Mangione, A., Imhoff, T.E., Lee, R.V., Shum, L.Y. and Jusko, W.J., Pharmacokinetics of theophylline in hepatic disease," *Chest*, 1978 (73):616- 622.

Glossary of Terms and Abbreviations

V	Volume of distribution
Cl	Clearance
$t_{1/2}$	Plasma half life
D_M	Maintenance dose
τ	Dosage interval
D_L	Loading dose
C_{ss}	Steady-state drug concentration after constant IV infusion
MEC	Minimum effective concentration
MTC	Minimum toxic concentration
F	Oral bioavailability
C_{ave}^{ss}	Average steady-state concentration after multiple dosing
C_{min}^{ss}	Minimum steady-state concentration after multiple dosing
C_{max}^{ss}	Maximum steady-state concentration after multiple dosing
τ_{max}	Maximum allowable dosage interval
K	Elimination rate constant
C_{max}^{1st}	Maximum plasma concentration after the first dose
C_{min}^{1st}	Minimum plasma concentration after the first dose
R	Accumulation factor

2

Bio-Degradable Polymers in Drug Delivery Systems

Introduction

Polymers are very important in the field of drug delivery. The pharmaceutical application of polymers not only include binding, suspending, emulsifying, viscosifying and flow controlling agent in tablet, suspension, emulsion and liquids but also can be used as film coatings to disguise the unpleasant taste of a drug, to enhance drug stability, to modify drug release characteristics and to act as carrier in drug targeting.

Recently, a number of novel drug delivery technologies have been developed, which include drug modification by chemical means, career based drug delivery and drug entrapment in polymeric matrices or within pumps that are placed in desired body compartments. These technical developments in drug delivery and drug targeting approaches improve the efficacy of drug therapy and patient compliance. Still there are many problem encountered by formulation scientists in designing the drug delivery systems, because of non-compatibility and biodegradability and elimination problem after delivering the drug from the polymer carrier *in-vivo*. There is a strong need to develop a proper delivery system to achieve the complete therapeutic effects of the existing drug molecules.

Use of polymeric materials in novel drug delivery approaches has attracted the pharmaceutical scientists, polymer chemists and chemical engineers in bringing out predictable, controlled delivery of bio active agents. The characterization of biocompatible polymers is more focused in the field of formulation development and drug delivery approaches etc. The biodegradable polymers have properties of degrading in biological fluids with progressive release of dissolved or dispersed drug. There is various novel drug delivery approaches developed in the field of polymer based drug delivery approaches. The bio-safety and biocompatibility are the important characteristics for the use of polymers in the field of pharmaceutical formulation development and in novel drug delivery systems. The main advantages of using biodegradable polymers include the enzymatic or chemical degradation of the polymers *in-vivo* resulting in the formation of biocompatible or non-toxic products in biological fluids, which are removed from the body through normal metabolic pathways and physiological mechanisms.

2.1 Classes of Biodegradable Polymers

A variety of natural, synthetic, and biosynthetic polymers are found to be biodegradable, which are classified basing on their chemical structures as:

2.1.1 Polyanhydrides

Polyanhydrides have been investigated for short-term controlled delivery of the bioactive agents, because of they exhibit rapid degradation *in-vivo* and have limited mechanical properties. Polyanhydrides are the hydrophobic polymers with hydrolytically labile anhydride linkages, in which the degradation rate can be manipulated by varying polymer composition; numerous di-acids are available for the design of the polymer with desired physicochemical properties. In general, this class of polymers shows minimal inflammatory reaction *in-vivo* and degrades into monomeric acids as non-mutagenic and non-cytotoxic products. Polyanhydrides are known to undergo surface erosion, a desired property to attain near zero-order drug release profile. Degradation of polyanhydrides depends on the rate of the water uptake, determined by hydrophilicity and crystallinity of the polymer. Recently, due to their biocompatibility and biodegradability, the Food and Drug Administration has approved the use of the polyanhydride derived from sebacic acid and 1, 3-bis (p-carboxyphenoxy) propane as the carrier of antitumor agents for the treatment of brain cancer. One drawback of polyanhydrides is that most of them have to be stored at frozen state under the anhydrous condition because of the hydrolytic instability of the anhydride bond. Some of the commercial products are cited below as examples:

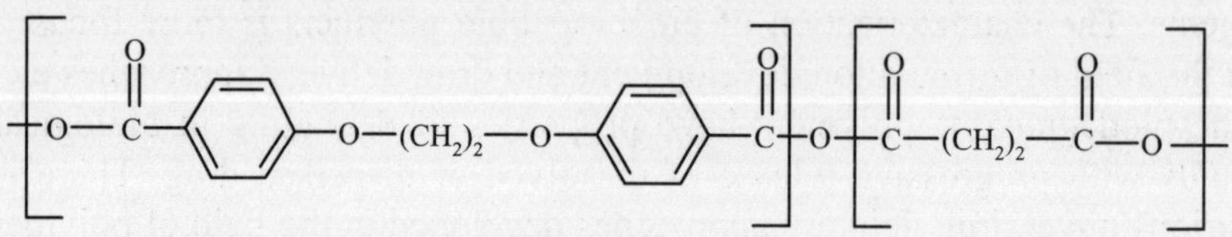

Poly [bis (p-carboxyphenoxy) propane-co-sebacic acid

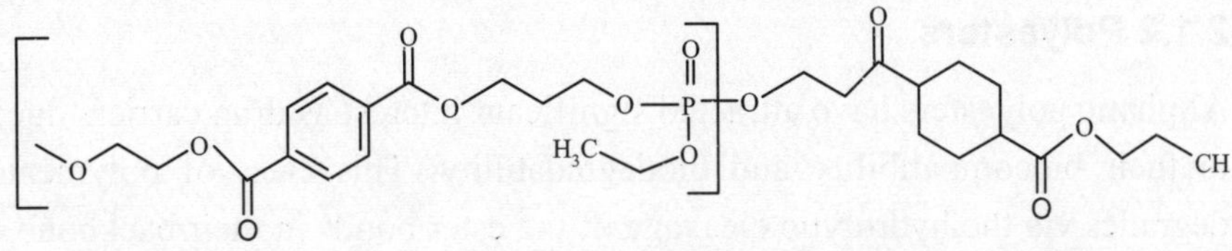

Poly [1, 4-bis (hydroxyethyl) terephthalate-alt-ethyloxyphosphate]

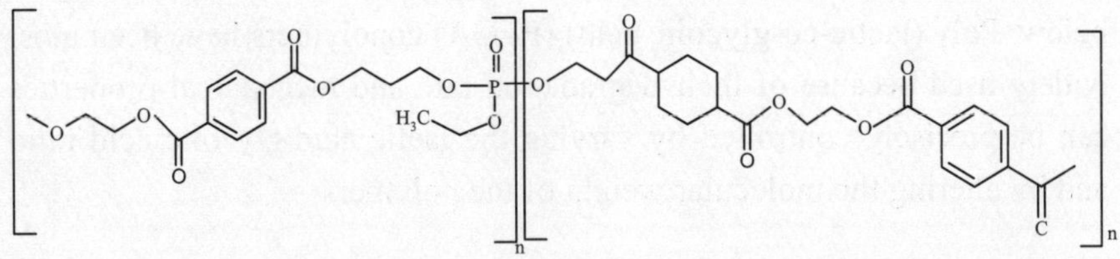

Poly [1, 4-bis (hydroxyethyl) terephthalate-alt-ethyloxyphosphate]-co-1, 4-bis (hydroxyethyl)terephthalate-co-terephthalate

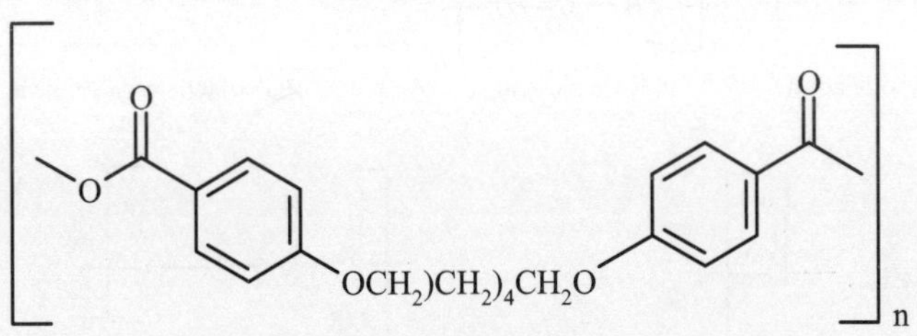

Poly [1, 6-bis (p-carboxyphenoxy) hexane]

Polymers with increasing hydrophobicity can be made from aromatic monomers including phthalic acid and various carboxyphenoxyalkanes such as poly [1-bis (p-carboxyphenoxy) methane] CPM, poly [1, 3-bis (p-carboxyphenoxy) propane] CPP and poly [1, 6-bis (p-carboxyphenoxy) hexane] CPH. High-molecular-weight polyanhydrides are usually synthesized by first converting the dicarboxylic acid monomer to mixed anhydride pre-polymers using acetic anhydride followed by polymerization of pre-polymers using poly-condensation reaction in melt. Polyanhydrides are often prepared as copolymers of aliphatic and/or aromatic monomers. The most common copolymers under investigation in drug delivery include poly-[fatty acid dimmer (FAD)-sebacic acid (SA)] and poly (CPP-SA).

(Figure ref: www.sigmaaldrich.com/catalog/search/.......)

2.1.2 Polyesters

Aliphatic polyesters have attracted significant interest as drug carriers due to their biocompatibility and biodegradability. This class of polymers degrades via the hydrolytic cleavage of the ester bonds in their backbone, whereas the role of enzymatic involvement in biodegradation is unclear. Chemical structures of the representative polyesters are shown in Figure below. Poly (lactic-co-glycolic acid) (PLGA) copolymers have been most widely used because of their degradation rate and mechanical properties can be precisely controlled by varying the lactic acid/glycolic acid ratio and by altering the molecular weight of the polymers

(Figure ref: www.sigmaaldrich.com/catalog/search/.......).

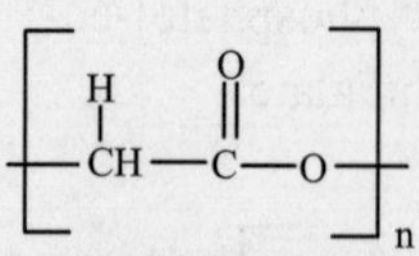

Poly (glycolic acid)

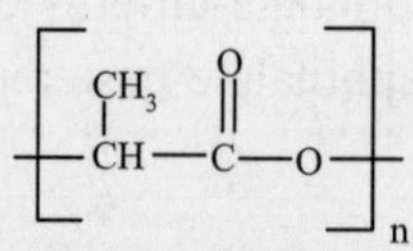

Poly (lactic acid)

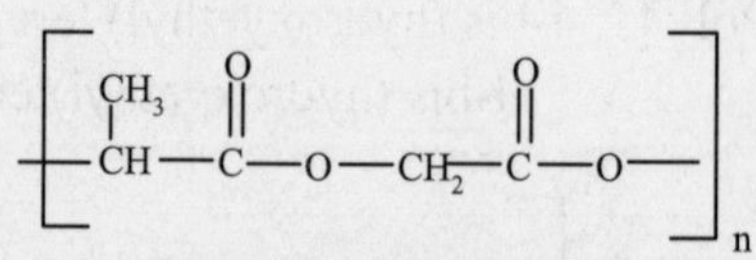

Poly (lactic-co-glycolic acid)

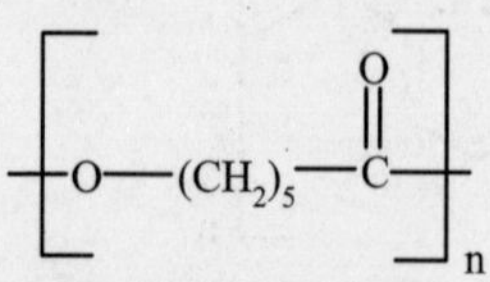

Poly (ε-carprolactone)

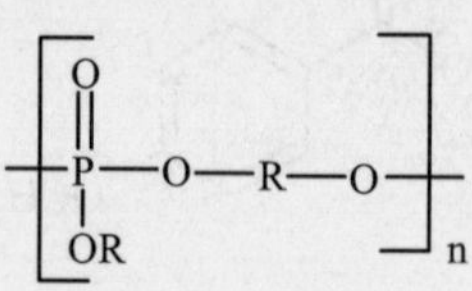

Poly (phosphoesters)

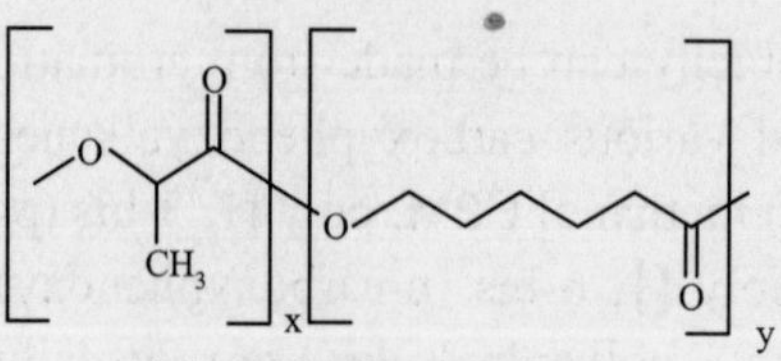

Poly (DL-lactide-co-caprolactone)

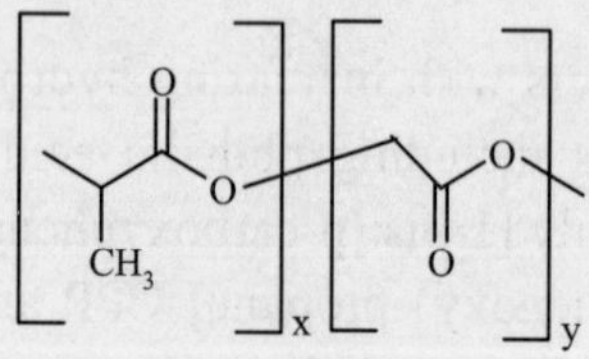

Poly(DL-lactide-co-glycolide)

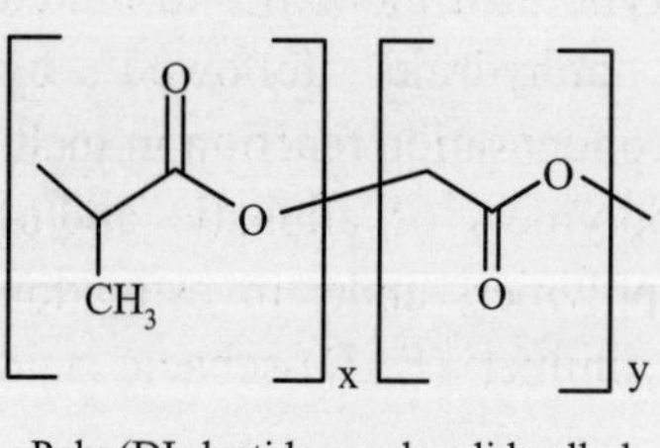

Poly (DL-lactide-co-glycolide alkyl glycolide) acid terminated

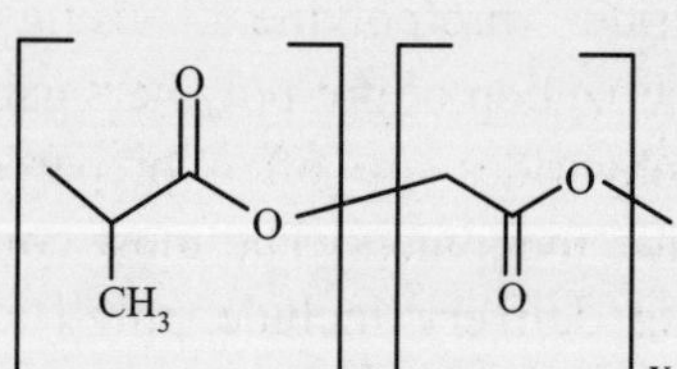

Poly (DL-lactide-co-terminated

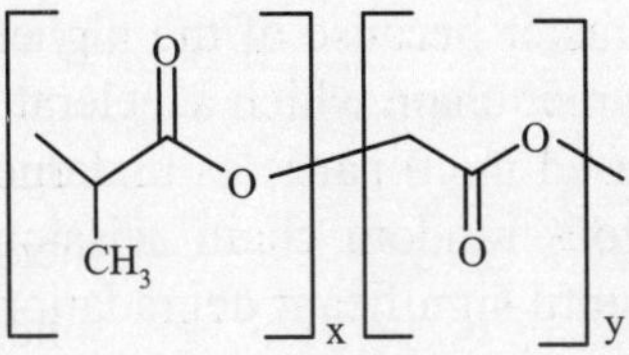

Poly (DL-lactide-co-glycolide alkyl ether terminated

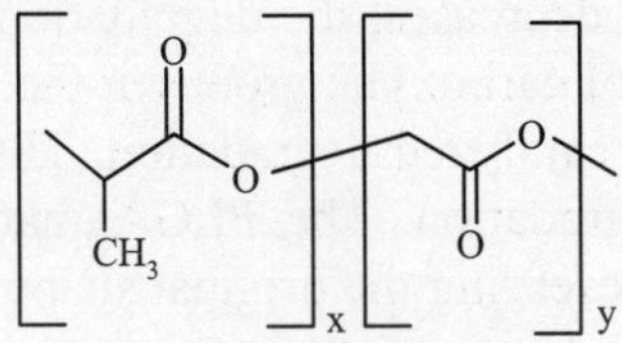

Poly (DL-lactide-co-glycolide ester terminated)

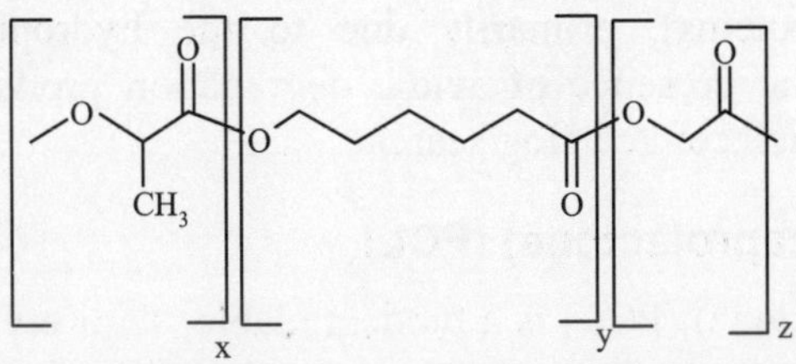

Poly(L-lactide-co-caprolactone-co-glycolide) L-lactide

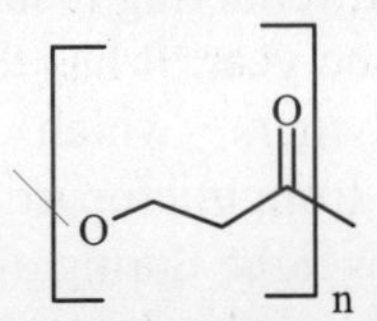

Poly(dioxanone)

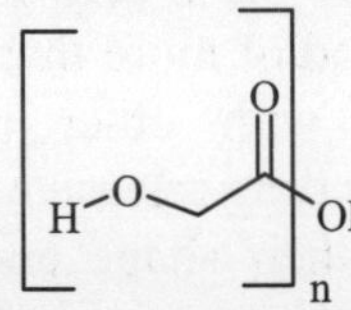

Polyglycolic acid

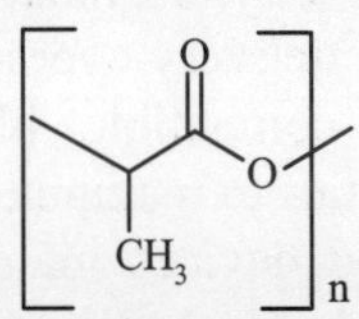

Polylactic acid

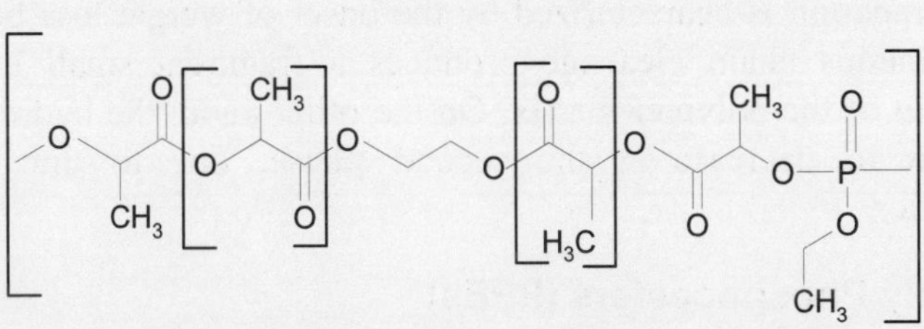

Poly[(lactide-co-ethylene glycol)-co-ethyloxyphosphate]

(Chemical Structures of polyesters)

PLGA polymers are cleaved into monomeric acids (i.e., lactic and glycolic acids), which are subsequently eliminated from the body as carbon dioxide and water. The degradation rate of PLGA is critical for determining the release rate of the encapsulated drug and depends on the crystallinity, hydrophobicity, and molecular weight of the polymer. In general, glycolic acid-rich PLGA copolymers (up to 70%) are amorphous in nature and degrade more rapidly. As the molecular weight of the polymer decreases, the degradation becomes faster because of the higher content of carboxylic groups at the end of polymer chain which accelerate the acid catalyzed degradation. The PLGA-based microparticles undergo bulk degradation. The PLGA matrix undergoes random chain scission while preserving the original shape and mass until significant degradation (~ 90%) has occurred. In spite of these promising characteristics of biodegradation, recent studies have demonstrated that PLGA copolymers significantly affect the stability and biological activity of the drugs (e.g., peptide and proteins), primarily due to the hydrophobicity of the polymers and the presence of acidic degradation products (Figure ref: www.sigmaaldrich.com/catalog/search/.......)

2.1.3 Poly (ε-caprolactone) (PCL)

Poly (ε-caprolactone) (PCL) is a biodegradable, semi-crystalline polymer having a low glass transition temperature (~60 °C). A number of drugs have been encapsulated using PCL. Due to its crystallinity and hydrophobicity, degradation of PCL is very slow, rendering it suitable for long-term delivery over a period of more than one year. It has the ability to form compatible blends with other polymers, which provides opportunities to manipulate the drug release rate from microparticles. The PCL-based devices maintain their shape and weight during the initial phase of biodegradation, where the molecular weight decreases up to 5000 through bulk hydrolysis of the ester bonds. The second phase of PCL degradation is characterized by the onset of weight loss because of the continuous chain cleavage produces a fragment small enough to diffuse out of the polymer matrix. On the other hand, the hydrolysis rate is known to decrease at the second phase, due to the increased crystallinity.

2.1.4 Poly Phosphoesters (PPEs)

Poly-phospho-esters (PPEs) have been used recently for delivery of low molecular weight drugs as well as high molecular weight proteins and DNA. This type of polymer degrades under the physiological conditions via hydrolysis or enzymatic cleavage of the phosphate bonds in the backbone. It is possible to obtain PPEs with a wide range of physicochemical properties; in particular by choosing biocompatible building blocks of the polymer, degradation products of PPEs can have minimal toxic effects and good biocompatibility. The degradation rate of PPEs is controllable by the percentage of the phosphate content in the backbone. The degradation rate increases with increasing the phosphate content of the polymer. In contrast to other polyesters, PPEs are known to degrade by a combined mechanism of surface erosion and bulk degradation. Recent studies have demonstrated that PPE based microparticles are promising for protein delivery because they don't generate acidic environments. Few examples are:

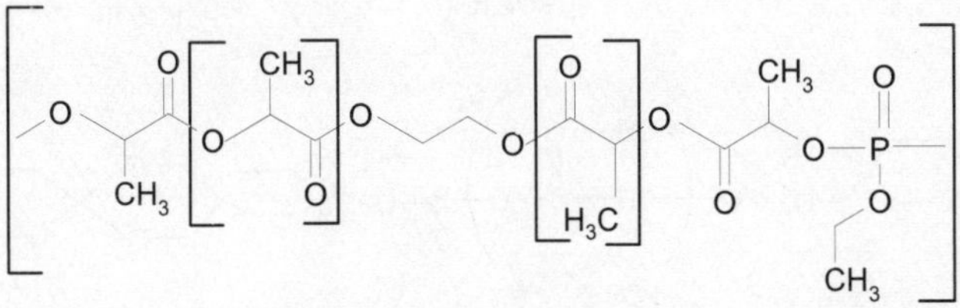

Poly[(lactide-co-ethylene glycol)-co-ethyloxyphosphate]

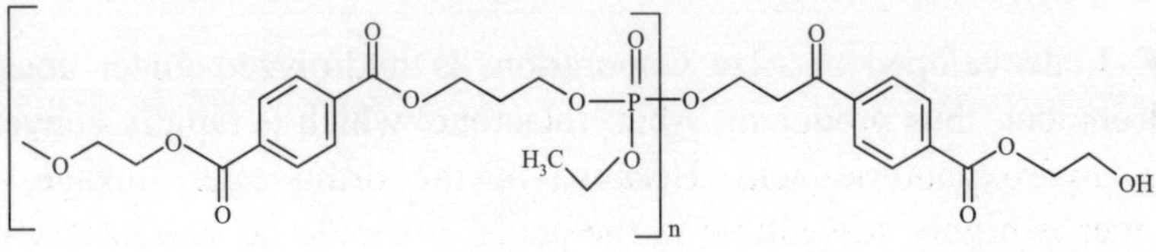

Poly[1,4-bis(hydroxyethyl)terephthalate-alt-ethyloxyphosphate]

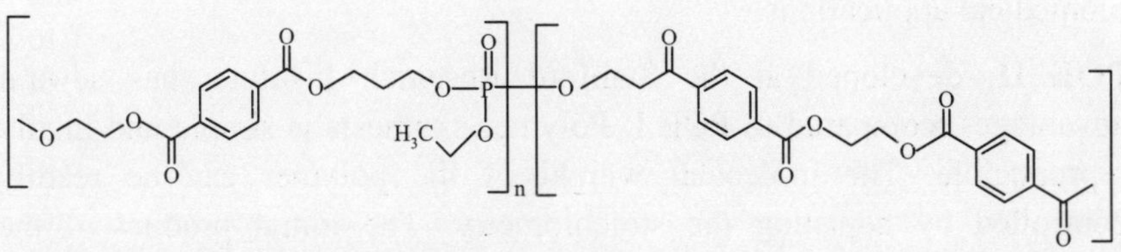

Poly-[1, 4-bis (hydroxyethyl)terephthalate-alt-ethyloxyphosphate]-co-1, 4-bis(hydroxyethyl)terephthalate-co-terephthalate

(Figure ref: www.sigmaaldrich.com/catalog/search/.......)

2.1.5 Poly-ortho Esters (POEs)

Poly-ortho-esters (POEs) categorized into four major classes of biodegradable polymers: POE I, POE II, POE III, and POE IV. POEs undergo surface erosion because of high hydrophobicity and water impermeability. Therefore, depending on the surface erosion rate of the POEs, the drug-loaded device may release the drug at a constant rate without significant burst release

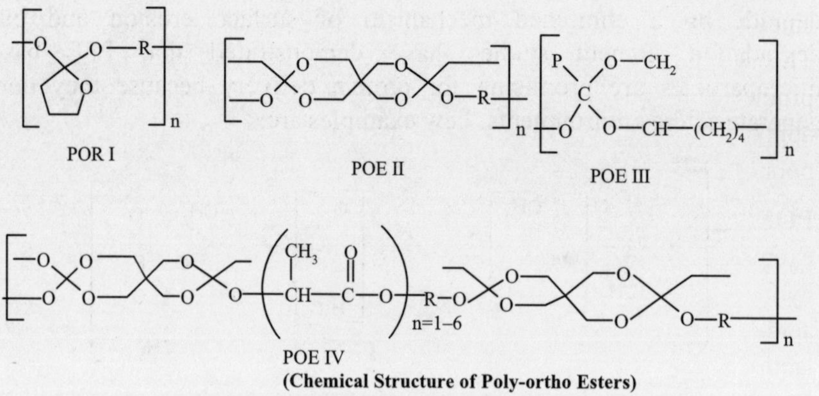

(Chemical Structure of Poly-ortho Esters)

POE I, deeveloped at Alza Coporation, is hydrolyzed under aqueous environment, thus producing γ-butyrolactone, which is rapidly converted to γ -hydroxybutyric acid. Because of the ortho ester linkage, this polymer is highly susceptible to the acids. It should be stabilized with a base such as Na_2CO_3 to prevent an uncontrolled, autocatalytic hydrolytic reaction. Such disadvantages of POE I, leads to limited application in biomedical applications.

POE II, developed at the Stanford Research Institute, has several advantages, compared to POE I. Polymer synthesis is simple and highly reproducible. The molecular weight of the polymer can be readily controlled by adjusting the stoichiometry. The initial product of the polymer hydrolysis is neutral, and thus it is not necessary to use a basic excipient. However, the polymers belonging to this family are extremely hydrophobic, which limits the access of water to the hydrolytically labile ortho-ester linkage. Hence, in order to achieve the increase in the surface erosion rate, it is necessary to incorporate acidic excipients into the polymer matrix. This limitation of POE II makes it difficult to design surface eroding devices because of the presence of acidic excipients, which may often accelerate the autocatalytic reaction.

POE III, developed at the Stanford Research Institute, is a semisolid material at room temperature. This semisolid material enables to prepare the injectable drug delivery system by simple mixing with the therapeutic agents without the need of using organic solvents or elevated temperatures. In addition, no autocatalysis occurs during degradation, since the initial hydrolysis generates one or more isomeric monoesters. The ortho ester bonds of the polymer are only sensitive to the acidic products. Despite these advantages of POE III, its biomedical applications have been limited due to difficulties in the synthesis and poor reproducibility of the synthesized polymers.

POE IV is a modified POE II, which can be used without acidic excipients. When POE IV, exposed to an aqueous environment, hydrolysis proceeds in three consecutive steps such as: First, carboxylic acid-terminated polymer fragments are produced by hydrolysis of the lactic acid or glycolic acid segment in the polymer backbone; Second, free-hydroxy acids are generated and they catalyze hydrolysis of the ortho ester linkages; and Third, the ortho esters are cleaved into the diol and pentaerythritol dipropionate, followed by ester hydrolysis to produce pentaerythritol and propionic acid. The erosion process of POE IV-based devices is primarily confined to the surface layers, and this property provides a number of advantages, one such example is to control the release behavior of the drug by manipulating the surface erosion rate of the polymer.

2.1.6 Polyphosphazenes

Polyphosphazenes are one of the most versatile and rapidly developing classes of biomedical Polymers. They are synthesized as linear polymers, composed of an inorganic backbone with nitrogen and phosphorous atoms. Since the synthetic method of the linear polydichlorophosphazene (PDPP) was found, a number of biodegradable polyphosphazenes have been developed by reaction of the highly reactive phosphorus-chlorine bonds of PDPP with alkoxide, primary (or secondary) amines, and organometallic reagents. Because of there are numerous substituents capable of being introduced into the backbone, a broad spectrum of polyphosphazenes can be synthesized by choosing the type and ratios of appropriate side groups. When exposed to an aqueous solution, these polymers are cleaved into nontoxic, low molecular weight products such as phosphates, ammonia and the corresponding side groups. They can degrade by both surface and bulk erosion, depending on the ability of the bond and hydrophobicity of the polymer. A few examples of synthesized molecules for biomedical applications are: polyposphazene-bearing amino acid ester, imidazole, glucosyl amino, glycolic acid ester, and lactic acid ester side groups. Some examples are:

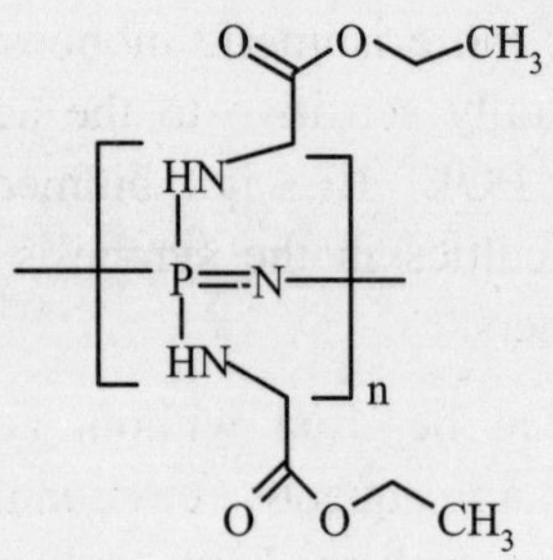

Poly (bis(1-(ethoxycarbonyl) methylamino)phosphazene)

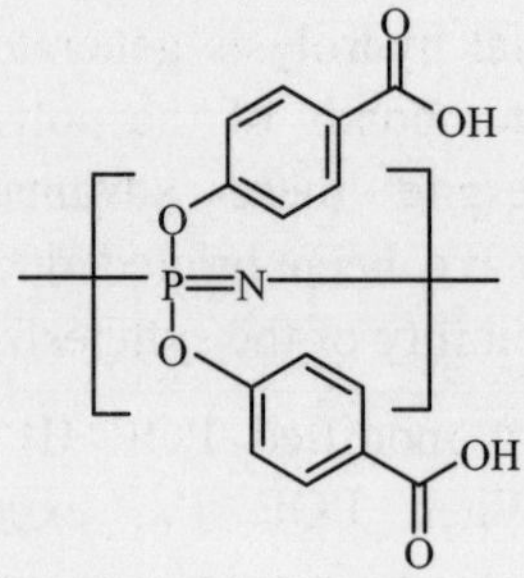

Poly-(bis(4-carboxyphenoxy) phosphazene)

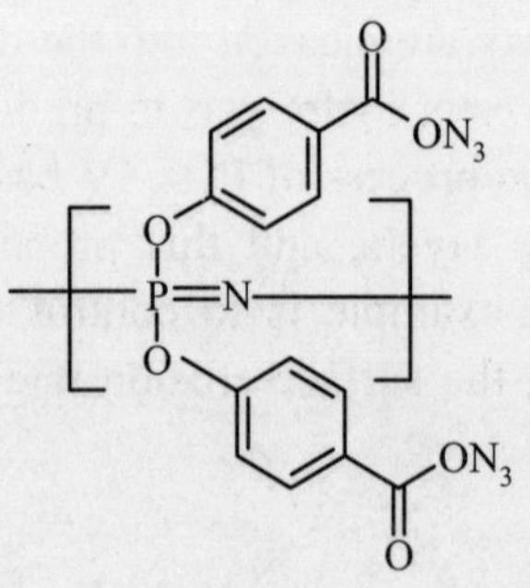

Poly-(bis(4-carboxyphenoxy) phosphazene) disodium salt

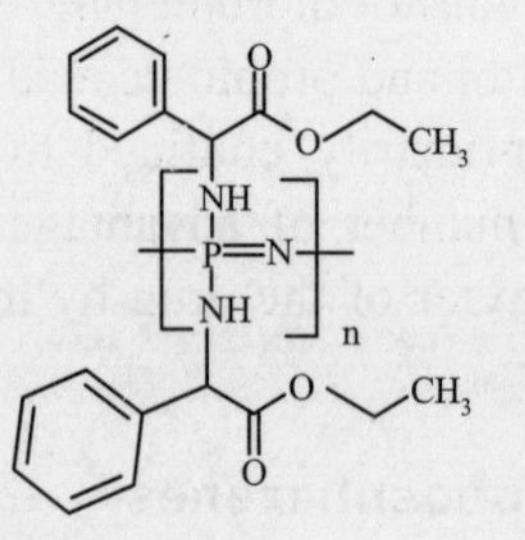

Poly[bis(1-(ethoxycarbonyl)-2-phenylethylamino)phosphazene]

(Chemical structure of Polyphosphazenes)

2.1.7 Block Copolymers

Multi-block Poly-(ether-ester)s based on poly-ethylene glycol (PEG), butylene terephthalate (BT) and butylene succinate (BS) have been developed as a new series of degradable polymers for controlled release applications. These poly-(ether-ester)s are the modification of poly-ethylene glycol-terephthalate(PEGT) / poly-butyleneterephthalate (PBT) copolymers, which have been successfully used as matrix in controlled release systems both *in-vitro* and *in-vivo*. However the degradation rate of some PEGT/PBT copolymers composition may be too slow. Substitution of the aromatic terephthalate units by aliphatic succinate units may be shown to increase the degradation rate of copolymers.

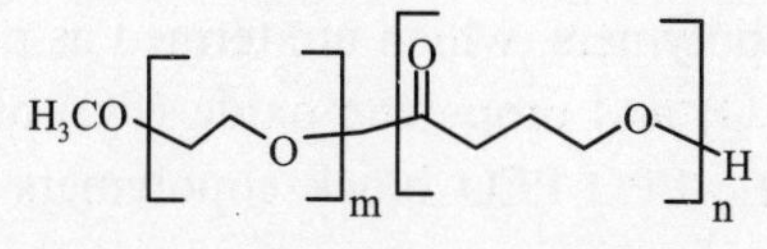

Poly-(ethylene glycol)-block-poly (ε−caprolactone) methyl ether PEG

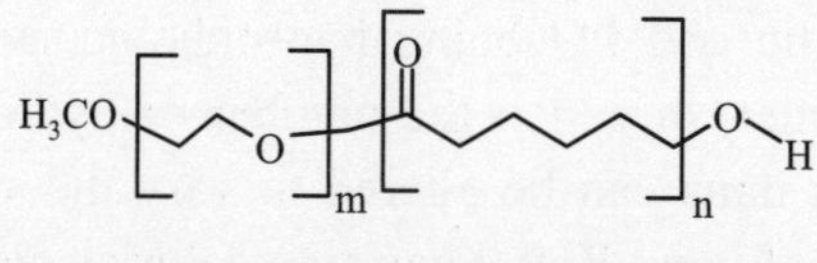

Poly-(ethylene glycol)-block-poly-lactide methyl ether PEG

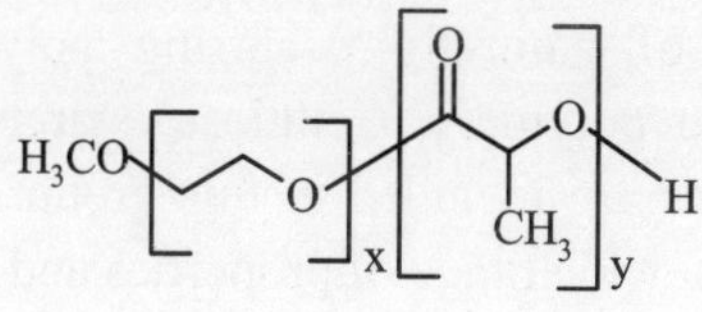

Polylactide-block-poly-(ethylene glycol)-block-polylactide PLA

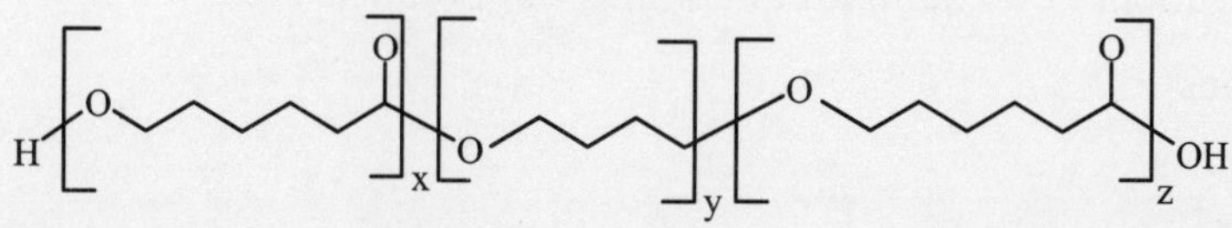

Polycaprolactone-block-polytetrahydrofuran -block-polycaprolactone

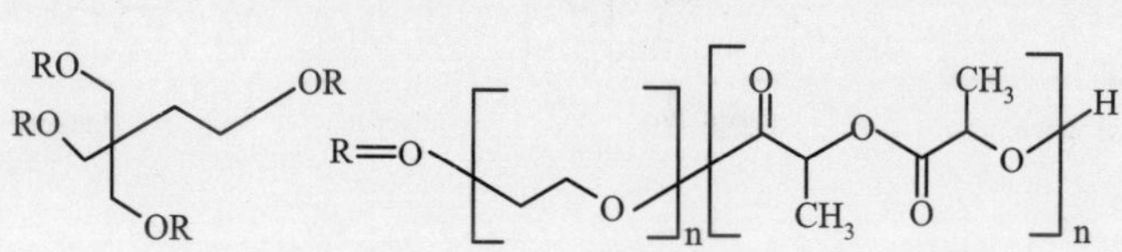

Poly-(ethylene oxide)-block-polylactide

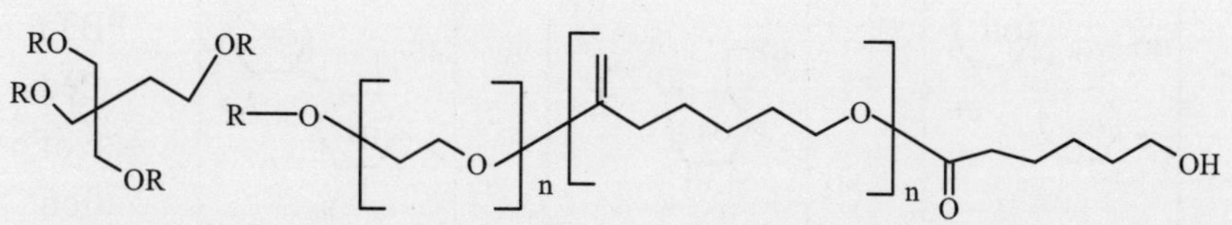

Poly-(ethylene oxide)-block-Polycaprolactone

Pluronic block copolymers, which are termed as poloxamers, consists of ethylene oxide (EO) and propylene oxide (PO) blocks arranged in a triblock structure. PEo-PPO-PEO block copolymers have been used in pharmaceutical formulations because of their ability to form self-aggregation, thereby forming micelles and liquid crystalline phases. Poloxamers consist of EO and PO blocks arranged in a triblock structure EO being hydrophilic and PO being hydrophobic, which results in an amphiphilic copolymer, in which the number of hydrophilic EO(X) and hydrophobic PO(Y) units can be altered to vary the size, hydrophilicity and hydrophobicity. (Figure Ref: www.sigmaaldrich.com/....)

2.1.8 Natural Polymers

This group consists of naturally occurring polymers and chemical modifications of these polymers. Cellulose, starch, lignin, chitin, and various polysaccharides are included in this group. These materials and their derivatives offer a wide range of properties and applications. Natural polymers tend to be readily biodegradable, although the rate of degradation is generally inversely proportional to the extent of chemical modification. Few examples of commercial products are:

Starches

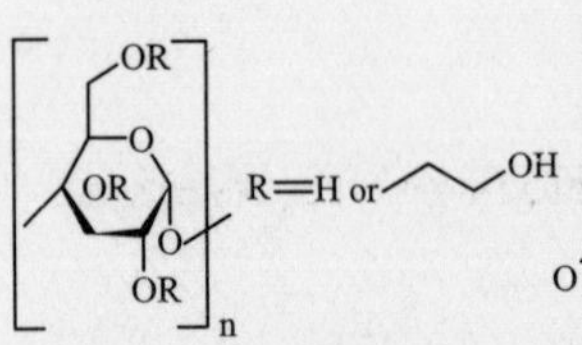

2-Hydroxyethyl starch

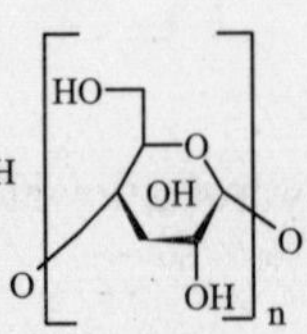

Amylose

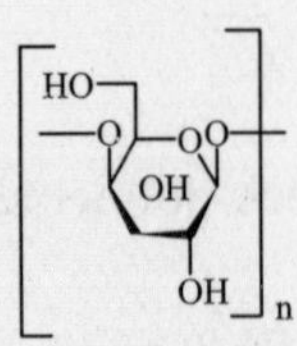

Galactan

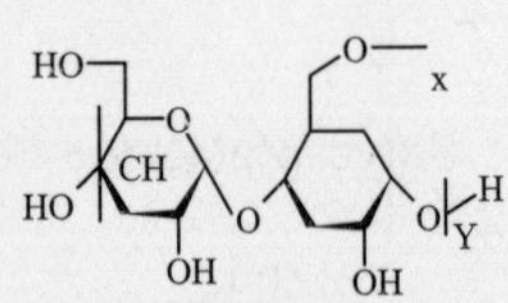

Maltodextrin

Celluloses

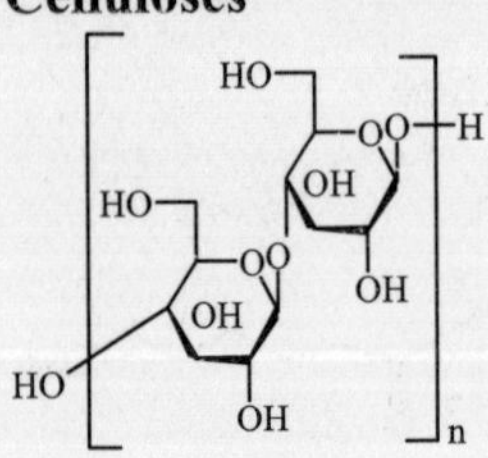

Cellulose Microcrystalline

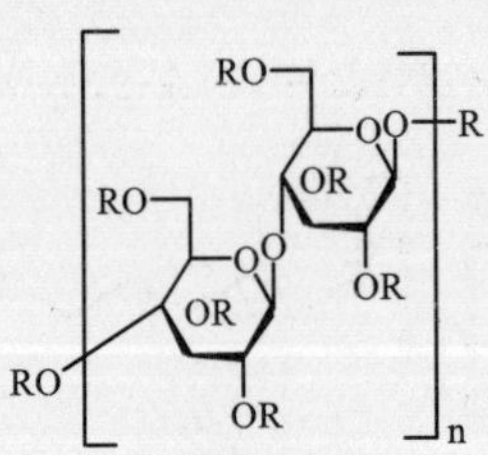

RH or CH

Methyl cellulose

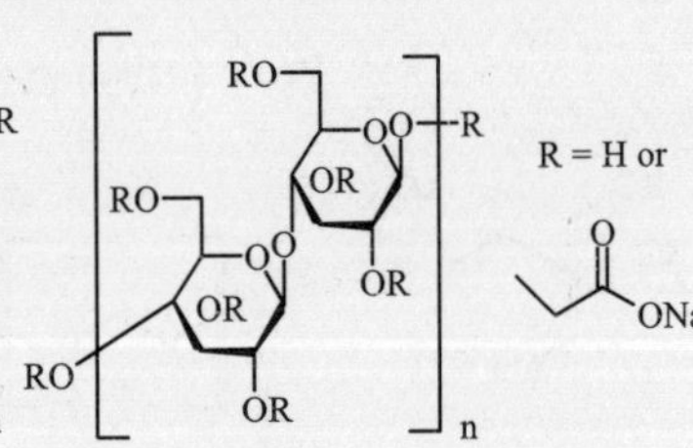

Sodium carboxymethyl cellulose

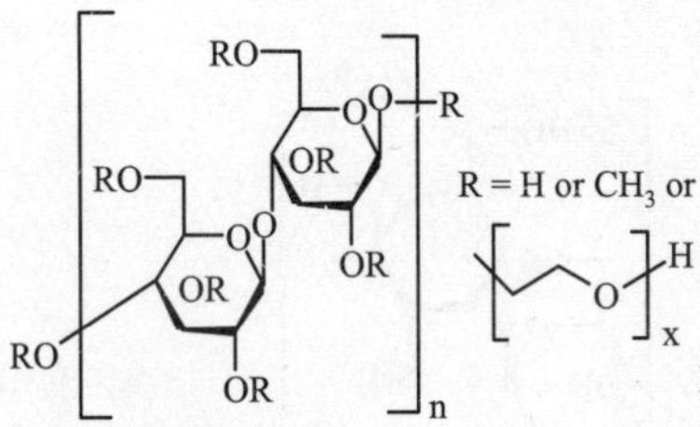

Methyl 2-hydroxyethyl Cellulose cellulose

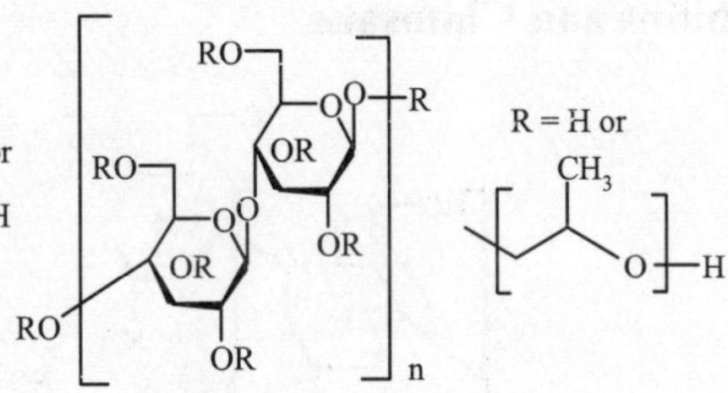

Hydroxypropyl ethoxylate

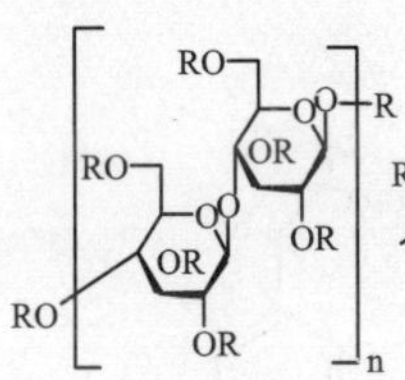

Ethyl cellulose

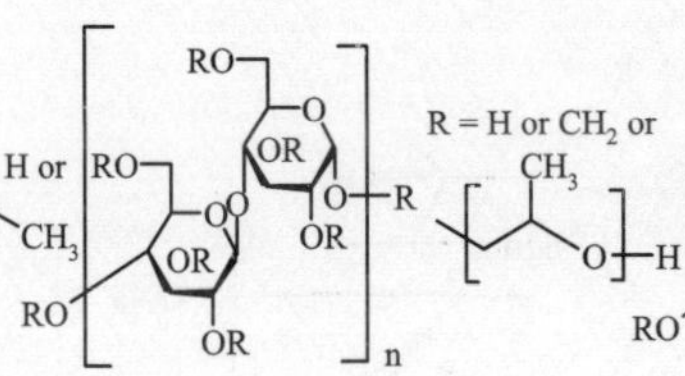

Hydroxy-propyl methyl Cellulose

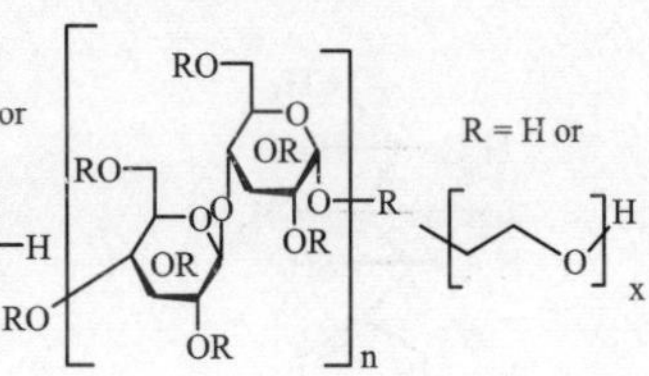

2-Hydrxy-ethyl cellulose

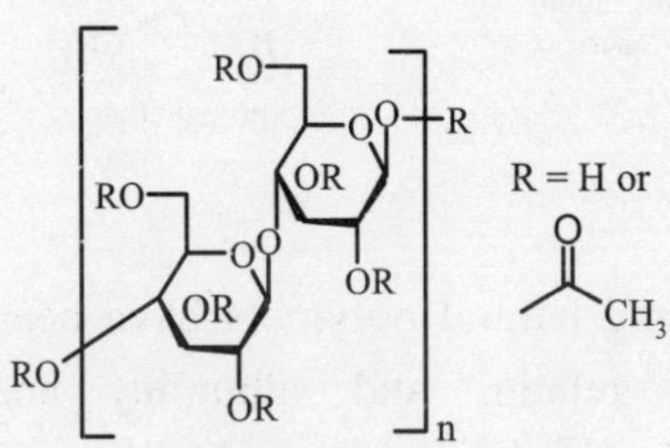

Cellulose acetate

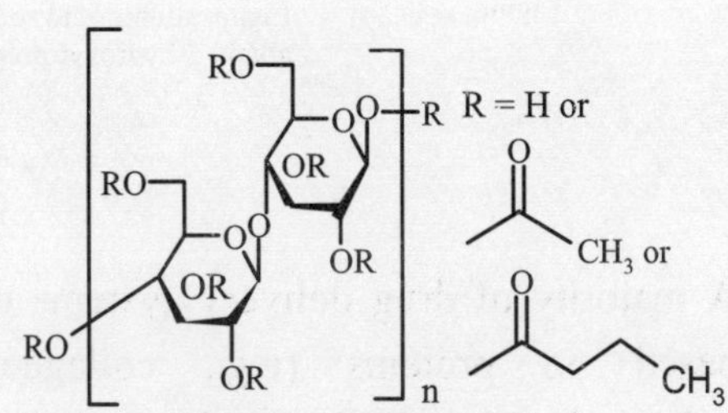

Cellulose acetate butyrate

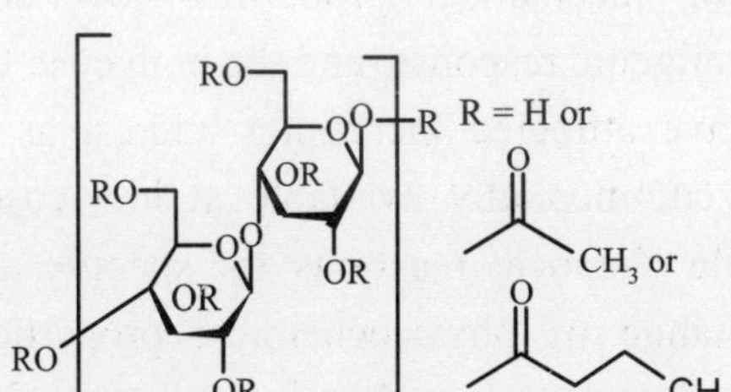

Cellulose acetate propionate

Chitins and Chitosans

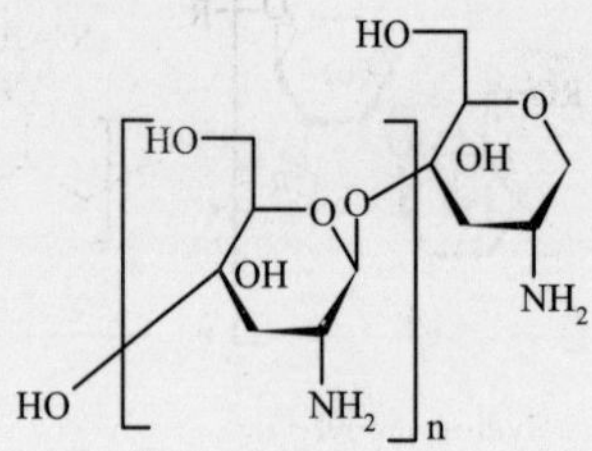

Chitosan

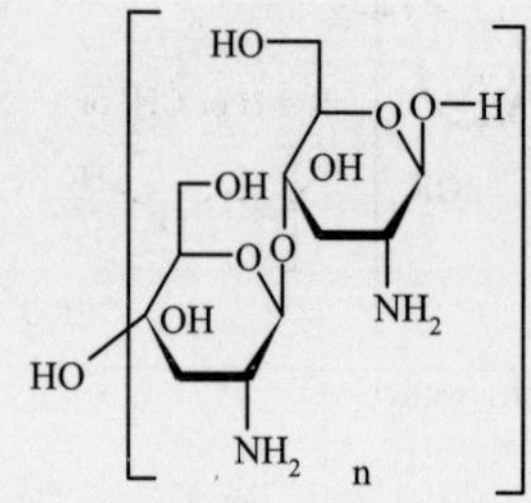

Chitosan Oligosaccharide lactate

Lignins

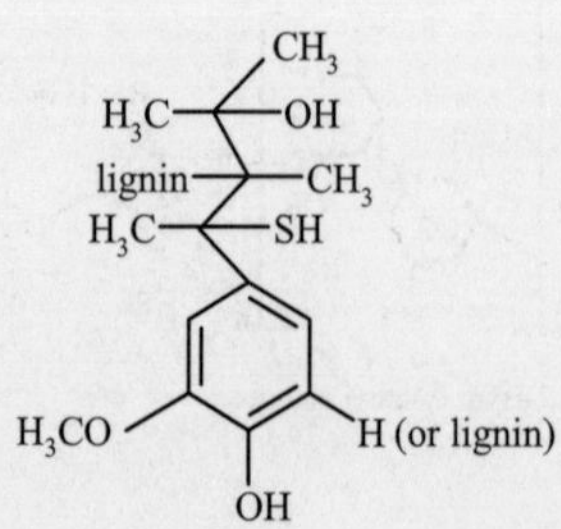

Lignin

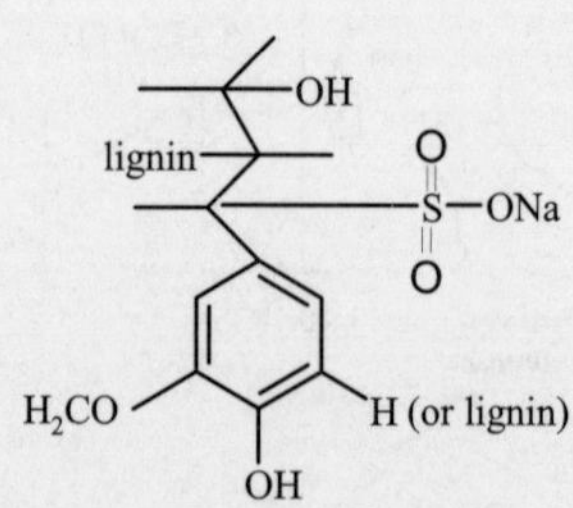

Lignosulfonic acid sodium Lignin, alkali, 2-hydroxyl-propyl eher

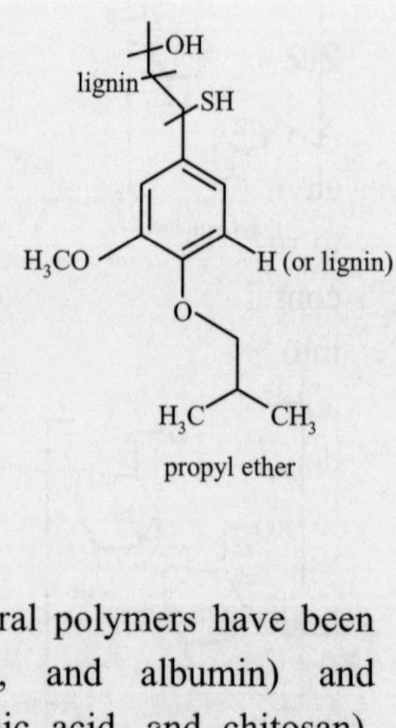

propyl ether

A majority of drug delivery systems using natural polymers have been based on proteins (e.g., collagen, gelatin, and albumin) and polysaccharides (e.g., starch, dextran, hyaluronic acid, and chitosan). Applications of proteins to delivery of protein drugs have been limited due to their poor mechanical properties, low elasticity, possible occurrence of an antigenic response, and the high cost. On the other hand, polysaccharides have attracted increasing interest as the drug carriers because they are commercially available at low cost and are readily modified by simple chemical reactions for specific applications. They exhibit a broad range of physicochemical properties. For example: chitosan, primarily composed of 2-amino-2-deoxy- β-D-glucopyranose (D- glucosamine), is obtained from chitin, which is the second most abundant natural polysaccharide. Chitosan and its derivatives have showed excellent biocompatibility, biodegradability, low immunogenicity, and biological activities. In particular, their biodegradability can be precisely controlled by modifying the structures with acetic anhydride. Since they contain primary amino groups in the main backbone that make their surfaces positive in the biological fluid, biodegradable microparticles can be readily prepared by treating them with a variety of biocompatible polyanionic substances such as sulfate, citrate, and tripolyphosphate. These unique features of chitosan have stimulated development of delivery systems for a wide range of biological agents. (Figure Ref: www.sigmaaldrich.com/catalog/....)

2.2 Bio-Degradation of Polymers

A variety of natural, synthetic, and biosynthetic polymers are bio and environmentally degradable. A polymer based on a C-C backbone tends to resist degradation, whereas heteroatom-containing polymer backbones confer biodegradability. Biodegradability can therefore be engineered into polymers by the judicious addition of chemical linkages such as anhydride, ester, or amide bonds, among others. The usual mechanism for degradation is by hydrolysis or enzymatic cleavage of the labile heteroatom bonds, resulting in a scission of the polymer backbone. Macroorganisms can eat, and sometimes digest polymers, and also initiate a mechanical, chemical, or enzymatic aging.

Biodegradable polymers with hydrolysable chemical bonds are researched extensively for biomedical, pharmaceutical, agricultural, and packaging applications. In order to be used in medical devices and controlled-drug-release applications, the biodegradable polymer must be biocompatible and meet other criteria to be qualified as processable, sterilizable and capable of controlled stability or degradation in response to biological conditions. The chemical natures of the degradation products, rather than that of the polymer itself, often critically influence biocompatibility. Poly (esters) based on polylactide (PLA), polyglycolide (PGA), polycaprolactone (PCL), and their copolymers have been extensively employed as biomaterials. Degradation of these materials yields the corresponding hydroxy acids, making them safe for *in-vivo* use.

Mechanism of Biodegradation

2.2.1 Hydrolysis

(Depend on main chain structure: anhydride > ester > carbonate)

A.

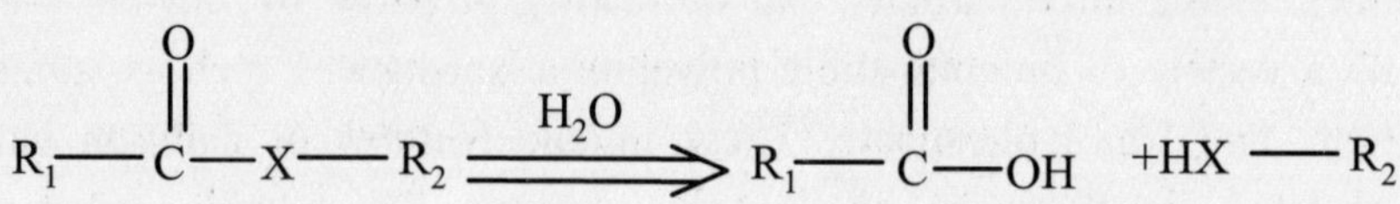

where, X = O, N, S

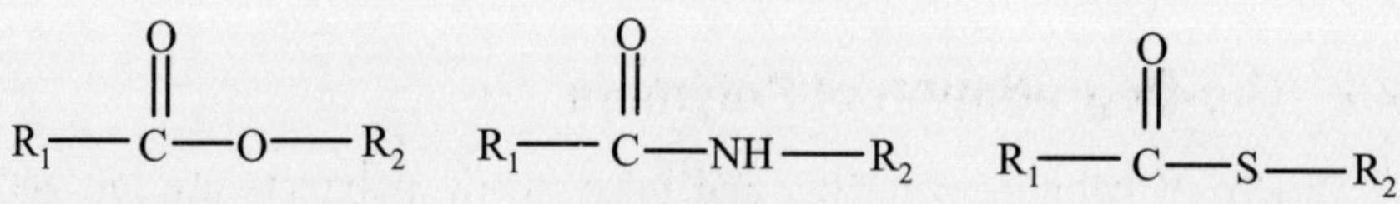

Ester Amide Thioester

B.

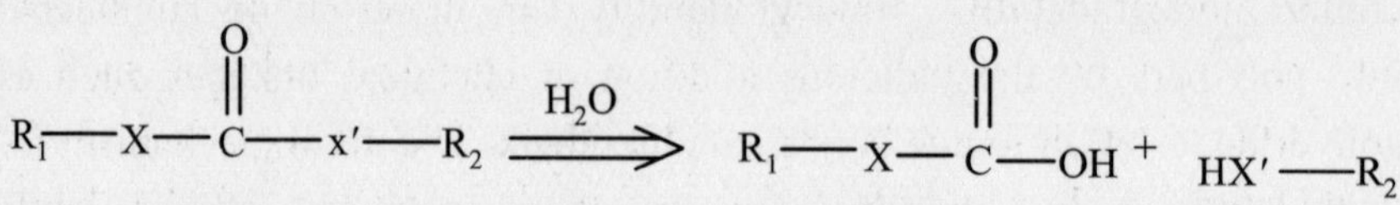

where, X and X' = O, N, S

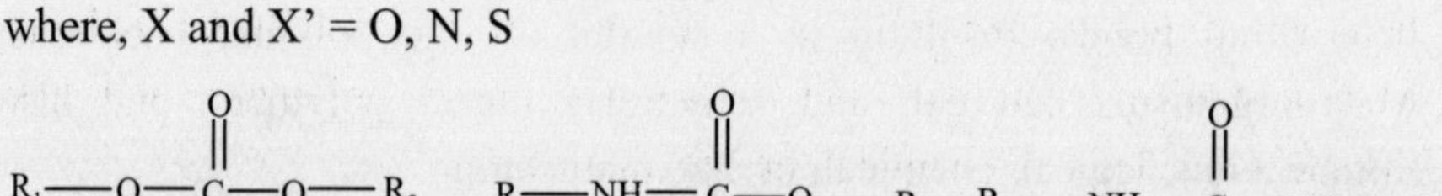

Carbonate Urethane Urea

C.

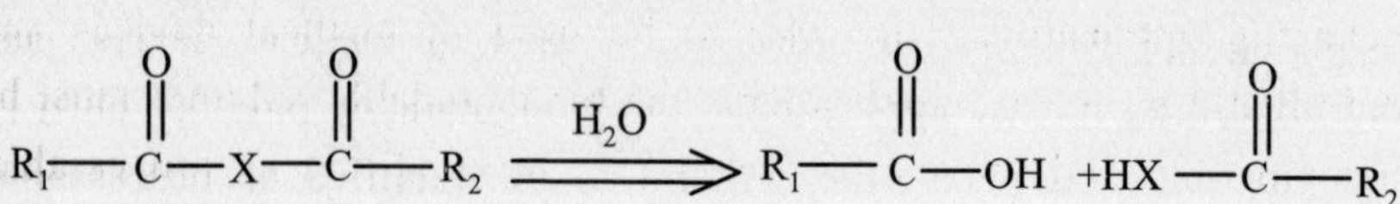

where, X and X' = O, N, S

Imide Anhydride

Examples:

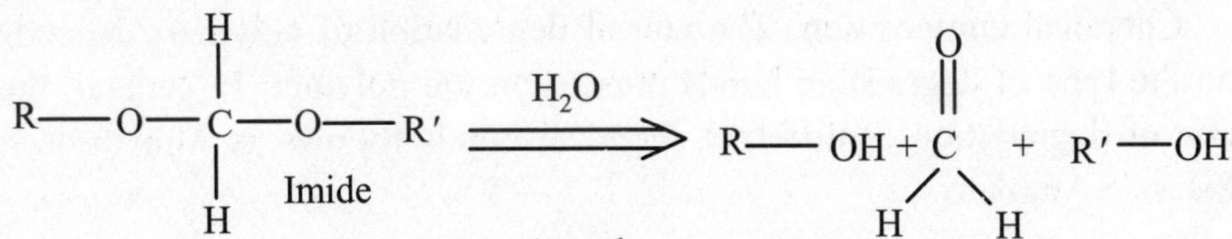

Acetals

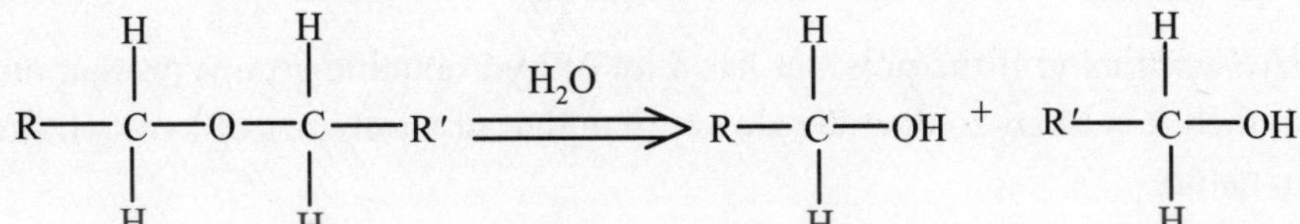

Ether

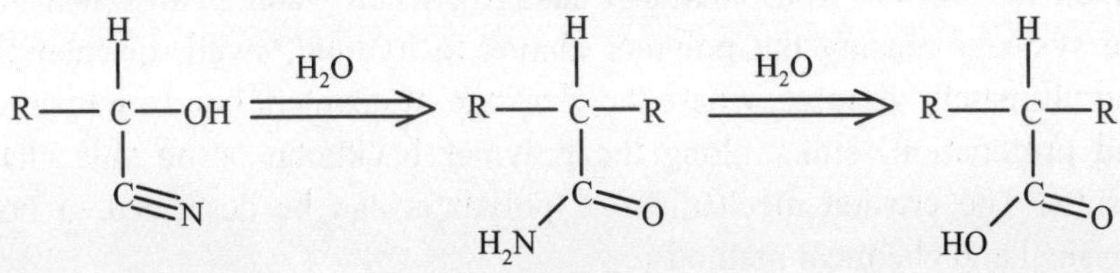

Nitrile

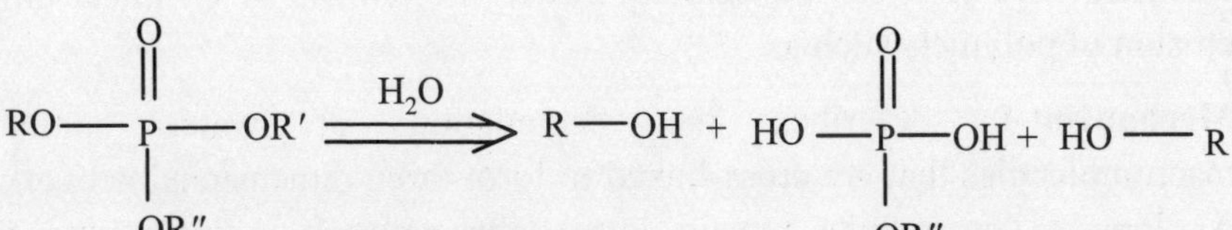

Phosphonate

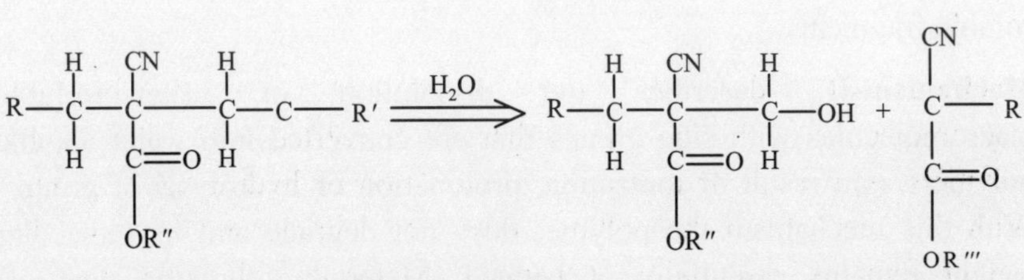

Polycyanoacrylate

Degradation of lactide based polymers and in general all hydrolytically degradable polymers, depends on the following properties:

Chemical composition: The rate of degradation of polymers depends on the type of degradable bonds present on the polymer. In general, the rate of degradation of different chemical bonds follows as Anhydride > Esters > Amides.

Crystallinity: Higher the crystallinity of a polymer, slower is its rate of degradation.

Hydrophilicity: If the polymer has a lot of hydrophobic groups present on it, then it is likely to degrade slower than that of a polymer of hydrophilic in nature.

2.2.2 Bioerosion

Bioerosion results from chemical changes, when water is absorbed into the systems causing the polymer chains to hydrate, swell, disentangle, and ultimately dissolve, where the cleavage of covalent bonds, ionization and protonation, either along the polymer backbone or on side chain results. The erosion mechanism of polymers can be described in both physical and chemical methods.

***Chemical Erosion*:** Heller, describes three mechanism of chemical bio-erosion of polymers such as:

Mechanism-I, describes the degradation of water soluble macromolecules that are cross-linked to form three-dimensional network. As long as cross-links remain intact, the network is intact and is insoluble. Degradation in these systems can occur either at crosslinks to form soluble backbone polymeric chain or at the main chain to form water soluble fragments, which results in high molecular weight, water soluble fragments.

Mechanism-II, describes the dissolution of water-insoluble macromolecules with side groups that are converted into water soluble polymers as a result of ionization, protonation or hydrolysis of groups. With this mechanism the polymer does not degrade and its molecular weight remains essentially unchanged. Materials following type –II erosion include cellulose acetates and partially esterified copolymers of maleic anhydride. These polymers become soluble by ionization of carboxylic groups.

Mechanism-III, describes the degradation of insoluble polymers with liable bonds. Hydrolysis of labile bonds causes scission of the polymer backbone, thereby forming low molecular weight, water soluble molecules. Polymers following type-III erosion include Poly lactic acid, Poly glycolic acid and their copolymers, polyorthoesters, polyamides, polycynoacrylates and polyanhydrides.

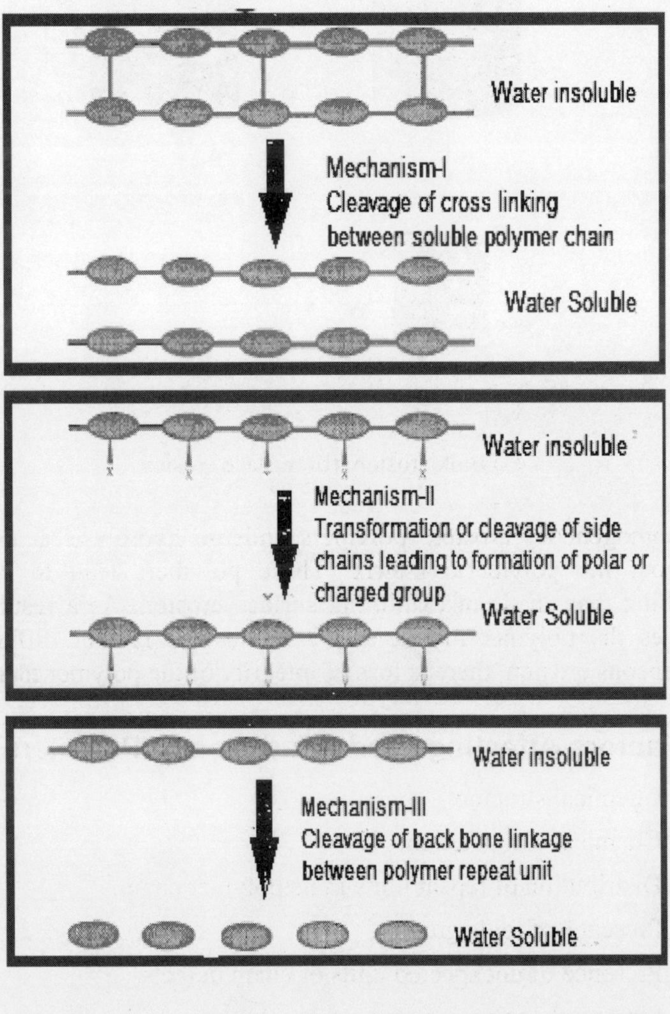

Physical Erosion: The physical erosion mechanisms can be characterized by heterogeneous or homogeneous type erosion.

In heterogeneous erosion, also called as surface erosion, the polymers erode at the surface, and maintained its physical integrity as its degradation proceeds. The mostly crystalline polymers undergo heterogeneous erosion following zero order release kinetics.

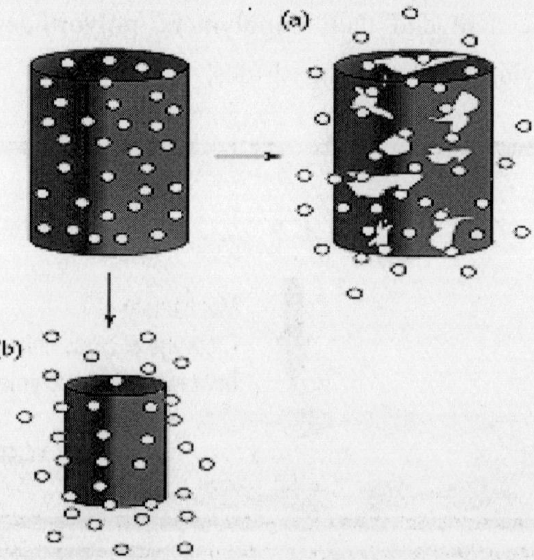

(a) Bulk erosion; (b) Surface erosion

In homogeneous erosion, polymers undergo hydrolysis at even rate throughout the polymeric matrix. These polymers tend to be more hydrophilic than those of exhibiting surface erosion. As a result, water penetrates the polymer matrix and increases the rate of diffusion. In homogeneous erosion, there is loss of integrity of the polymer matrix.

2.3 Factors Affecting Biodegradation of Polymers

- Chemical structure.
- Chemical composition.
- Distribution of repeat units in its polymer chain.
- Presence of ionic groups.
- Presence of unexpected units or chain defects.
- Structural configuration
- Molecular weight.

- Morphology (amorphous/semi crystalline, microstructures, residual stresses).
- Presence of low-molecular-weight compounds.
- Processing conditions.
- Annealing.
- Sterilization process.
- Storage history.
- Shape.
- Site of implantation.
- Adsorbed and absorbed compounds (water, lipids, ions, etc.).
- Physicochemical factors (ion exchange, ionic strength, pH).
- Physical factors (shape and size changes, variations of diffusion coefficients, mechanical stresses, stress- and solvent-induced cracking, etc.).
- Mechanism of hydrolysis (enzymes versus water).

2.4 Methods of Studying Polymer Degradation

- Morphological changes (swelling, deformation, bubbling and disappearance)
- Scanning Electron microscopy(SEM)
- Polymer Chain Reaction Study(PCR)
- Surface Characterization: X-ray Photoelectron Spectroscopy (XPS) and Secondary Ion Mass Spectrometry (SIMS)
- Weight loss
- Thermal behavior changes
- Differential Scanning Calorimetry (DSC)
- Thermo Gravimetric Analysis (TG)
- Thermo Mechanical Analysis(TMA)
- Molecular weight changes (Osmometry)
- Dilute solution viscosity (Viscometer)
- Size exclusion chromatography(SEC)

- Gel permeation chromatography(GPC)
- Mass spectroscopy
- Change in chemistry
- Infrared spectroscopy (IR)
- Nuclear Magnetic Resonance Spectroscopy (NMR)

Material selection

When investigating the selection of the polymer for biomedical applications, important criteria to consider are:

- The mechanical properties must match the application and remain sufficiently strong until the surrounding tissue has healed.
- The degradation time must match the time requirement.
- It does not invoke a toxic response.
- It is metabolized in the body after fulfilling its purpose.
- It is easily processible in the final product with an acceptable shelf life and if required easy sterility can be obtained.

Mechanical performance of a biodegradable polymer depends on various factors, which include:

The factors such as: selection of polymer, processing condition and the presence of additives, which may influence the polymers crystallinity, melt and glass transition temperatures and molecular weight. Each of these factors needs to be assessed on how they affect the biodegradation of the polymer. Biodegradation can be accomplished by synthesizing polymers with hydrolytically unstable linkages in the backbone. This is commonly achieved by the use of chemical functional groups such as esters, anhydrides, ortho-esters and amides. Most biodegradable polymers are synthesized by ring opening polymerization.

2.5 Biomedical Applications

Many opportunities exist for the application of synthetic biodegradable polymers in the biomedical area particularly in the fields of tissue engineering and controlled drug delivery Degradation is important in biomedicine for many reasons. Degradation of the polymeric implant means surgical intervention may not required for removal at the end of its functional life (eliminating don't need for a second surgery). In tissue engineering, biodegradable polymers can be designed as such to approximate tissues, providing a polymer scaffold that can withstand mechanical stresses, provide a suitable surface for cell attachment and growth, and degrade at a rate that allows the load to be transferred to the new tissue. In the field of controlled drug delivery, biodegradable polymers offer tremendous potential either as a drug delivery system alone or in conjunction to functioning as:

- Medical device
    - Wound management
    - Sutures
    - Staples
    - Clips
    - Adhesives
    - Surgical meshes
    - Orthopedic devices
    - Pins
    - Rods
    - Screws
    - Tacks
    - Ligaments
- Dental applications
    - Guided tissue regeneration Membrane
    - Void filler following tooth extraction
- Cardiovascular applications
    - Stents
- Intestinal applications
    - Anastomosis rings
- Drug delivery system
- Tissue engineering

An Over View of Different Technologies and Products based on Biodegradable Polymers:

Biodegradable Polymer-Metal Complexes for Gene and Drug Delivery: Hossein,et. Al., 2009, stated that a polymer- metal complex significantly enhanced the gene expression of cells in vivo treated with dextran or pullulan plasmid DNA complexes. Pullulan and Dextran are watersoluble polysaccharides with a repeated unit of maltotriose condensed through -1, 6 linkage and with multiple hydroxyl groups applicable to chemical modification, having low immunogenicity and a long history in clinical use as plasma expanders. Such complexes have been extensively used for pharmaceutical applications to demonstrate the feasibility of passive tumor targeting of drugs. They also demonstrated that chemical conjugation with this pullulan, enabled interferon (IFN) to target to the liver and induce the IFN-specific enzyme. However, the chemical conjugation often causes the denaturation and deactivation of IFN molecules. Therefore, metal coordination has been tried for conjugation of the IFN with pullulan. As a result, they have succeeded in targeting IFN to the liver through conjugation of pullulan with DTPA residues based on metal coordination and inducing the IFNspecific activity therein. This metal coordination was also effective in conjugating tumor necrotizing factor (TNF) with DTPA-introduced dextran and consequently targeting TNF to the tumor tissue for elucidation.

Biodegradable polymer nanoparticles that rapidly penetrate the human mucus barrier: Tang *et. al.*, 2009, synthesized PSA-PEG diblock copolymers (MW~18 kDa) via melt polycondensation of sebacic acid (SA) and methoxy-PEG prepolymers. They used 5 kDa PEG based on their previous observations that 2–5 kDa PEG provided a nonmucoadhesive coating on non-degradable latex beads, whereas coatings using 10 kDa PEG resulted in strong particle muco-adhesion.

Zoladex® (Astra Zeneca) is supplied as a sterile biodegradable product containing goserelin acetate equivalent to 3.6 mg of goserelin dispersed in a matrix of D, L-lactic and glycolic acids copolymer (13.3-14.3 mg/dose) designed to release over a 20 days period.

Lupron depot® (TAP Pharmaceutica Inc.) is a PLGA product, which is an emulsion consists of leuprorelin acetate in an aqueous solution containing gelatin dispersed in a solution of PLG in methylene chloride to form microsphere for sustained release.

Gliadel® wafer (Guilford Pharmaceuticals, Baltimore, MD) produced carmustine loaded glidel wafer fabricated from poly(carboxy phenoxy propane; sebacic acid) proved to be promising for the treatment of malignant glioma. Up to eight gliadel wafers are implanted in the cavity, where tumor resided.

Alzamer® depot Technology offers a non aqueous polymer solution for sustained delivery of biopharmaceuticals particularly therapeutic proteins for period of weeks to months.

Regel® depot Technology is one of the Macromed's proprietary drug delivery system based on triblock copolymer, composed of poly(lactide-co-glycolide)-poly(ethylenr glycol)-poly(lactide-co-glycolide). It is a family of thermally reversible gelling opolymers deloped for parenteral delivery. **Oncogel®** is a frozen formulation of paclitaxol in Regel. **Cytoryn®** is a novel, peritumoral, injectable depot formulation of interleukin-2 for cancer immunotherapy using Regel drug delivery system.

References

Jae Hyung Park, Mingli Ye and Kinam Park. Biodegradable Polymers for Microencapsulation of Drugs. *Molecules,* 2005(10):146-161

Hossein Hosseinkhani and Mohsen Hosseinkhani, Biodegradable Polymer-Metal Complexes for Gene and Drug Delivery. *Current Drug Safety*, 2009 (4): 79-83

Benjamin C. Tang, Michelle Dawson, Samuel K. Lai, Ying-Ying Wang, Jung Soo Suk, Ming Yang, Pamela Zeitlin, Michael P. Boyl, Jie Fu, and Justin Hanes, Biodegradable polymer nanoparticles that rapidly penetrate the human mucus barrier. *PNAS*, 2009, 106(46): 19268–19273

Zhao, Z.,Wang, J., Mao, H. Q., Leong, K. W. Polyphosphoesters in drug and gene delivery. *Adv. Drug Deliv. Rev*. 2003 (55): 483-499.

Zhao, Z., Wang, J., Mao, H. Q., Leong, K. W. Polyphosphoesters in drug and gene delivery. *Adv. Drug Deliv. Rev*. 2003 (55): 483-499.

V. B. Kotwal, M. Saifee, N. Inamdar and K. Bhsde. Biodegradable polymers: Which, When, Why? *Ind J Pharm Sci*, 2007 (5):616-625.

Zhang, Y., Chu, C. C. In vitro release behavior of insulin from biodegradable hybrid hydrogel networks of polysaccharide and synthetic biodegradable polyester. *J. Biomater. Appl.* 2002 (16): 305-325.

Lakshmi, S., Katti, D. S.; Laurencin, C. T. Biodegradable polyphosphazenes for drug delivery applications. *Adv. Drug. Deliv. Rev.* 2003 (55): 467-482.

Abraham, G. A.; Gallardo, A.; San Roman, J.; Fernandez-Mayoralas, A.; Zurita, M.; Vaquero, J. Polymeric matrices based on graft copolymers of PCL onto acrylic backbones for releasing antitumoral drugs. *J. Biomed. Mater. Res.* 2003 (64A): 638-647.

Chen, B. H., Lee, D. J. Slow release of drug through deformed coating film: effects of morphology and drug diffusivity in the coating film. *J. Pharm. Sci.* 2001(90):1478-1496.

Fulzele, S. V., Satturwar, P. M., Kasliwal, R. H., Dorle, A. K. Preparation and evaluation of microcapsules using polymerized rosin as a novel wall forming material. *J. Microencapsul.* 2004 (21): 83-89.

Jain, R. A. The manufacturing techniques of various drug loaded biodegradable poly (lactide-coglycolide) (PLGA) devices. *Biomaterial,* 2000 (21): 2475-2490.

Sinha, V. R., Trehan, A. Biodegradable microspheres for protein delivery. *J. Control. Release* 2003 (90): 261-280.

Kissel, T., Li, Y., Unger, F. ABA-triblock copolymers from biodegradable polyester A-blocks and hydrophilic poly(ethylene oxide) B-blocks as a candidate for in situ forming hydrogel delivery systems for proteins. *Adv. Drug. Deliv. Rev.* 2002 (54):99-134.

Tabata, Y., Gutta, S., Langer, R. Controlled delivery systems for proteins using polyanhydride microspheres. *Pharm. Res.* 1993 (10): 487-496.

Kipper, M. J., Shen, E.,; Determan, A. Narasimhan, B. Design of an injectable system based on bioerodible polyanhydride microspheres for sustained drug delivery. *Biomaterials,* 2002 (23):4405- 4412.

Wang, C., Ge, Q., Ting, D., Nguyen, D., Shen, H. R., Chen, J., Eisen, H. N., Heller, J., Langer, R., Putnam, D. Molecularly engineered poly(ortho ester) microspheres for enhanced delivery of DNA vaccines. *Nat. Mater.* 2004 (3): 190-196.

Jalil, R., Nixon, J. R. Biodegradable poly(lactic acid) and poly(lactide-co-glycolide) microcapsules: problems associated with preparative techniques and release properties. *J. Microencapsul.* 1990(7):297-325.

Sinha, V. R., Bansal, K., Kaushik, R., Kumria, R., Trehan, A. Poly-epsilon-caprolactone microspheres and nanospheres: an overview. *Int. J. Pharm.* 2004 (278):1-23.

Wang, J., Mao, H. Q., Leong, K. W. A novel biodegradable gene carrier based on polyphosphoester. *J. Am. Chem. Soc.* 2001(123): 9480-9481.

Dang, W., Daviau, T., Brem, H. Morphological characterization of polyanhydride biodegradable implant gliadel during in vitro and in vivo erosion using scanning electron microscopy. *Pharm. Res.* 1996 (13): 683-691.

Allcock, H. R., Kugel, R. L., Valan, K. J. Synthesis of high polymeric alkoxy and aryloxy phosphonitriles. *J. Am. Chem. Soc.* 1965 (87):4216-4217.

Allcock, H. R., Pucher, A. G., Scopelianos, A. G. Synthesis of poly(organophosphazenes) with glycolic acid ester and lactic acid ester side groups: prototypes for new bioerodible polymers. *Macromolecules*, 1994 (27): 1-4.

Hirano, S. Chitin and chitosan as novel biotechnological materials. *Polym. Int,* 1999 (48): 732-734.

A. shirwaikar, S. l. prabhu and G. A. Kumar, Herbal Excipients in Novel Drug delivery Systems. *Ind J Pharma Sci*, 2010 (4):415-422.

Park, J. H., Cho, Y. W., Chung, H., Kwon, I. C., Jeong, S. Y. Synthesis and characterization of sugar-bearing chitosan derivatives: aqueous solubility and biodegradability. *Biomacromolecules,* 2003 (4):1087-1091.

Hirano, S., Tsuchida, H., Nagao, N. N-acetylation in chitosan and the rate of its enzymic hydrolysis. *Biomaterials,* 1989 (10): 574-576.

Sinha, V. R.; Singla, A. K.; Wadhawan, S.; Kaushik, R.; Kumria, R.; Bansal, K.; Dhawan, S. Chitosan microspheres as a potential carrier for drugs. *Int. J. Pharm.* 2004 (274): 1-33.

D. S. Singhare; S. Khan, and P.G. Yeole. Poloxamer: Promising block Co-polymers in Drug Delivery. *Ind J Pharma Sci*, 2005 67(5):523-631.

Chandy, T., Rao, G. H., Wilson, R. F., Das, G. S. Development of poly(Lactic acid)/chitosan comatrix microspheres: controlled release of taxol-heparin for preventing restenosis. *Drug. Deliv*. 2001 (8): 77-86.

R. S. R. Murthy. N. K. Jain's Controlled and Novel Drug Delivery. 1st Edn., 1997:27-51.

J. Haller. Fundamentals of Polymer Science. Robinson's Controlled Drug Delivery. Special Edn., 2009, 20:140-166.

Hossin H. and Mohen H. Biodegradable polymer-metal complexes for Gene and Drug delivery. *Current drug safety*, 2009, 4(1):79-83.

3

Nanotechnology in Drug Delivery

Introduction

Nanotechnology is the design, characterization, production, and application of structures, devices and systems by controlling shape and size at the nano scale. Nanomedicine is the application of nanotechnology to health care systems. It has a potential impact on the easy and reliable diagnosis, monitoring and treatment of diseases. There are lot of research potentials in the field of nanoparticulate systems to be utilized in the drug delivery and drug targeting. Nanoparticulate systems have been used as a physical approach to alter and improve the pharmacokinetic and pharmaco-dynamic properties of various types of drug molecules. They have been used *in-vivo* to protect the drug entity in the systemic circulation, restrict access of the drug to the chosen sites and to deliver the drug at a controlled and sustained rate to the site of action. Various polymers have been used in the formulation of nanoparticles for drug delivery research to increase therapeutic benefit, while minimizing side effects.

Nanoparticles are defined as particulate dispersions or solid particles with a size in the range of 10-1000nm. The drug is dissolved, entrapped, encapsulated or attached to a nanoparticle matrix. Depending upon the method of preparation, nanoparticles, nanospheres or nanocapsules can be obtained. Nanocapsules are systems in which the drug is confined to a cavity surrounded by a unique polymer membrane, while nano-spheres are matrix systems in which the drug is physically and uniformly dispersed. In recent years, biodegradable polymeric nanoparticles, particularly those coated with hydrophilic polymer such as poly (ethylene glycol) (PEG) known as long-circulating particles, have been used as potential drug delivery devices because of their ability to circulate for a prolonged period of time targeting to a particular organ, as carriers of

DNA in gene therapy, and their ability to deliver proteins, peptides and genes.

The major goals in designing nanoparticles as a delivery system is to control particle size, surface properties and release of pharmacologically active agents in order to achieve the site-specific action of the drug at the therapeutically optimal rate and dose regimen. The advantages of using nanoparticles as a drug delivery system include the following:

1. Particle size and surface characteristics of nanoparticles can be easily manipulated to achieve both passive and active drug targeting after parenteral administration.
2. They can control and sustain release of the drug during the transportation and at the site of action, by altering organ distribution of the drug and subsequent clearance of the drug so as to achieve increase in drug therapeutic efficacy and reduction in side effects.
3. Controlled release and particle degradation characteristics can be readily modulated by the choice of matrix constituents. Drug loading is relatively high and drugs can be incorporated into the systems without any chemical reaction, which is an important factor for preserving the drug activity.
4. Site-specific targeting can be achieved by attaching targeting ligands to surface of particles or use of magnetic guidance.
5. The system can be used for various routes of administration including oral, nasal, parenteral, intra-ocular, etc.

In spite of these advantages, nanoparticles do have limitations. For example, their small size and large surface area can lead to particle-particle aggregation, making physical handling of nanoparticles difficult in liquid and dry forms. In addition, small particles size and large surface area readily result in limited drug loading and burst release. These practical problems have to be overcome before nanoparticles can be used clinically or made commercially available.

3.1 Preparation of Nanoparticles

3.1.1 Pearl/Ball-Milling Technology for the Production of Drug Nanocrystals

It has been found that running a pearl mill over a sufficiently long milling time can be able to produce drug nanosuspension. These mills consist of a milling container filled with fine milling pearls or larger-sized balls (made up of steel, glass, and zirconium oxide). The container can be static and the milling material can move by a stirrer; alternatively, the complete container is moved in a complex movement leading consequently to movement of the milling pearls. In the production process the coarse drug powder is dispersed by high-speed stirring in a surfactant/stabilizer (added for physical stability) solution to yield a nanosuspension. Adsorption onto the particle surface leads to high zeta potential values providing good physical stabilities. In case of parenteral drug nanocrystals, the choice is limited; e.g., for intravenous injection, accepted surfactants include: lecithins, Poloxamer 188, Tween 80, low molecular weight polyvinylpyrrolidone (PVP), sodium glycocholate (in combination with lecithin), etc.

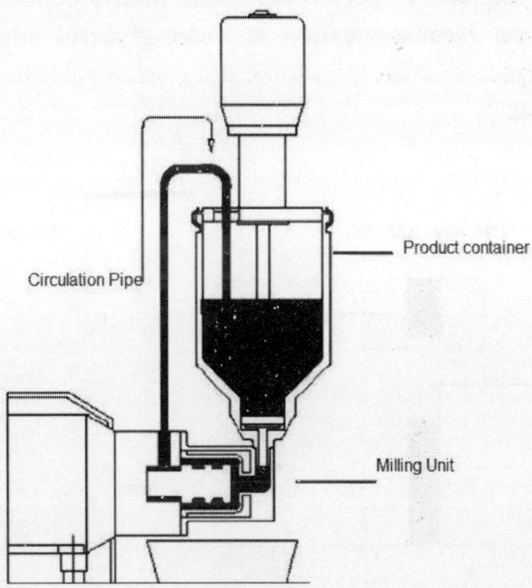

Fig. 3.1 DISPERMAT1 SL: schematic view of a bead mill using recirculation method.

3.1.2 Drug Nanocrystals Produced by High-Pressure Homogenization Followed by Spray Drying

High-pressure homogenization is a technology that has been applied for many years in various areas for the production of emulsions and suspensions. In the pharmaceutical industry parenteral emulsions are produced by this technology. Most of the homogenizers used are based on the piston-gap principle; an alternative is the jet-stream technology (Fig 3.2). It was found that similar efficient particle diminution can be achieved by homogenization in nonaqueous media such as oils and liquid polyethylene glycols (PEGs), which means media with low vapor pressure. In the case of low vapor pressure liquids, the cavitation in the homogenization gap is distinctly reduced or does not exist at all. As a consequent next step, after homogenization in water (100% water) and homogenization in nonaqueous media (0% water), homogenization was performed in mixtures containing different percentages of water (1–99% water). The dispersion media were mixed with water-miscible liquids (e.g., alcohols, glycerol). Preparation of drug nanosuspensions in water–ethanol mixtures is favorable for producing dry products, because later the spray drying can be performed under milder conditions when using such a mixture. Homogenization in water–glycerol mixtures (2.25% of water-free glycerol) leads to isotonic drug nanosuspensions for parenteral administration.

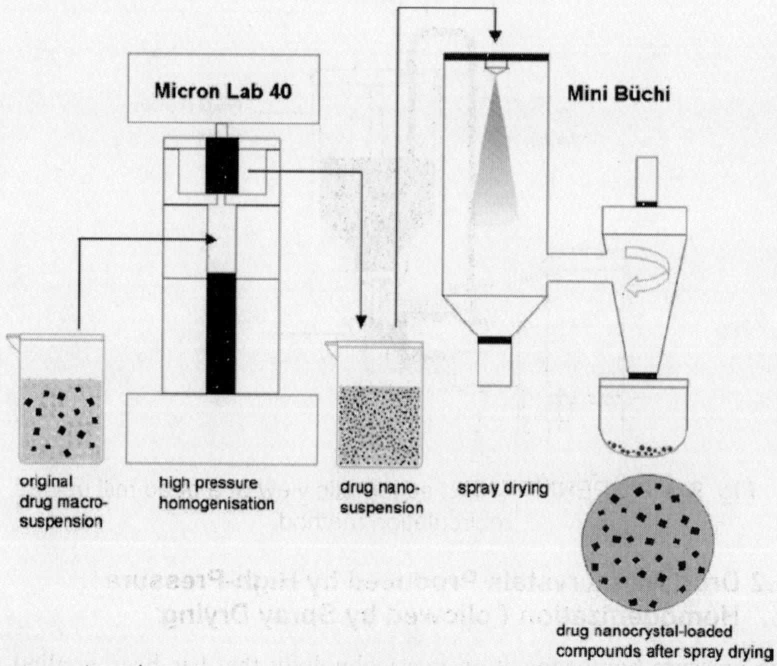

Fig. 3.2 Two-step process of the production of drug nanocrystal loaded compounds: the drug nano-suspension obtained by high pressure homogenization (Micron Lab 40) is further processed by spray drying using a Mini Bu¨ chi. Drug nanocrystals embedded in the matrix are obtained.

3.1.3 Dispersion of Preformed Polymers

Dispersion of preformed polymers is a common technique used to prepare biodegradable nanoparticles from poly (lactic acid) (PLA); poly (D,L-glycolide), PLG; poly (D, L lactide-co-glycolide) (PLGA) and poly (cyanoacrylate) (PCA). This technique can be used in various ways such as:

Emulsification Solvent evaporation method: In this method, the polymer is dissolved in an organic solvent such as dichloromethane, chloroform or ethyl acetate which is also used as the solvent for dissolving the hydrophobic drug. The mixture of polymer and drug solution is then emulsified in an aqueous solution containing a surfactant or emulsifying agent to form oil in water (o/w) emulsion. After the formation of stable emulsion, the organic solvent is evaporated either by reducing the pressure or by continuous stirring. Particle size was found to be influenced by the type and concentrations of stabilizer, homogenizer speed and polymer concentration. In order to produce small particle size, often a high speed homogenization or ultrasonication may be employed.

One such example of production of protein stabilized nanoparticles is to form a solid and stable layer of albumin onto drug nanoparticles, the protein needs to be cross-linked (or denatured) onto the particle surface. Typically, albumin crosslinking can be achieved by heat, use of cross-linker such as gluteraldehyde, or high shear. Fortunately, in the emulsification solvent evaporation process high shear is already in use, hence it can also be used for cross-linking protein stabilizers. High-shear cross-linking works for the protein-bearing sulfhydryl or disulfide groups (e.g., albumin). The high-shear conditions produce cavitation in the liquid, which causes tremendous local heating and results in the formation of hydroxyl radicals that are capable of cross-linking the polymer, for example, by oxidizing the sulfhydryl residues (and/or disrupting the existing disulfide bonds) to form new cross-linked disulfide bonds.

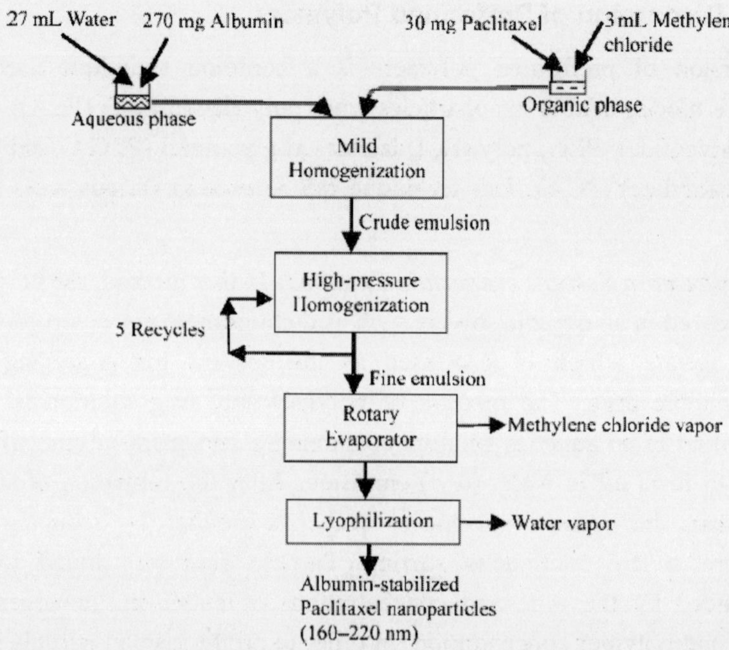

Fig. 3.3 Schematic of the protein-stabilized drug (Paclitaxol) nanoparticle formation.

Spontaneous emulsification or solvent diffusion method: This is a modified version of solvent evaporation method. In this method, the water-miscible solvent along with a small amount of the water immiscible organic solvent is used another phase. Due to the spontaneous diffusion of solvents, an interfacial turbulence may be created between the two phases leading to the formation of small particles. As the concentration of water miscible solvent increases, a decrease in the size of particle can be achieved.

For creating fine emulsion for obtaining nanoparticles, the use of a high amount of surface stabilizer is avoided to reduce the high load of the polymer exipients, as some of these exipients have shown toxicity. This leaves us to the use of high shear to generate fine emulsions for which sonication and homogenization techniques are required. Sonication generates emulsions through ultrasound-driven mechanical vibrations, which causes cavitation. Ambient pressure homogenizers use rotor–stator types of mixers, which can go to very high rotational speeds. High-pressure homogenization uses high pressure to force the fluid into microchannels of a special configuration and initiates emulsification via a combined mechanism of cavitations, shear, and impact, exhibiting excellent emulsification efficiency.

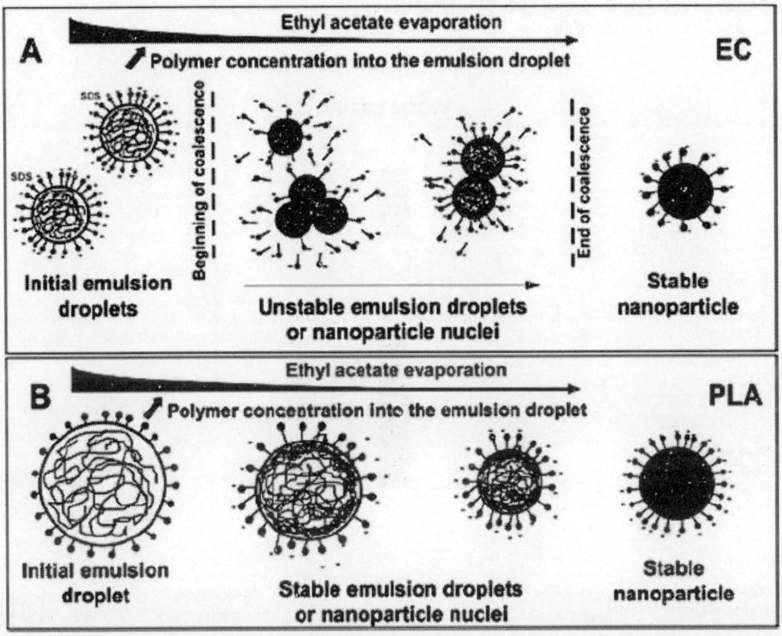

Fig. 3.4 The mechanisms of formation of the nanoparticles by emulsion solvent evaporation using solutions (A) of EC in ethyl acetate and (B) of PLA in ethyl acetate.

Abbreviation: PLA, poly-lactic acid.

Both solvent evaporation and solvent diffusion methods can be used for hydrophobic or hydrophilic drugs. In the case of hydrophilic drug, a multiple w/o/w emulsion needs to be formed with the drug dissolved in the internal aqueous phase.

3.1.4 Production by Hot-Melt Matrix Method

The production of drug nanocrystals in solid matrices can be done with high-pressure homogenization in hot melts. It offers advantages over production in aqueous solution and subsequent spray drying. The process is completely anhydrous, avoiding possible drug degradation or instabilities. The production can directly be performed by hot high pressure homogenization in melted material. The homogenizers equipped with temperature control jackets placed around the sample/product containers. Working temperatures up to 100^{0}C (heated with water) or higher (heated with silicon oil) can be selected depending on the melting temperature of the used matrix material.

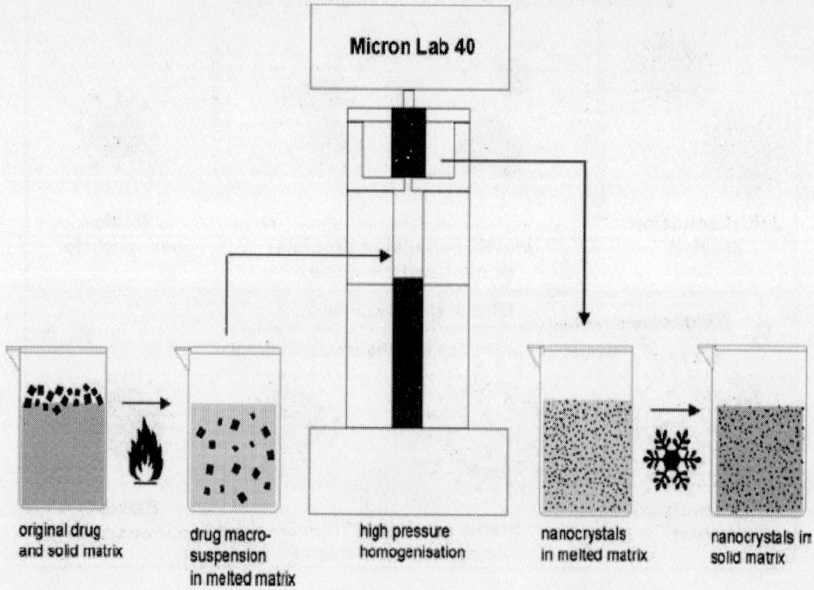

Fig. 3.5 Schematic of the process utilizing melted matrices: the coarse drug material is added to the solid matrix material, which is then melted for dispersing the drug. The nano-suspension is obtained by high pressure homogenization. Subsequent cooling leads to drug nanocrystals embedded in a solid matrix.

3.1.5 Production using Supercritical Fluid Technology

A fluid is supercritical when it is compressed beyond its critical pressure (Pc) and heated beyond its critical temperature (Tc). SCF technology has emerged as an important technique for particle manufacturing. In many industrial applications, it is poised to replace the conventional recrystallization and milling processes, mainly because of the quality and the purity of the final particles and environmental benefits. There are a variety of SCFs available as listed in Table 3.1.

Table 3.1 Critical Constants and Safety Data for Various Supercritical Solvents

SCF	T_c (°C)	P_c (bar)	Safety hazard
Ethylene	9.3	50.3	Flammable gas
Trifluoromethane (fluoroform)	25.9	47.5	
Chlorotrifluoromethane	28.9	39.2	
Ethane	32.3	48.8	Flammable gas
Carbon dioxide	31.1	73.7	
Dinitrogen monoxide (laughing gas)	36.5	72.6	Not combustible but enhances combustion of other substances
Sulfur hexafluoride	45.5	37.6	
Chlorodifluoromethane (HCFC 22; R 22)	96.4	49.1	Combustible under specific conditions
Propane	96.8	43.0	Extremely flammable
Ammonia	132.4	112.7	Flammable and toxic
Dimethyl ether (wood ether)	126.8	52.4	Extremely flammable
Trichlorofluoromethane (CFC 11, R 11)	198.0	44.1	
Isopropanol	235.2	47.6	Highly flammable
Cyclohexane	280.3	40.7	Highly flammable
Toluene	318.6	41.1	Highly flammable
Water	374.0	220.5	

Abbreviation: SCF, supercritical fluid.

Supercritical CO_2 (SC CO_2) is the most widely used supercritical fluid because of its mild critical conditions (Tc = 31.1 °C, Pc = 73.8 bars), non-toxicity, non-flammability, and low price. The most common processing techniques involving supercritical fluids are supercritical anti-solvent (SAS) and rapid expansion of super critical solution (RESS). The process of SAS employs a liquid solvent, e.g. methanol, which is completely miscible with the supercritical fluid (SC CO_2), to dissolve the solute to be micronized; at the process conditions, because the solute is insoluble in the supercritical fluid, the extract of the liquid solvent by supercritical fluid leads to the instantaneous precipitation of the solute, resulting the formation of nanoparticles. Thote and Gupta (2005) reported the use of a modified SAS method for formation of hydrophilic drug dexamethasone phosphate for nano-encapsulation purpose.

RESS differs from the SAS process in that its solute is dissolved in a supercritical fluid (such as supercritical methanol) and then the solution is rapidly expanded through a small nozzle into a region of lower pressure. Thus the solvent power of supercritical fluids dramatically decreases and the solute eventually precipitates. This technique is clean because the precipitate is basically solvent free.

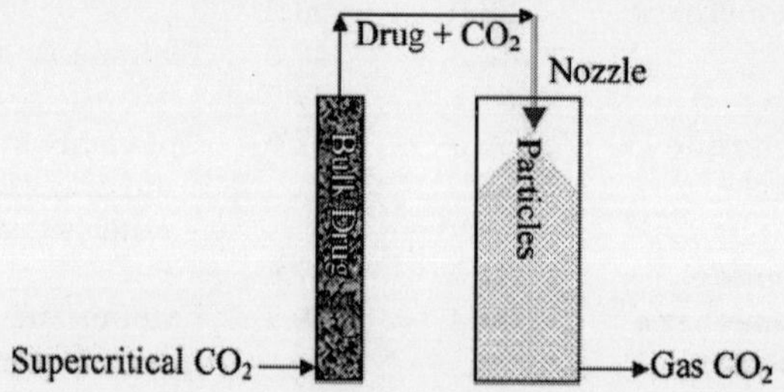

Fig. 3.6 Schematic of RESS process.
Abbreviations: RESS, rapid expansion of supercritical solution; CO_2, carbon dioxide.

The rapid expansion of supercritical CO_2 can produce nuclei of 5–10nm in diameter. These nanonized drugs include: 100nm lidocaine, 200nm griseofulvin and 200nm b-sitosterol, etc.

3.1.6 Polymerization Method

In this method, monomers are polymerized to form nanoparticles in an aqueous solution. Drug is incorporated either by being dissolved in the polymerization medium or by adsorption onto the nanoparticles after polymerization completed. The nanoparticle suspension is then purified to remove various stabilizers and surfactants employed for polymerization by ultracentrifugation and re-suspending the particles in an isotonic surfactant-free medium. This technique has been reported for making polybutylcyanoacrylate or poly (alkylcyanoacrylate) nanoparticles. Nanocapsule formation and their particle size depend on the concentration of the surfactants and stabilizers used.

3.1.7 Coacervation or Ionic Gelation Method

Much research has been focused on the preparation of nanoparticles using biodegradable hydrophilic polymers such as chitosan, gelatin and sodium alginate. Calvo et al., developed a method for preparing hydrophilic chitosan nanoparticles by ionic gelation. The method involves a mixture of two aqueous phases, of which one is the polymer chitosan, a di-block co-polymer ethylene oxide or propylene oxide (PEO-PPO) and the other is a polyanion sodium tripolyphosphate. In this method, positively charged amino group of chitosan interacts with negative charged tripolyphosphate to form coacervates with a size in the range of nanometer. Coacervates are formed as a result of electrostatic interaction between two aqueous phases, whereas, ionic gelation involves the material undergoing transition from liquid to gel due to ionic interaction at room temperature.

3.2 Characterization of Nanoparticles

3.2.1 Particle Size

Particle size plays major role in determining the *in vivo* distribution, biological fate, toxicity and the targeting ability of nanoparticle systems. In addition, they can also influence the drug loading, drug release and stability of nanoparticles. Because of the comparable size of the components in the human cells and if one needs to deliver or penetrate the drug inside, nanoparticles are of great interest in drug delivery. The unique qualities and performance of nanoparticles as devices of drug delivery arise directly from their physicochemical properties. Hence, determining such characteristics is essential in achieving a mechanistic understanding of their behavior. A good understanding allows prediction of in vivo performance as well as allowing particle design, formulation development, and process troubleshooting to be carried out in a rational fashion. So it is important to study the size and surface characteristics of the drug particle and/or the drug loaded carrier to determine the drug delivery at the target site.

Table 3.2 Typical Size of Various Objects

Object	Size (nm)
Carbon atom	0.1
DNA double helix (diameter)	3
Ribosome	10
Virus	100
Bacterium	1,000
Red blood cell	5,000
Human hair (diameter)	50,000
Resolution of unaided human eyes	100,000

The physical characterization of nanoparticle can be determined by the following methods:

3.2.1.1 Dynamic Light Scattering (DLS)

DLS, also known as photon correlation spectroscopy (PCS) or quasi-elastic light scattering (QELS) records the variation in the intensity of scattered light on the microsecond time scale. This variation results from interference of light scattered by individual particles under the influence of Brownian motion, and is quantified by compilation of an autocorrelation function. This function is fit to an exponential, or some combination or modification thereof, with the corresponding decay constant(s) being related to the diffusion coefficient(s). Using standard assumptions of spherical size, low concentration, and known viscosity of the suspending medium, particle size is calculated from this coefficient.

Table 3.3 Methods for Assessing the Properties of Nanoparticles

Properties	Relevant Analytical Method(s)
Size	Dark field optical microscopy, Dynamic light scattering, Static light scattering, Ultrasonic spectroscopy, Turbidimetry, NMR, Single particle optical sensing, FFF Hydrodynamic fractionation, Filtration
Morphology	TEM, SEM, Atomic force microscopy
Surface charge	Electrophoretic light scattering, U-tube electrophoresis, Electrostatic-FFF, Zeta meter
Surface hydrophobicity	Hydrophobic interaction chromatography
Surface adsorbents	Electrophoresis
Density	Isopycnic centrifugation, sedimentation-FFF
Interior structure	Freeze-fracture SEM, DSC, X-ray diffraction, NMR
Abbreviations: DSC, differential scanning calorimetry; FFF, field fractionation; NMR, nuclear magnetic resonance; SEM, scanning electron microscopy; TEM, transmission electron microscopy.	

3.2.1.2 Turbidimetry

For non-absorbing particles, turbidity is the complement to light scattering because it represents the amount of incident radiation not reaching a detector, that is, light lost to scattering. Hence the turbidity spectrum is also described by Mie theory and thus can be used to determine particle size as long as the data are normalized for concentration. This approach requires tiny amounts of sample and can be easily executed using a spectrophotometer.

3.2.1.3 Single-Particle Optical Sensing (SPOS)

SPOS, which is also known as optical particle counting, involves recording the obscuration or scattering of a beam of light that results from the passage of individual particles through a sensor. Signal magnitude is translated to the size of the particle via use of a previously determined calibration curve using standards approximating the sample in terms of shape and optical properties.

3.2.1.4 Optical Microscopy

Microscopy is still useful to get an estimate of size and crystallinity of starting materials, however the dark field techniques, in which particles are observed indirectly as bright spots on a dark background because of their scattering under oblique illumination, is extremely valuable in assessing the presence and numbers of nanoparticles.

3.2.1.5 Electron Microscopy

Scanning and transmission electron microscopy, SEM and TEM, respectively, provide a way to directly observe nanoparticles, with the former method being better for morphological examination. TEM has a smaller size limit of detection, is a good validation for other methods, and affords structural information via electron diffraction, but staining is usually required, and one must be cognizant of the statistically small sample size and the effect that vacuum can have on the particles. Very detailed images data can result from freeze-fracture approaches in which a cast is made of the original sample.

3.2.1.6 Isopycnic Centrifugation

It is a bio-analytical method applied to nanoparticles is centrifugation of analyte using a sucrose gradient as the suspending media. Under the influence of Stokes' laws, sedimenting particles will settle until they reach a point where their density matches that of the gradient. This self-focusing separation allows nanoparticle density to be determined, which along with particle size and bulk substitute concentration can in turn be used to calculate a number concentration.

3.2.1.7 Zeta Potential

Zeta potential is used as a surrogate for surface change, and is often measured by observing the oscillations in signal that result from light scattered by particles located in an electric field.

3.2.1.8 X-Ray Diffraction (Powder X-ray Diffraction / Small-Angle Neutron Scattering / Small-Angle X-ray Scattering)

The geometric scattering of radiation from crystal planes within a solid allow the presence or absence of the former to be determined thus permitting the degree of crystallinity to be assessed.

3.2.1.9 Differential Scanning Calorimetry (DSC)

DSC can be used to determine the nature and speciation of crystallinity within nanoparticles through the measurement of glass and melting point temperatures and their associated enthalpies

3.2.1.10 Electrophoresis

This process will determine the clearance and bio-distribution of the colloid, so evaluating the exact nature of the surface coverage, which is required to achieve a useful understanding. The small size of nanoparticles allows their electrophoretic behavior to be observed using bioanalytical tools such as isoelectric focusing electrophoresis and 2-D polyacrylamide gel electrophoresis (PAGE).

3.2.1.11 Hydrophobic Interaction Chromatography

In this method the analyte is first adsorbed onto a chromatographic stationary phase using a high concentration of an antichaotropic salt. Elution occurs using a gradient in which the salt concentration is decreased, so that those materials eluting first are the least hydrophobic because the salt concentration did not need to be decreased much before the analyte desorbed. Hydrophobic interaction chromatography has been used as a means of characterizing the hydrophobicity of nanoparticle surfaces, a property influenced by the choice of surfactant and/or polymer and also a key parameter in determining their *in-vivo* fate

3.2.2 Nanoparticle Suspension and Settling

Because of the small size of the nanoparticles, it is easy to keep them suspended in a liquid. Large microparticles precipitate out more easily because of gravitational force, whereas the gravitational force is much smaller on a nanoparticle. Particle settling velocity, v, is given by Stokes' law as

$$v = \frac{d^2 g(\rho_s - \rho_l)}{18\mu_l}$$

Where, g is gravitation acceleration (9.8 m/sec at sea level), ρ_l is liquid density (997 kg/m3 for water at 25^{0}C), μ_l is viscosity (0.00089 Pa/sec for water at 25 ^{0}C). For various particles sizes, settling velocities are calculated in Table 3. 4.

Table 3.4 Particle Settling Velocities

Particle Size in nm	Settling Velocity in nm/Sec
1	0.00043
10	0.043
100	4.30
1000	430
10000	43,005

Thermal (Brownian) fluctuations resist the particle settlement. According to Einstein's fluctuation–dissipation theory, average Brownian displacement x in time t is given as:

$$x = \sqrt{\frac{2k_B Tt}{\pi \mu d}}$$

Where, k_B is the Boltzman constant (1.38 10^{-23} J/K), and T is temperature in Kelvin.

3.2.3 Magnetic and Optical Properties

Small nanoparticles also exhibit unique magnetic and optical properties. For example, ferromagnetic materials become super-paramagnetic below about 20 nm, i.e., the particles do not retain the magnetization because of the lack of magnetic domains; however, they do experience force in the magnetic field. Such materials are useful for targeted delivery of drugs and heat.

For example, interaction of electromagnetic pulses with nanoparticles can be utilized for enhancement of drug delivery in solid tumors. The particles can be attached to antibodies directed against antigens in tumor vasculature and selectively delivered to tumor blood vessel walls. The local heating of the particles by pulsed electromagnetic radiation results in perforation of tumor blood vessels, *microconvection* in the *interstitium*, and perforation of cancer cell membrane, and therefore provides enhanced delivery of drugs from the blood into cancer cells with minimal thermal and mechanical damage to the normal tissues.

Gold and silver nanoparticles show size-dependent optical properties. The intrinsic color of nanoparticles changes with size because of surface plasmon resonance. Such nanoparticles are useful for molecular sensing, diagnostic, and imaging applications.

3.2.4 Drug Loading

A successful nanoparticulate system should have a high drug-loading capacity thereby reduce the quantity of matrix materials for administration. Drug loading and entrapment efficiency very much depend on the solid-state drug solubility in matrix material or polymer (solid dissolution or dispersion), which is related to the polymer composition, the molecular weight, the drug polymer interaction and the presence of an end functional groups (ester or carboxyl). The PEG moiety has no or little effect on drug loading. The macromolecule or protein shows greatest loading efficiency when it is loaded at or near its isoelectric point when it has minimum solubility and maximum adsorption.

3.2.5 Drug Release

For a successful nanoparticulate system, both drug release and polymer biodegradation are important factors to be considered. In general, drug release rate depends on: (1) solubility of drug; (2) desorption of the surface-bound/ adsorbed drug; (3) drug diffusion through the nanoparticle matrix; (4) nanoparticle matrix erosion/degradation; and (5) combination of erosion/diffusion process. Thus solubility, diffusion and biodegradation of the matrix materials govern the release process.

In-vitro release from the nanoparticulate systems can be studied by the methods such as: (1) side-by-side diffusion cells with artificial or biological membranes; (2) dialysis bag diffusion technique; (3) reverse dialysis bag technique; (4) agitation followed by ultracentrifugation/ centrifugation; (5) Ultra-filtration or centrifugal ultra-filtration techniques. Usually the release study is carried out by controlled agitation followed by centrifugation.

3.3 Biomedical Application of Nanoparticles

Nanoparticles are devices or systems can be made variety of materials including polymers (polymeric nanoparticles, micelles, or dendrimers), lipids (liposomes), viruses (viral nanoparticles), and even organometallic compound (nanotubes).

Nanotechnology is expected to have great impact on many areas in medical technology, especially:

- Surgery
- Cancer diagnosis
- Cancer therapy
- Biosensors
- Molecular imaging
- Implant technology
- Tissue engineering
- Drug delivery
- Gene delivery

Nanotechnology is foreseen to change health care in a fundamental way, which includes:

- Novel methods for disease diagnosis and therapy
- New and effective tools for disease prevention
- Point to point care, fast testing and daily screening of health
- Therapeutic selection tailored to the patient's profile

3.3.1 Polymer based drug carriers

The drug is either physically entrapped in or covalently bound to the polymer matrix. The resulting compounds may have the structure of capsules (polymeric nanoparticles), amphiphilic core/shell (polymeric micelles), or hyperbranched macromolecules (dendrimers; Fig. 3.7). Polymers used as drug conjugates can be divided into two groups of natural and synthetic polymers.

Polymeric nanoparticles (polymer-drug conjugates)

Polymers such as albumin, chitosan, and heparin occur naturally and have been a material of choice for the delivery of oligonucleotides, DNA, and protein, as well as drugs. Recently, a nanoparticle formulation of paclitaxel, in which serum albumin is included as a carrier [nanometer-sized albuminbound paclitaxel (Abraxane); Fig. 3.7A], has been applied in the clinic for the treatment of metastatic breast cancer. Among synthetic polymers such as N-(2-hydroxypropyl)-methacrylamide copolymer (HPMA), polystyrene-maleic anhydride copolymer, polyethylene glycol (PEG), and poly-L-glutamic acid (PGA), PGA was the first biodegradable polymer to be used for conjugate synthesis.

Several representative chemotherapeutics that are used widely in the clinic have been tested as conjugates with PGA *in-vitro* and *in-vivo* and showed encouraging abilities to circumvent the shortcomings of their free drug counterparts. Among them, Xyotax (PGApaclitaxel) and CT-2106 (PGA-camptothecin) are now in clinical trials.

Polymeric micelles (amphiphilic block copolymers)

The functional properties of micelles are based on amphiphilic block copolymers, which assemble to form a nanosized core/shell structure in aqueous media (Fig. 3.7B). The hydrophobic core region serves as a reservoir for hydrophobic drugs, whereas the hydrophilic shell region stabilizes the hydrophobic core and renders the polymers water-soluble, making the particle an appropriate candidate for i.v. administration. The drug can be loaded into a polymeric micelle in two ways: physical encapsulation or chemical covalent attachment. The first polymeric micelle formulation of paclitaxel, Genexol-PM (PEG-poly(D,L lactide)-paclitaxel), is a cremophor-free polymeric micelle-formulated paclitaxel.

3.3.2 Dendrimers

A dendrimer is a synthetic polymeric macromolecule of nanometer dimensions, composed of multiple highly branched monomers that emerge radially from the central core (Fig. 3.7C). Properties associated with these dendrimers such as their monodisperse size, modifiable surface functionality, multivalency, water solubility, and available internal cavity make them attractive for drug delivery. Polyamidoamine dendrimer, the dendrimer most widely used as a scaffold was conjugated with cisplatin.

3.3.3 Lipid-based Drug Carriers

Liposomes

Liposomes are self-assembling closed colloidal structures composed of lipid bilayers and have a spherical shape in which an outer lipid bilayer surrounds a central aqueous space (Fig. 3.7D). Currently, several kinds of cancerdrugs have been applied to this lipid-based system using a variety of preparation methods. Among them, liposomal formulations of the anthracyclines doxorubicin (Doxil®, Myocet®) and daunorubicin (DaunoXome®) are approved for the treatment of metastatic breast cancer and AIDS-related Kaposi's sarcoma. Besides these approved agents, many liposomal chemotherapeutics are currently being evaluated in clinical trials.

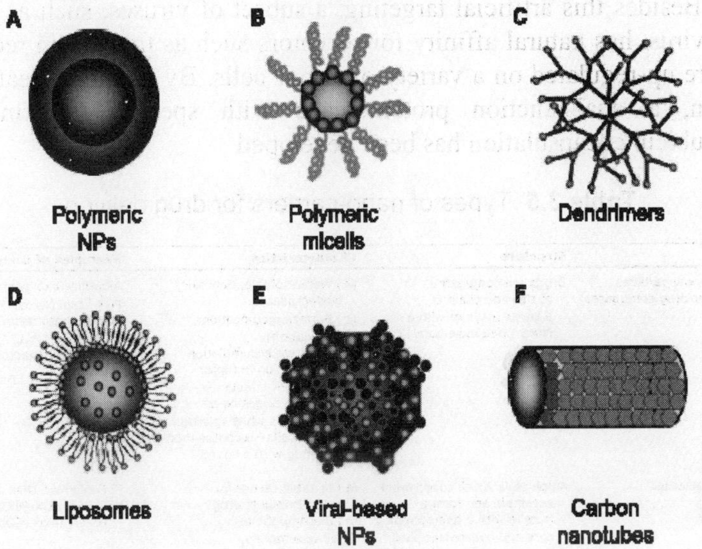

[Fig. 3.7 Types of nanocarriers for drug delivery. A, polymeric nanoparticles: polymeric nanoparticles in which drugs are conjugated to or encapsulated in polymers. B, polymeric micelles: amphiphilic block copolymers that form to nanosized core/ shell structure in aqueous solution. The hydrophobic core region serves as a reservoir for hydrophobic drugs, whereas hydrophilic shell region stabilizes the hydrophobic core and renders the polymer to be water-soluble. C, dendrimers: synthetic polymeric macromolecule of nanometer dimensions, which is composed of multiple highly branched monomers that emerge radially from the central core. D, liposomes: self-assembling structures composed of lipid bilayers in which an aqueous volume is entirely enclosed by a membranous lipid bilayer. E, viral-based nanoparticles: in general structure are the protein cages, which are multivalent, self-assembles structures. F, carbon nanotubes: carbon cylinders composed of benzene rings.]

3.3.4 Viral Nanoparticles

A variety of viruses including cowpea mosaic virus, cowpea chlorotic mottle virus, canine parvovirus, and bacteriophages have been developed for biomedical and nanotechnology applications that include tissue targeting and drug delivery (Fig. 3.7E). A number of targeting molecules and peptides can be displayed in a biologically functional form on their capsid surface using chemical or genetic means. Therefore, several ligands or antibodies including transferrin, folic acid, and single-chain antibodies have been conjugated to viruses for specific tumor targeting in vivo. Besides this artificial targeting, a subset of viruses, such as canine parvovirus, has natural affinity for receptors such as transferrin receptors that are up-regulated on a variety of tumor cells. By targeting heat shock protein, a dual-function protein cage with specific targeting and doxorubicin encapsulation has been developed

Table 3.5 Types of nano-carriers for drug delivery

System	Structure	Characteristics	Examples of compounds
Polymeric nanoparticles (polymer-drug conjugates)	Drugs are conjugated to the side chain of a linear polymer with a linker (cleavable bond)	(*a*) Water-soluble, nontoxic, biodegradable (*b*) Surface modification (pegylation) (*c*) Selective accumulation and retention in tumor tissue (EPR effect) (*d*) Specific targeting of cancer cells while sparing normal cells—receptor-mediated targeting with a ligand	Albumin-Taxol (Abraxane) PGA-Taxol (Xyotax) PGA-Camptothecin (CT-2106) HPMA-DOX (PK1) HPMA-DOX-galactosamine (PK2)
Polymeric micelles	Amphiphilic block copolymers assemble and form a micelle with a hydrophobic core and hydrophilic shell	(*a*) Suitable carrier for water-insoluble drug (*b*) Biocompatible, self-assembling, biodegradable (*c*) Ease of functional modification (*d*) Targeting potential	PEG-pluronic-DOX PEG-PAA-DOX (NK911) PEG-PLA-Taxol (Genexol-PM)
Dendrimers	Radially emerging hyperbranched synthetic polymer with regular pattern and repeated units	(*a*) Biodistribution and PK can be tuned (*b*) High structural and chemical homogeneity (*c*) Ease of functionalization, high ligand density (*d*) Controlled degradation (e) Multifunctionality	PAMAM-MTX PAMAM-platinate
Liposomes	Self-assembling closed colloidal structures composed of lipid bilayers	(*a*) Amphiphilic, biocompatible (*b*) Ease of modification (*c*) Targeting potential	Pegylated liposomal DOX (Doxil) Non-pegylated liposomal DOX (Myocet) Liposomal daunorubicin (DaunoXome)
Viral nanoparticles	Protein cages, which are multivalent, self-assembled structures	(*a*) Surface modification by mutagenesis or bioconjugation—multivalency (*b*) Specific tumor targeting, multifunctionality (*c*) Defined geometry and remarkable uniformity (*d*) Biological compatibility and inert nature	HSP-DOX CPMV-DOX
Carbon nanotubes	Carbon cylinders composed of benzene ring	(*a*) Water-soluble and biocompatible through chemical modification (organic functionalization) (*b*) Multifunctionality	CNT-MTX CNT-amphotericin B

[Abbreviations: PGA, poly-(L-glutamate); HPMA, N-(2-hydroxypropyl)-methacrylamide copolymer; PEG, polyethylene glycol; PAA, poly-(Laspartate); PLA, poly-(L-lactide); PAMAM, poly(amidoamine); DOX, doxorubicin; MTX, methotrexate; PK, pharmacokinetics; EPR, enhanced permeability and retention; CNT, carbon nanotube; HSP, heat shock protein; CPMV, cowpea mosaic virus.]

3.3.5 Carbon Nanotubes

Carbon nanotubes are carbon cylinders composed of benzene rings (Fig. 3.7F) that have been applied in biology as sensors for detecting DNA and protein, diagnostic devices for the discrimination of different proteins from serum samples, and carriers to deliver vaccine or protein. Carbon nanotubes are completely insoluble in all solvents, generating some health concerns and toxicity problems. However, the introduction of chemical modification to carbon nanotubes can render them water-soluble and functionalized so that they can be linked to a wide variety of active molecules such as peptides, proteins, nucleic acids, and therapeutic agents. Antifungal agents (amphotericin B) or anticancer drugs (methotrexate) have been covalently linked to carbon nanotubes with a fluorescent agent (FITC), and these drugs bound to carbon nanotubes were shown to be more effectively internalized into cells compared with free drug alone and to have potent antifungal activity *in-vivo*.

3.3.6 Magnetic Nanoparticles for Biomedical Applications

It is seen that because of size and composition there is an exponential growth in activities associated with the potential use of magnetic nanoparticles in biomedical applications. Thus, for applications in angiography and tumour permeability, ultra-small superparamagnetic iron oxide particles (USPIO) are preferred. However, for liver imaging superparamagnetic particles (SPIO) with intense macrophage uptake are preferred. For hyperthermia treatment, particles with sizes around the monodomain–multidomain transition, i.e. particles below 50 nm in diameter, have been found to produce the maximum specific absorption rate (SAR). It has been reported that the SAR of 35 nm magnetite particles is twice than that of 10 nm particles. Particles around the monodomain–multidomain size exhibit larger SAR values in the allowed frequency range of 10^5–10^6 Hz, which makes them useful for hyperthermia treatment. This is also the ideal size range for tomography imaging using the non-linear response of the magnetic particles. This technique, called MPI from 'magnetic particle imaging', has shown important advantages in resolution time and sensitivity.

Iron oxide–carbon systems are becoming popular due to their high chemical stability. Mirkin's group described an original method for the large scale synthesis of core– shell iron–carbon nanoparticles with diameters smaller than 5 nm from the residues produced in the formation of carbon nanotubes. Magnetic composites of Fe-based nanoparticles encapsulated in carbon/silica (C/SiO_2–Fe) or carbon (C–Fe) matrixes have been prepared by laser-induced pyrolysis of aerosols. Carbon nanocomposites were formed by amorphous carbon nanoparticles of 50–100 nm diameters in which isolated iron based nanoparticles of 3–10 nm in size are located. The powders were dispersed in aqueous solutions at pH-7 resulting in biocompatible colloidal dispersions with potential high resistance to biodegradation. Structural and magnetic properties and the suitability of aqueous dispersions as contrast agents for MRI were analyzed. The results of these characterizations and the NMR relativity data are very encouraging for applications of laser pyrolysis products in living tissues

3.3.7 Therapeutic Application of Antibody-Conjugated Nanoparticles

Gupta and Torchilin 2007, described the synthesis and efficacy of the monoclonal anticancer antibody 2C5 conjugated to commercially available PEGylated liposomes loaded with doxorubicin (a DNA interacting drug widely used in chemotherapy), in an intracranial model in nude mice. The treatment with the antibody-conjugated liposomes provided a significant therapeutic benefit over controls, with a pronounced reduction in the tumour size.

Poly-(d,l-lactide-co-glycolide)/montmorillonite nanoparticles (PLGA/MMT NPs) have been decorated with human epidermal growth factor receptor-2 (HER2) antibody, Trastuzumab (Herceptin_), for targeted breast cancer chemotherapy with paclitaxel as a model anticancer drug. According to the authors, their *in vitro* studies revealed that the therapeutic effects of the drug formulated in the conjugated nanoparticles could be 12.7 times higher than that of the bare nanoparticles, and 13.1 times higher than Taxol. The same humanised IgG1 monoclonal antibody has been covalently attached to human serum albumin-based nanoparticles, and the resulting carrier showed specific targeting to HER_2-positive breast cancer cells.

Paclitaxel-loaded poly(lactide-co-glycolide) (PLGA) nanoparticles coated with cationic SM5-1 singlechain antibody (scFv), containing a polylysine (SMFv-polylys), were synthesized and effectively tested *in-vitro*. The purpose of tagging the antibody with the cationic polypeptide polylysine is to achieve electrostatic attraction with the negatively charged nanoparticles. When compared to nontargeted paclitaxel-loaded PLGA nanoparticles, the antibodyconjugated nanoparticles showed enhanced *in-vitro* cytotoxicity against human hepatocellular carcinoma cell lines.

Tyner et al. 2004, have recently reported the surface functionalization of inorganic nanoparticles (made of magnesium–aluminium layered double hydroxides) with disuccinimidyl carbonate (DSC), which were then loaded with a huA33 antibody and a blood plasma protein (serum albumin). The biological *in vitro* tests showed that LDH-DSC-huA33 nanobiohybrids had an activity against human A33 antigen, which is 30 times higher than that of LDH-DSC-albumin.

3.3.8 Antibody-Conjugated Nanoparticles in Diagnosis

Cell Sorting: Cell sorting and bioseparation are found to be main applications for the antibody-conjugated nanoparticles. Kandzia et al. 1981, described the use of a monoclonal HLA-BW6 antibody coupled to albumin coated magnetite microspheres via surface-incorporated *Staphylococcus aureus* Protein-A. The mixture of HLABW6 and -BW4 human peripheral blood lymphocytes was incubated with these immuno-micropheres and applied to a glass column located in a magnetic field. Only HLA-BW4 lymphocytes passed through the column and were collected. The recovered cells were 97% viable. Protein A (protein isolated from the cell wall of *Staphyloccocus aureus*) was chosen, because it interacts specifically with the Fc fragment of the antibody.

Sensing: In a biosensor, a ligand and a receptor binds together in a reaction, which is collected as a signal to a transductor using different methods, including: optical, magnetic, electrochemical, radioactive, piezoelectric, mechanical, mass spectrometric methods, etc. Like any sensor, a biosensor should be cheap, compact, selective, sensitive, portable, reusable, and have a fast readout. Nanoparticles offer their physical properties as that of the biosensor. In some cases, nanoparticles are used simply as carriers of antibodies to recognize them by association in biosensors. For example, in 1996, surface plasmon resonance (SPR) was used, through the BIACORE® biosensor, to demonstrate the specific interaction between an anti-CD4 monoclonal antibody (IOT4a), adsorbed on poly(methylidene malonate) nanoparticles, and the CD4 molecule. Nanoparticles have been used for the sensing of inorganic phosphate, by using the fluorescence resonance energy transfer (FRET). The determination of inorganic phosphate is necessary to elucidate metabolic processes. Anti-HER2 and anti-IgG gold-coated silica nanoparticle conjugates have been used to measure the number of targeted plasmonic nanoparticles that bind to cell, due to their strong light scattering properties. The ability to quantify this is crucial for improved diagnostic and therapeutic efficacy.

Gold nanoparticles are widely used in optical sensing, due to their large light absorption and scattering cross section in the surface plasmon resonance wavelength region.

Gold-coated silver nanoparticles, covalently attached to goat anti-mouse IgG antibodies and to a reporter, were used in biosensing applications by employing surface enhanced Raman spectroscopy (SER)

Fluorescent reporter proteins embedded in polyacrylamide nanoparticles have been used for the sensing of inorganic phosphate, by using the fluorescence resonance energy transfer (FRET). The determination of inorganic phosphate is necessary to elucidate metabolic processes

Anti-prostate specific antibodies have been conjugated to gold nanoparticles and used to quantify the amount of prostate specific antigen (an FDA-approved biomarker for prostate cancer diagnosis) by utilizing dynamic light scattering.

3.3.9 Nanoparticles for oral Delivery of Therapeutics Including Peptides and Proteins

Significant advances in biotechnology and biochemistry have led to the discovery of a large number of bioactive molecules and vaccines based on peptides and proteins. Development of suitable carriers remains a challenge due to the fact that bioavailability of these molecules is limited by the epithelial barriers of the gastrointestinal tract and their susceptibility to gastrointestinal degradation by digestive enzymes. Polymeric nanoparticles allow encapsulation of bioactive molecules and protect them against enzymatic and hydrolytic degradation. For instance, it has been found that insulin-loaded nanoparticles have preserved insulin activity and produced blood glucose reduction in diabetic rats for up to 14 days following the oral administration.

Looking to the excellent biocompatibility, the biodegradable polyester called poly (D, L-lactide-co- glycolide) PLGA most frequently used biomaterial for a variety of drug delivery systems (blends, films, matrices, micro-spheres nanoparticles, pallets, etc.), which include antipsychotics, anaesthetics, antibiotics, antiparasites, antitumors, hormones, proteins, etc. Polymeric nanoparticles are investigated especially in drug delivery systems for drug targeting because of their particle size ranging from 10-1000nm and long circulation in the blood.

Examples of such systems include: Triclosan/PLGA for periodontal application, Paclitaxel/PLGA a bioadhesive system for oral delivery of paclitaxel, Ellagic acid-loaded PLGA nanoparticles for oral administration for more radical scavenging, PLGA nanoparticles encapsulating streptomycin targeted for oral delivery for targeting to maintain the plasma level for 4-7 days following single oral administration, Estradiol loaded PLGA nanoparticulate for better oral bioavailability, Cyclosporin loaded PLGA nanoparticles found to a potential career for oral administration, Cisplatin loaded PLGA-PEG nanoparticles found to be effective in cancer therapy.

3.3.10 Nanoparticles for Gene Delivery

Polynucleotide vaccines work by delivering genes encoding relevant antigens to host cells, where they are expressed, producing the antigenic protein within the vicinity of professional antigen presenting cells to initiate immune response. Such vaccines produce both humoral and cell-mediated immunity because intracellular production of protein, as opposed to extracellular deposition, stimulates both arms of the immune system. The key ingredient of polynucleotide vaccines, DNA, can be produced cheaply and has much better storage and handling properties than the ingredients of the majority of protein-based vaccines. Hence, polynucleotide vaccines are set to supersede many conventional vaccines particularly for immunotherapy. Nanoparticles loaded with plasmid DNA could also serve as an efficient sustained release gene delivery system due to their rapid escape from the degradative endo-lysosomal compartment to the cytoplasmic compartment.

3.3.11 Nanoparticles for Drug Delivery into the Brain

The BBB is characterized by relatively impermeable endothelial cells with tight junctions, enzymatic activity and active efflux transport systems. It effectively prevents the passage of water-soluble molecules from the blood circulation into the CNS. Consequently, the BBB only permits selective transport of molecules that are essential for brain function. Strategies for nanoparticle targeting to the brain rely on the presence of and nanoparticle interaction with specific receptor-mediated transport systems in the BBB. For example polysorbate 80/LDL, transferrin receptor binding antibody (such as OX26), lactoferrin, cell penetrating peptides and melanotransferrin have been shown capable of delivery of a self non transportable drug into the brain through the chimeric construct that can undergo receptor-mediated transcytosis. It has been reported poly-(butylcyanoacrylate) nanoparticles was able to deliver hexapeptide dalargin, doxorubicin and other agents into the brain which is significant because of the great difficulty for drugs to cross the BBB.

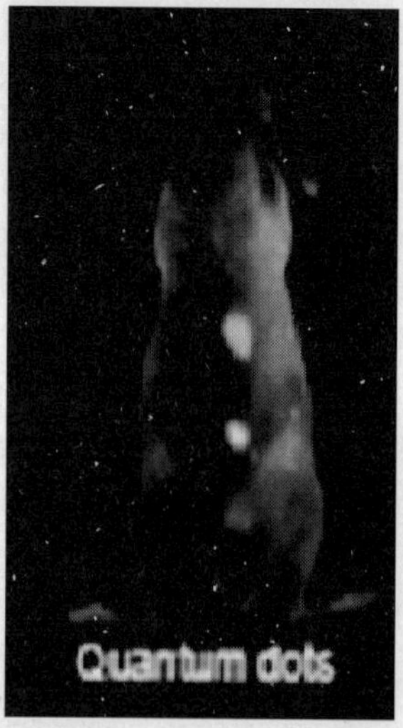

Source: JAMMA, Vol-292, 15: 1944-45, 2004

3.3.12 Nanoparticles and Cancer Therapy: Quantum Dots

Semiconductor QDs are rapidly emerging as popular luminescence probes for many biological and biomedical applications owing to their extremely small size (approximately 10 nm in diameter), high photostability, tunable optical properties, and multimodality. Such inorganic–organic composite nanomaterials have shown extreme efficiency in cancer diagnosis *in-vivo*, with their small size which facilitates unimpeded systemic circulation and attached targeting molecules, allowing for specific 'honing in' at neoplastic sites. Similar to other nanoparticles, QDs can be modified via conjugation of various surface molecules for targeted delivery. QDs also provide sufficient surface area to attach therapeutic agents for simultaneous drug delivery and *in-vivo* imaging as well as for tissue engineering. *In-vivo* cancer targeting and imaging in living animals by QDs was demonstrated by Gao and colleagues, where both subcutaneous injection of QD-tagged prostate cancer cells and systemic injection of multifunctional QD probes were used to achieve sensitive and multicolor fluorescence imaging of cancer cells. In a recent study, Bagalkot and colleagues used QD–apatamer (Apt)–doxorubicin (Dox) conjugate for targeted cancer therapy, imaging, and sensing. It was shown that this multifunctional nanoparticle system can deliver doxorubicin to the targeted prostate cancer cells and sense the delivery of doxorubicin by activating the fluorescence of QD, while allowing for simultaneous imaging of the cancer cells. Other such biomarkers for cancer diagnosis include (Nanoware and Cantilever; Fig 3.8 and Fig 3.9):

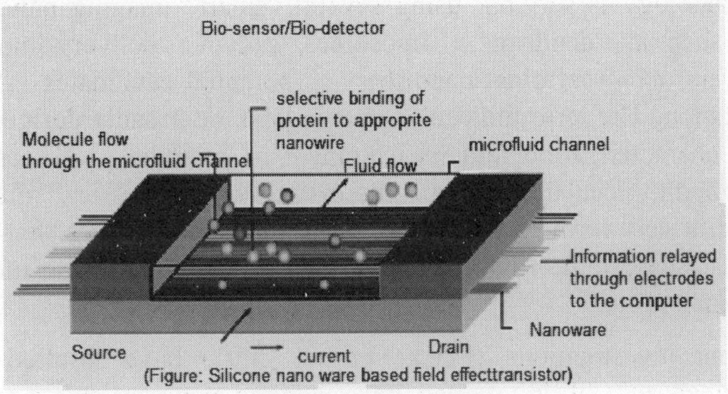

Fig. 3.8 Nanoware: Nano technology based Biomarkers for cancer diagnosis

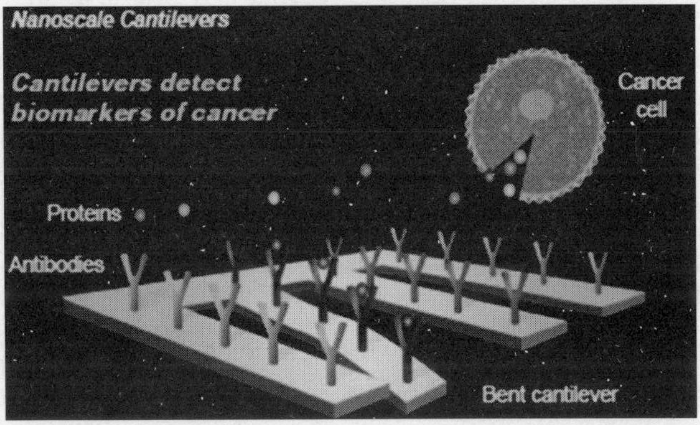

Fig. 3.9 Cantilevers: Nano technology based Biomarkers for cancer diagnosis

3.3.13 Nano particles in Delivery of Drugs to Cardiovascular Disease

Nanotechnology offers four areas, in which cardiovascular diseases can be better combated with immediate impact, *Targeted therapeutics*: delivering drugs where they are needed. *Tissue engineering*: building new tissues to replace defective valves, damaged heart muscle, clogged blood vessels, and so forth, *Molecular imaging*: using "smart" imaging agents that identify disease more specifically and *Biosensors and diagnostics*: improved diagnostic devices for the lab, and implantable sensors to detect problems inside the body.

Cardiovascular molecular imaging is another area where nanotechnology is serving, using several "smart" imaging nanoparticle agents such as: dendrimers, liposomes, polymer delivery molecules, cantilevers, nanoscaffolds, nanofibers as potential candidates in cardiac visualization. The term molecular imaging can be broadly defined as the *in-vivo* characterization and measurement of biologic processes at the cellular and molecular level. In contradistinction to "classical" diagnostic imaging, it sets forth to probe the molecular abnormalities that are the basis of disease rather than to image the end effects of these molecular alterations.

Recent investigations (Flacke *et al.,* 2001) have resulted in the development of a fibrin targeted, paramagnetic, nanoparticulate magnetic resonance (MR) contrast agent that allows enhanced sensitive detection and quantification of occult microthrombi within the intimal surface of atherosclerotic vessels. This agent is a ligand-directed, lipidencapsulated, liquid perfluorocarbon nanoparticles (250- nm diameter); it has a prolonged systemic half-life and can carry high gadolinium-based–agent payloads. Other investigators (Zhao *et al.,*2001) have also introduced a site-targeted MR contrast agent by conjugating a C2- glutathione S-transferase fusion protein to super-paramagnetic iron oxide nanoparticles, which allows noninvasive detection of apoptosis at MR imaging.

References

V. J. Moharajan and Y. Chen. Nanoparticles- A Review. *Tropical J Pharma Res*, 2006, 5(1):561-573.

Rainer H. Muller, Jan M. W., and Faris N. Rushrab. Manufacturing of nanoparticles by Milling and Homogenization Technologies. A. B. Guptas Nanoparticle Technology in Drug Delivery, by Taylor & Francis, New York, 2006 (152):21-52.

Ram B. Gupta. Polymer or protein stabilized nanoparticles from Emulsions. Nanoparticle Technology in Drug Delivery, by Taylor & Francis, New York, 2006 (159):85-102.

Ram B. Gupta. Supercritical Fluid Technology. Nanoparticle Technology in Drug Delivery, by Taylor & Francis, New York, 2006 (159):53-84.

Dhruba J Bharali, Marianne K, Mujgan G., Tessa M. S., Shaker A Mousa. Nanoparticles and cancer Therapy: A concise review with emphasis on dendrimers. *Int J of Nanomedicine*, 2009 (4):1-7.

V. Gajbhiye, P. Vijayaraj K., A. Sharma, A. Agarwal, A. Asthana and N. K. Jain. Dendrimeric Nanoarchitectures Mediated Transdermal and Oral Delivery of Bioactivities. *Ind J Pharma Sci*, 2008 (4):431-439.

M. Saeed Arayne, Najma Sultana and Faiza Qureshi. Nanoparticles in delivery of cardiovascular drugs. *Pak. J. Pharm. Sci.*, 2007, 20(4): 340-348

S. Mukherjee, S. Ray and R.S. Thakur. Solid Lipid Nanoparticles: A Modern Formulation Approach in Drug Delivery System. *Ind J Pharma Sci*, 2009 (July-Aug): 349-358.

M. S. Muthu. Nanoparticles based on PLGA and its co-polymer: An overview. *Asian J Pharma Sci*, 2009 (Oct-Dec): 266-273.

A G Roca, R Costo, A F Rebolledo, S Veintemillas-Verdaguer, P Tartaj, T Gonz´alez-Carre ˜ no, M P Morales and C J Serna. Progress in the preparation of Magnetic Nanoparticles for applications in Biomedicine. *J. Phys. D: Appl. Phys.,* 2009, 42: (224002):1-24.

Manuel Arruebo, M´ onica Valladares, and ´ Africa Gonz´alez-Fern´andez. Antibody-Conjugated Nanoparticles for Biomedical Applications. *Journal of Nanomaterials*, 2009, Article ID 439389: 1-24.

Flacke S, Fischer S and Scott M. Novel MRI contrast agent for molecular imaging of fibrin: implications for detecting vulnerable plaque. *Circulation*, 2001 (4): 1280-1285.

Zhao M, Beauregard D, Loizou L, Davletov B and rindle K . Non-invasive detection of apoptosis sing magnetic resonance imaging and a targeted ontrast agent. *Nat. Med.*, 2001: 241-1244.

Desai M. P., Labhasetwar V., Amidon G. L. , Levy R. J. Gastrointestinal uptake of biodegradable microparticles: effect of particle size. *Pharm Res,* 1996, 13(suppl 12):1838–1845.

Chow TS. Size-dependent adhesion of nanoparticles on rough substrates. *J Phys: Condens Matter,* 2003,15(suppl 2):L83–L87.

Lamprecht A, Schafer U, Lehr C-M. Size-dependent bioadhesion of micro- and nanoparticulate carriers to the inflamed colonic mucosa. *Pharm Res* 2001, 18(suppl 6):788–793.

Esenaliev RO. Radiation and nanoparticles for enhancement of drug delivery in solid tumors. *PCT Int Appl,* 2000 (WO 2000002590).

Kelly L, Coronado E, Zhao LL, Schatz GC. The optical properties of metal nanoparticles: the influence of size, shape, and dielectric environment. *J Phys Chem B*, 2003, (107):668–677.

Gaur U, Sahoo SK, De TK, Ghosh PC, Maitra A, Ghosh PK. Biodistribution of fluoresceinated dextran using novel nanoparticles evading reticuloendothelial system. *Int J Pharm*, 2000, 202(suppl 1–2):1–10.

Hillyer, JF, Albrecht RM. Gastrointestinal persorption and tissue distribution of differently sized colloidal gold nanoparticles. *J Pharm Sci*, 2001, 90(suppl 12):1927–1936.

Prabha S, Zhou W-Z, Panyam J, Labhasetwar V. Sizedependency of nanoparticle-mediated gene transfection: studies with fractionated nanoparticles. *Int J Pharm*, 2002, 244(suppl 1–2):105–115.

Uekama K. Design and evaluation of cyclodextrin-based drug formulation. *Chem Pharm Bull*, 2004, 52(8):900–915.

Mu¨ ller RH, Bo¨hm BHL. Nanosuspensions. In: Mu¨ ller RH, Benita S, Bo¨hm, B, eds. Emulsions and Nanosuspensions for the Formulation of Poorly Soluble Drugs. *Stuttgart: Medpharm,* 1998:149–174.

Liversidge E, Wei L. Inventor stabilization of chemical compounds using nanoparticulate formulations. U.S. Patent 2003054042 A1, 2003.

Chen X, Young TJ, Sarkari M, Williams RO, 3rd, Johnston KP. Preparation of cyclosporine: A nanoparticles by evaporative precipitation into aqueous solution. *Int J Pharm*, 2002, 242(1–2): 3–14.

Muller RH, Keck CM. Challenges and solutions for the delivery of biotech drugs—a review of drug nanocrystal technology and lipid nanoparticles. *J Biotechnol,* 2004, 113(1–3):151–170.

Merisko-Liversidge E, Sarpotdar P, Bruno J, et al. Formulation and antitumor activity evaluation of nanocrystalline suspensions of poorly soluble anticancer drugs. *Pharm Res*, 1996, 13(2):272–278.

Muller RH, Peters K. Nanosuspensions for the formulation of poorly soluble drugs: Preparation by a size-reduction technique. *Int J Pharm*, 1998, 160(2):229–237.

Muller RH, Dingler A, Schneppe T, Gohla S. Large scale production of solid lipid nanoparticles (SLNTM) and nanosuspensions (DissoCubesTM). In: Wise D, ed. Handbook of Pharmaceutical Controlled Release Technology. 2000:359–376.

Freitas C, Muller RH. Spray-drying of solid lipid nanoparticles (SLNTM). *Eur J Pharm Biopharm*, 1998, 46(2):145–151.

Bushrab FN, Mu¨ ller, RH. Drug nanocrystals: Amphotericin B-containing capsules for oral delivery. Philadelphia: *AAPS*, 2004.

Krause KP, Muller RH. Production and characterisation of highly concentrated nanosuspensions by high pressure homogenisation. *Int J Pharm,* 2001, 214(1–2):21–24.

Moschwitzer J, Achleitner G, Pomper H, Muller RH. Development of an intravenously injectable chemically stable aqueous omeprazole formulation using nanosuspension technology. *Eur J Pharm Biopharm*, 2004, 58(3):615–619.

Vergote GJ, Vervaet C, Van Driessche I. In vivo evaluation of matrix pellets containing nanocrystalline ketoprofen. *Int J Pharm*, 2002, 240(1–2):79–84.

Muller RH, Jacobs C. Buparvaquone mucoadhesive nanosuspension: preparation, optimisation and long-term stability. *Int J Pharm*, 2002, 237(1–2):151–161.

Muller RH, Keck CM. Drug delivery to the brain—realization by novel drug carriers. *J Nanosci Nanotechnol*, 2004, 4(5):471–483.

K. M. Tyner, S. R. Schi.man, and E. P. Giannelis, "Nanobiohybrids as delivery vehicles for camptothecin," *Journal of Controlled Release*, 2004, 95(3): 501–514.

O'Donnell PB, McGinity JW. Preparation of microspheres by the solvent evaporation technique. *Adv Drug Deliv Rev*, 1997, 28(1):25–42.

Bala I, Hariharan S, Kumar MNVR. PLGA nanoparticles in drug delivery: the state of the art. Crit Rev Ther Drug Carrier Syst, 2004, 21(5):387–422.

Bibette J, Calderon FL, Poulin P. Emulsions: basic principles. *Rep Prog Phys*, 1999, 62(6):969–1033.

Quintanar-Guerrero D, Fessi H, Allemann E, Doelker E. Influence of stabilizing agents and preparative variables on the formation of poly(D,L-lactic acid) nanoparticles by an emulsificationdiffusion technique. *Int J Pharm*, 1996, 143(2):133–141.

Maa Y-F, Hsu CC. Performance of sonication and microfluidization for liquid-liquid emulsification. *Pharm Dev Technol*, 1999, 4(2): 233–240.

Maa YF, Hsu C. Liquid-liquid emulsification by rotor/stator homogenization. *J Contr Rel*, 1996 (38):219–228.

Cegnar M, Kos J, Kristl, J. Cystatin incorporated in poly(lactide- co-glycolide) nanoparticles: development and fundamental studies on preservation of its activity. *Eur J Pharm Sci*, 2004, 22(5):357–364.

Yoncheva K, Vandervoort J, Ludwig A. Influence of process parameters of high-pressure emulsification method on the properties of pilocarpine-loaded nanoparticles. *J Microencapsul*, 2003, 20(4): 449–458.

Vandervoort Jo, Yoncheva K, Ludwig A. Influence of the homogenisation procedure on the physicochemical properties of PLGA nanoparticles. *Chem Pharm Bull,* 2004, 52(11): 1273–1279.

Desgouilles S, Vauthier C, Bazile D, et al. The design of nanoparticles obtained by solvent evaporation: a comprehensive study. *Langmuir*, 2003, 19(22):9504–9510.

Chattopadhyay P, Shekunov BY, Seitzinger JS, Huff RW. Particles from supercritical fluid extraction of emulsion. *PCT Int Appl*, 2004, 61: wo2004004862A1.

Sahoo SK, Panyam J, Prabha S, Labhasetwar V. Residual polyvinyl alcohol associated with poly (D,L-lactide-coglycolide) nanoparticles affects their physical properties and cellular uptake. *J Contr Rel,* 2002, 82(1):105–114.

Leucuta SE, Risca R, Daicoviciu D, Porutiu D. Albumin microspheres as a drug delivery system for epirubicin: pharmaceutical, pharmacokinetics and biological aspects. *Int J Pharm*, 1988, 41(3):213–217.

Pecora R. Dynamic light scattering measurement of nanometer particles in liquids. *J Nanoparticle Res,* 2000, (2):123–131.

Chu B, Liu T. Characterization of nanoparticles by scattering techniques. *J Nanoparticle Res,* 2000 (2):29–41.

McClements DJ. Principles of ultrasonic droplet size determination in emulsions. *Langmiur,* 1996(12):3454–3461.

Irache JM, Durrer C, Ponchel G, Duchene D. Determination of particle concentration in latexes by turbidimetry. *Int J Pharmaceutics,* 1993, (90):R9–R12.

Westesen K, Bunjes H, Koch MHJ. Physicochemical characterization of lipid nanoparticles and evaluation of their drug loading capacity and sustained release potential. *J Controlled Release,*1997, (48):223–236.

Jenning V, Mader K, Gohla SH. Solid lipid nanoparticles (SLN) based on binary mixtures of liquid and solid lipids: a 1H-NMR study. *Int J Pharmaceutics*, 2000 (205):15–21.

Mayer C. Nuclear magnetic resonance on dispersed nanoparticles. *Progress in nuclear magnetic resonance spectroscopy*, 2002 (40):307–366.

Valentini M, Vaccaro A, Rehor A, Napoli A, Hubbell JA, Tirelli N. Diffusion NMR Spectroscopy for the characterization of the size and interactions of colloidal matter: The case of vesicles and nanoparticles. *J Am Chem Soc*, 2004 (126):2142–2147.

Lieberman A. Particle characterization in liquids. In: Knapp JZ, Barber TA, Lieberman A, eds. Liquid- and Surface-Borne Particle Measurement Handbook. New York: Marcel-Dekker, 1996:1–28.

Ito T, Sun L, Bevan MA, Crooks RA. Comparison of nanoparticle size and electrophoretic mobility measurements using a carbon-nanotube-based coulter counter, dynamic light scattering, transmission electron microscopy, and phase analysis light scattering. *Langmuir*, 2004 (20): 6940–6945.

Jores K, Mehnert W, Drechsler M, Bunjes H, Johann C, Maeder K. Investigations on the structure of solid lipid nanoparticles (SLN) and oil-loaded solid lipid nanoparticles by photon correlation spectroscopy, field-flow fractionation and transmission electron microscopy. *J Controlled Release,* 2004 (95):217–227.

Tobio M, Gref R, Sanchez A, Langer R, Alonso MJ. Stealth PLA-PEG nanoparticles as protein carriers for nasal administration. *Pharm Res* 1998 (15):270–275.

Zur Mu¨ hlen A, zur Mu¨ hlen E, Niehus H, Mehnert W. Atomic force microscopy studies of solid lipid nanoparticles. *Pharm Res*, 1996 (13):1411–1416.

Shi HG, Farber L, Michaels JN, et al. Characterization of crystalline drug nanoparticles using atomic force microscopy and complementary techniques. *Pharm Res*, 2003 (20):479–484.

Liu J. Scanning transmission electron microscopy of nanoparticles. In: Wang ZL, ed. Characterization of Nanophase Materials. Weinheim: Wiley-VCH Verlag, 2000:81–132.

Hitchcock AP, Morin C, Heng YM, Cornelius RM, Brash JL. Towards practical soft X-ray spectromicroscopy of biomaterials. *J Biomaterials Sci Polymer Edition*, 2002 (13):919–937.

De Serio M, Zenobi R, Deckert V. Looking at the nanoscale: scanning near-field optical microscopy. TrAC, *Trends Anal Chem*, 2003 (22):70–77.

De Campos AM, Diebold Y, Carvalho ELS, Sanchez A, Alonso MJ. Chitosan nanoparticles as new ocular drug delivery systems: in vitro stability, in vivo fate, and cellular toxicity. *Pharm Res,* 2004 (21):803–810.

Colfen H, Antonietti M. Field-flow fractionation techniques for polymer and colloid analysis. *Adv Polym Sci,* 2000, 150(New Developments in Polymer Analytics I):67–187.

Tan JS, Butterfield DE, Voycheck CL, Caldwell KD, Li JT. Surface modification of nanoparticles by PEO/PPOblock copolymers to minimize interactions with blood components and prolong blood circulation in rats. *Biomaterials*, 1993 (14):823–833.

Williams A, Varela E, Meehan E, Tribe K. Characterization of nanoparticulate systems by hydrodynamic chromatography. *Int J Pharmaceutics,* 2002 (242):295–299.

DosRamos JG. Recent developments on resolution and applicability of capillary hydrodynamic fractionation (CHDF). *Polymeric Mat Sci Eng.,* 2002 (87):338.

Blom MT, Chmela E, Oosterbroek RE, Tijssen R, Van den Berg A. On-chip hydrodynamic chromatography separation and detection of nanoparticles and biomolecules. *Anal Chem*, 2003 (75):6761–6768.

Westesen K, Siekmann B, Koch MHJ. Investigations on the physical state of lipid nanoparticles by synchrotron radiation X-ray diffraction. *Int J Pharmaceutics*, 1993 (93):189–199.

Hunter RJ, ed. Colloid Science: Zeta Potential in Colloid Science: Principles and Applications. London: Academic Press, 1981.

McNeil-Watson F, Tscharnuter W, Miller J. A new instrument for the measurement of very small electrophoretic mobilities using phase analysis light scattering (PALS). Colloids Surf, A: Physicochemical and Engineering Aspects 1998 (140):53–57.

Mu¨ller RH. Hydrophobic interaction chromatography (HIC) for determination of the surface hydrophobicity of particulates. In: Particle and Surface Characterisation Methods, Based on the Invited Lectures presented at the Colloidal Drug Carriers Expert Meeting, 2nd, Mainz, Mar. 6, 1997. Mu¨ ller RH, Mehnert W, Hildebrand GE, eds. 1997.

Goeppert TM, Mu¨ ller RH. Alternative sample preparation prior to two-dimensional electrophoresis protein analysis on solid lipid nanoparticles. *Electrophoresis*, 2004 (25):134–140.

Vauthier C, Schmidt C, Couvreur P. Measurement of the density of polymeric nanoparticulate drug carriers by isopycnic centrifugation. *J Nanoparticle Res* 1999, (1):411–418.

Huve P, Verrecchia T, Bazile D, Vauthier C, Couvreur P. Simultaneous use of size-exclusion chromatography and photon correlation spectroscopy for the characterization of poly(lactic acid) nanoparticles. *J Chromatogra*, 1994, (675):129–139.

Khlebtsov BN, Kovler LA, Bogatyrev VA, Khlebtsov NG, Shchyogolev SY. Studies of phosphatidylcholine vesicles by spectroturbidimetric and dynamic light scattering methods. *J Quant Spectr Rad Transfer*, 2003 (79–80):825–838.

Teipel U. Problems in characterizing transparent particles by laser light diffraction spectrometry. *Chem Eng Technol*, 2002 (25):13–21.

Kommareddy S, Tiwari SB, Amiji MM. Long-circulating polymeric nanovectors for tumor-selective gene delivery. *Technol Cancer Res Treat*, 2005 (4): 615-25.

Mu L, Feng SS. A novel controlled release formulation for the anticancer drug paclitaxel (Taxol(R)): PLGA nanoparticles containing vitamin E TPGS. *J Control Release,* 2003 (86): 33-48.

Reverchon E, Adami R. Nanomaterials and supercritical fluids. The *Journal of Supercritical Fluids,* 2006, (37): 1-22.

Florence AT, Hussain, N. Transcytosis of nanoparticle and dendrimer delivery systems: evolving vistas. *Adv Drug Deliv Rev*, 2001, 50(S1):69-89.

4
Aerosol Technology in Drug Delivery

Introduction

Pulmonary drug delivery is the preferred route of administration of aerosolized drugs in the treatment of chronic respiratory diseases such as asthma and chronic obstructive pulmonary disease (COPD) as well as in the treatment of some non-respiratory diseases. This route has several advantages over other delivery routes such as: drug delivery by inhalation utilizes the extensive surface area of the alveoli (approximately 100 m^2), which are in close proximity to blood flow; it avoids hepatic first-pass metabolism and are effective at lower doses as compared to that of oral doses and there are fewer associated side-effects and enables non-invasive administration of drugs.

Aerosol: A product, which is packaged under pressure and containing therapeutically active ingredients that are released upon activation of an appropriate valve system; it is intended for topical application to the skin as well as local application into the nose (nasal aerosols), mouth (lingual aerosols), or lungs (inhalation aerosols). The drugs, delivery by aerosols are deposited in the airways by: gravitational sedimentation, inertial impaction, and diffusion. Mostly larger drug particles are deposited by first two mechanisms in the airways, while the smaller particles get their way into the peripheral region of the lungs by following diffusion.

The term aerosol describes a nebulized solution consists of very fine particles carried by a gas (usually air) to the site of therapeutic application. When the site of application is the alveoli and small bronchioles, the medicament must be dispersed as droplets of roughly 5 micron diameter. When the target is the nasal and pharyngeal region, larger droplets are appropriate.

Other devices for generating aerosols employ compressed gases, usually hydro-fluorocarbons and chloro-fluorocarbons, which are mixed with the medicament and other necessary excipients in a pressurized container.

Pulmonary delivery of drugs can be achieved by using pressurized metered dose inhalers (pMDIs), dry powder inhalers (DPIs), and nebulizers.

Aerosol-Foam: A dosage form containing one or more active ingredients, surfactants, aqueous or non-aqueous liquids, and the propellants; if the propellant is in the internal (discontinuous) phase (i.e., oil-in-water type), a stable foam can be discharged and if the propellant is in the external (continuous) phase (i.e., water-in-oil type), a spray or a quick-breaking foam can be discharged.

Aerosol-Metered: A pressurized dosage form consists of a metered dose valve, which allows the delivery of a uniform quantity in the form of spray up on each actuation.

Aerosol-Powder: A product, which is packaged under pressure and contains therapeutically active ingredients in the form of a powder that are released upon activation of an appropriate valve system.

Aerosol-Spray: An aerosol product which utilizes a compressed gas as the propellant to provide the force necessary to expel the product as a wet spray; it is applicable to solutions of medicinal agents in aqueous solvents.

Current Research in Pulmonary Drug Delivery

Other than the drugs such as decongestants, steroids, estradiol, butorphenol, nicotine, desmopresin, metoclopramide, which are well established in the form of aerosol delivery the new drugs such as: peptides, proteins, and gene-based therapies for treating local and systemic diseases are the current focus of research in pulmonary drug delivery. The following products are of major interest: insulin, interferon, erythropoietin, α_1-antitrypsin, calcitonin, and factor VIII.

The delivery of drugs through the pulmonary route is a potentially effective form of therapy for patients with chronic disease, which include: the debilitating hereditary disease, cystic fibrosis, and type-I diabetes (insulin is absorbed well through the lungs). Pulmonary delivery of aerosolized insulin may ultimately have a significant impact on quality of life by reducing the number of daily insulin injections for patients with type-I diabetes.

4.1 Formulation of Aerosol

The aerosol product essentially consists of four components such as: a) product concentrate, b) Propellant, c) Container, and d) Valve and actuator.

4.1.1 Product Concentrate

The Product concentrate consists of active ingredients or mixture of active ingredients and other such necessary agents (excipients) such as solvents, antioxidants and surfactants, etc.

4.1.2 Propellant

The propellant may be single propellant or a blend of various propellants. This is equivalent to other vehicles used in pharmaceutical formulations. Just as a blend of solvents is used to achieve desired solubility characteristics or various surfactants are mixed to get the desired HLB value for an emulsion system, the propellant is selected to give the desired vapor pressure, solubility and particle size and help in delivering the drugs to the desired site in desired forms. Propellants can be combined with active ingredients in many different ways producing products with varying characteristics. Depending on the type of aerosol system utilized, the pharmaceutical aerosol may be dispensed as a fine mist, wet spray, and quick–breaking foam, stable foam, semi-solid or solid. The type of system selected depends on many factors such as: physical, chemical and pharmacological properties of active ingredients and site of application. Various types of propellants used are:

1. Fluorinated Hydrocarbons such as Trichloromonofluoromethane (propellant-11), Dichlorodifluoromethane (Propellant-12), and Dichlorotetrafluoroethane (propellant-114) find wide application in the oral, inhalation and topical route. A range of pressures can be obtained by mixing of different blends of fluorocarbons. The chemical inertness, lack of toxicity, lack of inflammability and explosiveness make the fluorocarbon ideal for use, but because of ozone depletion in environment the use is confined to drugs only unless and until other such propellants are compatible with drug concentrate.
2. Hydrocarbons such as Propane, Butane, and Isobutane alone as well as their blends are also utilized for the aerosols
3. Compressed gases such as Nitrogen, Carbon dioxide, and Nitrous oxide are also utilized for topical applications.

4.1.3 Containers

Various materials have been utilized for the manufacturing of aerosol containers, which must with stand pressure as high as 140 to 180 psig at 130^0F such as:

Metal

1. Tinplated steel
   (a) Side seam (three-Piece)
   (b) Two piece or drawn
   (c) Tin free steel
2. Aluminum
   (a) Two piece
   (b) One piece (extruded or drawn)
3. Stainless steel
4. Glass
   (a) Uncoated glass
   (b) Plastic coated glass
5. Plastics

The containers must be chosen looking to the compatibility with the product concentrate.

4.1.4 Valves and Actuators

A typical Aerosol Valve (Continuous spray Valve) is made up from several components as follows:

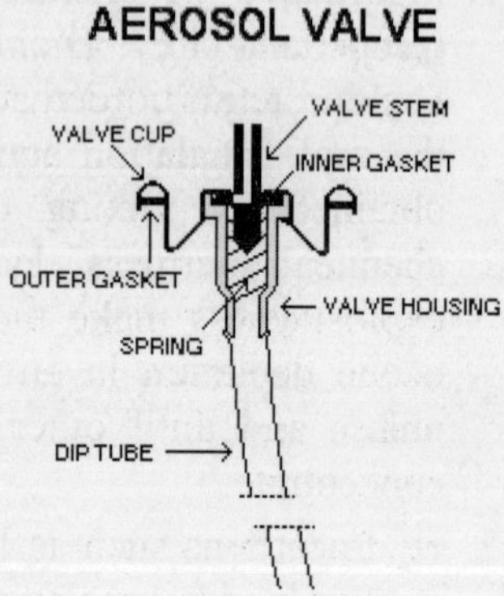

- *Valve Cup*: typically constructed from Tin-plated steel or aluminium.
- *Outer Gasket*: this is the seal between the valve cup and the aerosol can.
- *Valve Housing*: contains the valve stem, spring and inner gasket.
- *Valve Stem*: in effect, the tap through which the product flows.
- *Inner Gasket*: covers the hole in the valve stem.
- *Valve Spring*: usually stainless steel.
- *Dip Tube*: allows the liquid to enter the valve.
- *Actuator (not shown)*: fits onto the valve stem.

The valve constructed as shown in the figure 4.1, to which the actuator may also be fitted. The valve stem is fitted with a small hole, through which the product flows. Some valves may contain two or even four holes, depending on the nature of the product to be dispensed. The holes are very small, with diameters as low as 0.30mm, and as high as 1.00mm. In the closed position the hole(s) is covered by the inner gasket. When the actuator is depressed it pushes the valve stem through the inner gasket, and the hole(s) is uncovered, allowing liquid to pass through the valve and into the actuator.

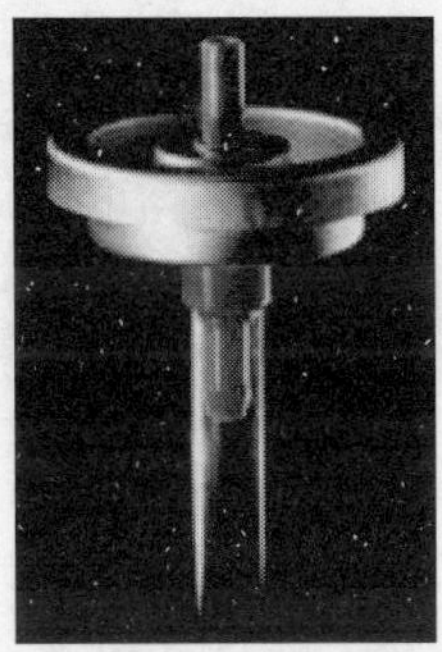

Fig. 4.1 Continuous spray valve

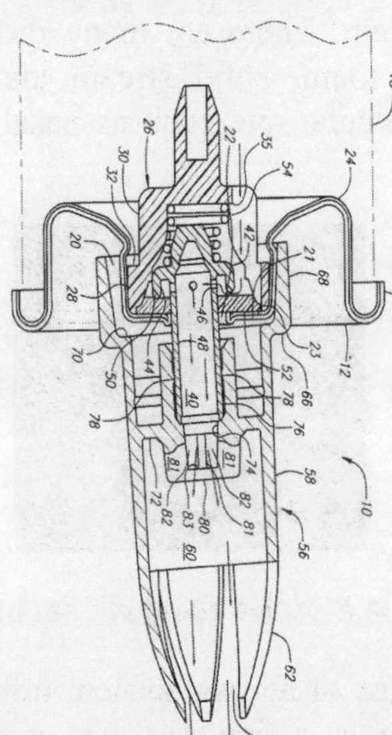

Metering Valve: 14-Mounting Ferrule, 78-Metering Chamber, 64-Dip Tube, 48-Stainless Steel Stem, 26-Sealing Gasket

Fig. 4.2 Metering Valve

Metering Valves: Metering valves operates on the principle of a chamber, whose size determines the amount of medicament dispensed after each time actuation (Figure-4.2).

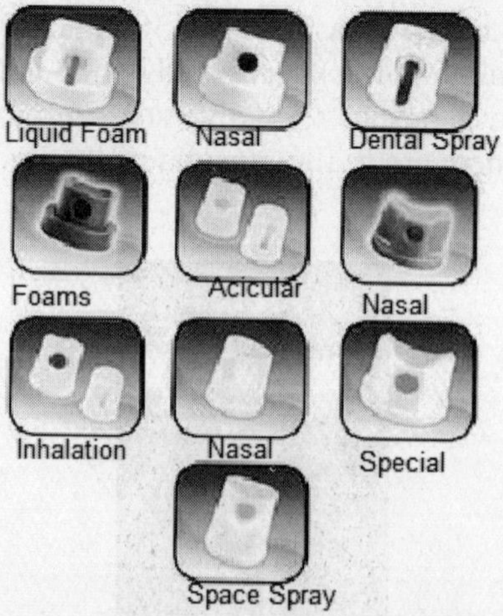

Fig. 4.3 Different types of actuators.

Actuator: The actuator allows for easy opening and closing valve, which is fitted to the valve stem. There are many different types of actuators which produce spray, foam, solid stream and special application to deliver the drug at particular site such as nasal, pharyngeal, inhalation, etc. (Fig 4.3).

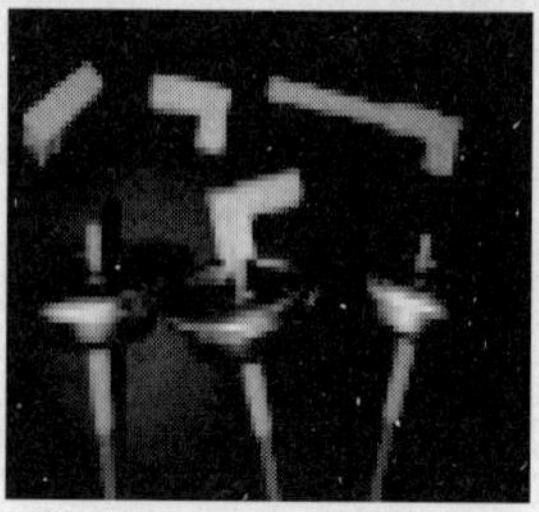

Fig. 4.4 Valve fitted with actuator.

Depending on the type of aerosol system utilized, the pharmaceutical aerosol may be dispensed as a fine mist, wet spray, and quick – breaking foam, stable foam, semi-solid or solid. The type of system selected depends on many factors such as: physical, chemical and pharmacological properties of active ingredients and site of application

The different types of aerosol formulation can be classified as follows:

1. Solution system or two phase system
2. Water-based system or three phase system
3. Suspension or dispersion systems
4. Foam Systems
   (a) Stable foam
   (b) Aqueous stable foam
   (c) Non aqueous stable foam
   (d) Quick breaking foam
5. Intranasal aerosol.

Solution system: A large number of aerosols can be formulated as a solution system which consists of two phases: a vapor phase and a liquid phase. Here the active ingredient is soluble in the propellant and thus forms a solution of the drug or active ingredient. Depending on the type of spray required, the propellant many consist of:

1. Propellant 12 or A-70 (which procedure very fine particles because of higher vapor pressure)
2. A mixture of propellant 12 and other propellants.

If the other propellants have lower vapor pressure, the pressure of the final system decreases resulting in larger particles. Also lowering of the vapor pressure is produced through the addition of less volatile solvents such as ethyl alcohol, propylene glycol, ethyl acetate, glycerin and acetone. These are solvents added to increase the solubility of drug in the propellant. The amount of propellant used may vary from 5% for foams to 95% for inhalation products of the entire formulation. Spray systems with large particles are also useful for topical preparations, since they tend to coat the affected area with a film of active ingredients.

For Examples:

Aerosol intended for inhalation or for local activity in the respiratory system in the treatment of asthma may be formulated as follows:

Ingredients	Weight %ge
Isoproterenol HCl	0.25
Ascorbic Acid	0.10
Ethanol	35.75
Propellant-12	63.90

Water based systems: These are three phase system contain large amounts of water added to replace all or part of the non aqueous solvent used in aerosol. Depending on the formulation, they are emitted as a spray or foam to produce a spray, the formulation must consist of dispersed active ingredients and other solvents in an emulsion system in which the propellant is in the external phase. In this way, when the product is dispensed, the propellant vaporizes and dispenses the active ingredients into minute particles. Since propellant and water are not miscible, a three phase aerosol forms (propellant phase, water phase and vapor phase)

Suspension or Dispersion Systems: Various methods have been used to overcome the difficulties arises due to addition of a co-solvent to the propellant i.e., especially alcohol, which is flammable and causes chilling sensation, which is used to suspend or disperse the drug in a propellant or mixture of propellants. To prevent or decrease the rates of settling of disperse particles, various surfactants or suspending agents are added to the system. These systems used for oral inhalation.

For Example:

Ingredients	**Weight %ge**
Epinephrine bitartrate (1-5 micron)	0.50
Sorbitan Trioleate	0.50
Propellant-114	49.50
Propellant-12	49.50

Foam Systems: Foam systems are mainly used for external preparations and not used for pulmonary delivery. However, they form an essential part of the aerosol product and has therefore been described in detail below. Emulsion and foam aerosols consist of active ingredients, aqueous or non-aqueous vehicle, surfactant and propellant and are dispensed as stable or quick breaking foam depending on the nature of the ingredients and the formulation. The liquefied propellant is emulsified and is generally found in the internal phase. Non-aerosol emulsions are usually dispersed in lotion or viscous liquid form, but aerosol emulsions are dispensed as foams and can be advantageous for various applications involving irritating ingredient or when the material is applied to a limited area.

Aqueous stable foams: As the name indicates, aqueous stable foams contain water as the vehicle. In order to impart stability to the formulation, a large concentration of lipids is incorporated as oil in water emulsion. The general formula for the foam, the different types of lipids used and their concentration are given below:

For Example:

Ingredients	Weight %ge
Active ingredients	2
Emulsion Base (O/W Type)	95
Hydrocarbon Propellant	3

Non aqueous stable foams: Non-aqueous stable foams are formulated by the use of various glycols such as polyethylene glycol. A general formula is:

For Example:

Ingredients	Weight %ge
Active ingredients	2
Glycerol	95
Hydrocarbon Propellant	3

Quick – Breaking Foams: Here, the propellant is present in low concentration and in the external phase. When the product is dispensed it is emitted as foam which immediately collapses into liquid, which is applicable to topical medication, applied to limited or large areas without the use of a mechanical force to dispense the active ingredients.

For Example:

Ingredients	Weight %ge
Ethyl alcohol	62
Surfactant	5
Water	30
Hydrocarbon Propellant	3

Intranasal Aerosols: Drug delivery systems intended for the deposition of medication into the nasal route is being utilized as an effective means of administering drugs, intended to produce either local or systemic effect. Traditionally, the modes of administering intranasal preparation have been limited to nasal drops, nasal sprays (non-pressurized), inhalants, intranasal gels (jellies), cream and ointments. A new popular means of delivering the drugs to the intranasal route is pressurized metered nasal aerosols. Intranasal aerosol offers various advantages such as: 1. Delivery of a measured dose of drug; 2. Excellent depth of penetration into the nasal passage with minimal inadvertent penetration into the lungs; 3. Reduced droplet or particle size; 4. Lower dosage; 5. Maintenance of sterility from dose to dose; 6. Greater patient compliance;

7. Decreased mucosal irritability and 8. Greater flexibility in product formulation. The basic formula for intranasal aerosol suspension is:

For Example:

Ingredients	Weight %ge
Active ingredients (micronized)	1
Solvents (dispensing agent)	1
Propellant-12/11 (60:40)	98

Manufacturing of Aerosols

The steps involved:

- Formulation of product concentrate
- Filling of product concentrate to the container
- Replacement of valve and fitting to the neck of the container
- Filling of Propellants to the container

The filling of propellants can be done by

1. **Pressure Filling** in which a pressure burette capable of metering small volumes of liquefied gas under pressure fills in to the container.
2. **Cold Filling** in which an insulated box fitted with copper tubing that has been coiled to increase the area exposed to cooling surface, filled with dry ice/acetone prior to use. This system can be used with metered valves as well as with non-metered valves to fill the hydrocarbon propellants to the container.
3. **Compressed Gas Filling** in which the compressed gas under pressure utilizing a pressure reducing valve through the delivery gauge fitted with a filling head and flow indicator used to fill the container.

Quality Control Tests of Aerosols

The aerosols can be evaluated by the series of Physical, chemical, and biological tests, which include:

A. Flammability and combustibility

- Flash point (i.e. measurement of ignition temperature; applicable for hydrocarbon propellant)
- Flame extension, including flashback (extension of flame on flushing of spray to the flame)

B. Physicochemical properties:
- Vapor Pressure (measured by pressure gauze)
- Density (measured by Hydrometer of Pycnometer)
- Moisture content (Karl Fischer apparatus)
- Identification of propellant (Gas Chromatography)
- Concentrate-Propellant ratio (Gas Chromatography)

C. Performance
- Aerosol valve discharge rate (by reweighing the container after each actuation)
- Spray pattern(impingement of a spray on a piece of paper)
- Dosage with metered valve (assay of content of medicament from each actuation)
- Net content (reweighing the contents in the filling line)
- Foam Stability (visual observation)
- Particle size determination(Coulter counter)
- Leakage

D. Biological Testing (determination of therapeutic activity *in-vivo*)

4.2 New Horizons in the Pulmonary Drug Delivery

Advances in device technology led to the development of more efficient delivery systems capable of delivering larger doses and finer particles into the lung. As more efficient pulmonary delivery devices and sophisticated formulations become available, physicians and health professionals have a choice of a wide variety of device and formulation combinations that will target specific cells or regions of the lung, avoid the lung's clearance mechanisms and be retained within the lung for longer periods. The development of an inhalant therapy that is efficacious and safe depends not only on a pharmacologically active molecule, but also on a well-designed delivery system and formulation. It is the optimization of the whole system (drug, drug formulation and device) that is necessary for the successful development of inhalation therapies, both new and old, for the treatment of local and systemic diseases. Drug–device combinations must aerosolize the drug in the appropriate particle size distribution and concentration to ensure optimal deposition and dose in the desired region of the lung. The development of modern inhalation devices can be divided into three different categories, the refinement of the nebulizer and the evolution of two types of compact portable devices, the metered-dose inhaler (MDI) and the dry powder inhaler (DPI).

4.2.1 Nebulizers

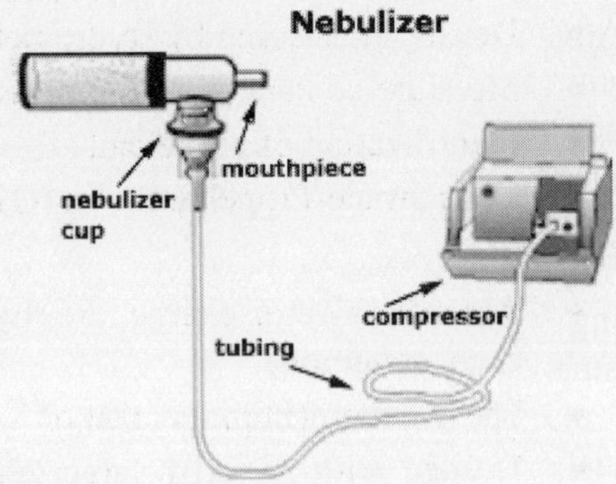

Nebulizers have been used for many years to treat asthma and other respiratory diseases. There are two basic types of nebulizer available such as: jet and ultrasonic nebulizers. The jet nebulizer functions by the Bernoulli's principle by which compressed gas (air or oxygen) passes through a narrow orifice creating an area of low pressure at the outlet of the adjacent liquid feed tube. This results in drug solution being drawn up from the fluid reservoir and shattered into droplets in the gas stream. The ultrasonic nebulizer uses a piezoelectric crystal, vibrating at a high frequency (usually 1–3 MHz) to generate a fountain of liquid in the nebulizer chamber; the higher the frequency, the smaller the droplets produced.

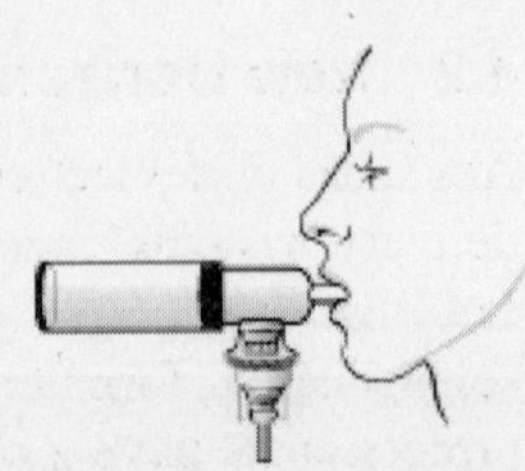

Constant output jet nebulizers can aerosolize most drug solutions and provide large doses with very little patient co-ordination or skill. Treatments using these nebulizers can be time-consuming but are also inefficient, with large amounts of drug wastage (50% loss) with continuously operated nebulizers. Most of the prescribed drug never reaches the lung with nebulization. The majority of the drug is either retained within the nebulizer (referred to as dead volume) or released into the environment during expiration. On average, only 10% of the dose placed in the nebulizer is actually deposited in the lungs.

Limitations

The physical properties of drug formulations and particle size may have an effect on nebulization rates. The viscosity, ionic strength, osmolarity, pH and surface tension may prevent the nebulization of some formulations. If the pH is too low, or if the solution is hyper- or hypo-osmolar, the aerosol may induce bronchoconstriction, coughing and irritation of the lung mucosa. As well, high drug concentrations may decrease the drug output with some nebulizers; colomycin at concentrations >75 mg ml^{-1} foams in all nebulizers, especially ultrasonic ones, making aerosolization of the drug very inefficient.

Advances in technology have led to the recent development of novel nebulizers that reduce drug wastage and improve delivery efficiency. Enhanced delivery designs (Pari LC Star, Pari, Germany) increase aerosol output by directing auxiliary air, entrained during inspiration, through the nebulizer, causing generation of aerosol to be swept out of the nebulizer and available for inhalation. Drug wastage during exhalation is reduced to the amount of aerosol produced by the jet air flow rate that exceeds the storage volume of the nebulizer. Adaptive aerosol delivery (Halolite; Medic-Aid, Bognor Regis, UK) monitors a patient's breathing pattern in the first three breaths and then targets the aerosol delivery into the first 50% of each inhalation. This ensures that the aerosol is delivered to the patient during inspiration only, thereby eliminating drug loss during expiration that occurs with continuous output nebulizers. The results of a radiolabelled deposition study showed that 60% of the emitted dose from the Halolite device is deposited in the lungs, 37% in the oropharynx and stomach and only 3% is lost to the environment. A number of metered-dose liquid inhalers, including AERx (Aradigm, Hayward, CA, USA), AeroDose (AeroGen, Sunnyvale, CA, USA) and Respimat (Boehringer Ingelheim, Ingelheim, Germany), have been developed that produce a fine aerosol in the respirable range by forcing the drug solution through an array of nozzles with 30–75% of the emitted dose being deposited in the lungs.

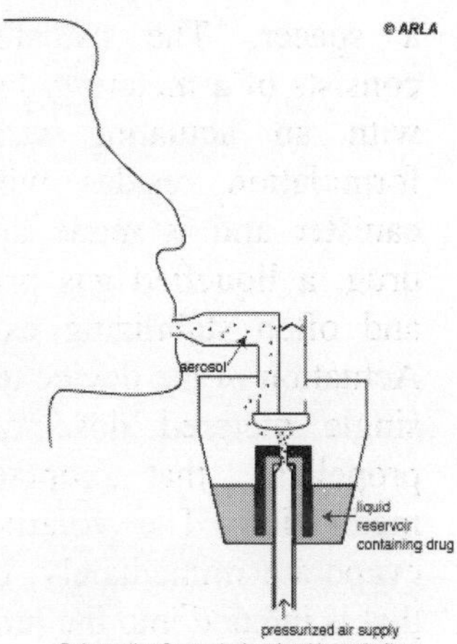

Schematic of a vented, valved jet nebulizer.

Examples of some commercial products:

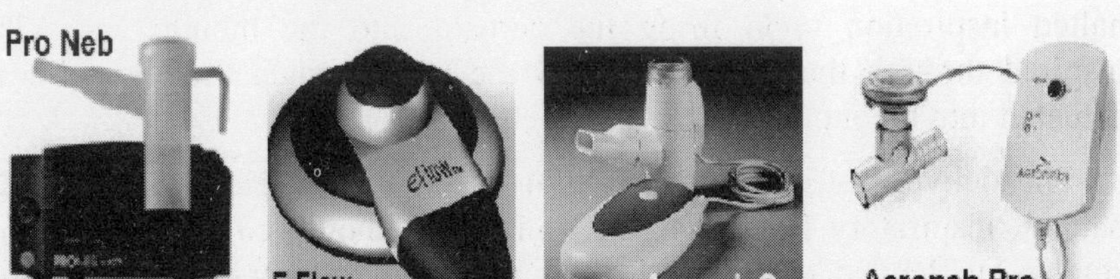

(Air-Jet) (Ultrasonic Mesh) (Ultrasonic) (Inspiration Sensing)

(Nebulizers)

4.2.2 Metered-dose Inhalers (MDI)

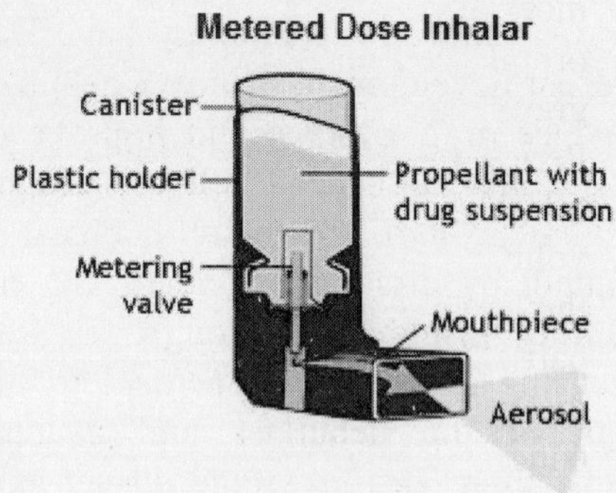

The MDI device consists of a canister, actuator, and sometimes a spacer. The canister itself consists of a metering dose valve with an actuating stem. The formulation resides within the canister and is made up of the drug, a liquefied gas propellant, and often stabilizing excipients. Actuation of the device releases a single metered dose of liquid propellant that contains the medication. The volatile propellant breaks up into droplets, which evaporate immediately, creating an aerosol containing micronized drug that is inhaled into the lungs.

The MDI was a revolutionary invention that overcame the problems of the hand-bulb nebulizer, as the first portable outpatient inhalation device and is the most widely used aerosol delivery device today.

The MDI emits a drug aerosol driven by propellants, such as chlorofluorocarbons (CFC) and more recently, hydrofluoroalkanes (HFAs) through a nozzle at high velocity ($> 30 \text{ m s}^{-1}$).

Limitations

MDIs deliver only a small fraction of the drug dose to the lung. Typically, only 10–20% of the emitted dose is deposited in the lung. The high velocity and large particle size of the spray causes approximately 50–80% of the drug aerosol to impact in the oropharyngeal region. Hand–mouth dis-coordination is another obstacle in the optimal use of the MDI. Crompton et al., found that 51% of patients experienced problems coordinating actuation of the device with inhalation; 24% of patients halted inspiration upon firing the aerosol into the mouth; and 12% inspired through the nose instead of the mouth when the aerosol was actuated into the mouth.

The delivery efficiency of a MDI depends on a patient's breathing pattern, inspiratory flow rate (IFR) and hand–mouth co-ordination. The studies by Bennett *et al.* and Dolovich *et al.* demonstrated that for any particle size between 1 and 5 μm mass median aerodynamic diameter (MMAD), deposition was more dependent on IFR than any other variable. Increases in IFR result in decreases in total lung dose deposition and penetration into the peripheral airways. Fast inhalations (> 60 l min^{-1}) result in a reduced peripheral deposition because the aerosol is more readily deposited by inertial impaction in the conducting airway and oropharyngeal regions. When aerosols are inhaled slowly, deposition by gravitational sedimentation in peripheral regions of the lung is enhanced. Peripheral deposition has also been shown to increase with an increase in tidal volume and a decrease in respiratory frequency. As the inhaled volume is increased, aerosols are able to penetrate more peripherally into the lungs. A period of breath holding on completion of inhalation enables particles which penetrate the periphery to be deposited in that region, instead of being exhaled during the expiratory phase. Thus, the optimal conditions for inhaling MDI aerosols are from a starting volume equivalent to the functional residual capacity, actuation of the device at the start of inhalation, IFR of <60 l min^{-1} followed by a 10-s breath-hold at the end of inspiration.

An assortment of different spacer tubes, valve holding chambers and mouthpiece extensions have been developed to eliminate co-ordination requirements, reduce the 'cold Freon®' effect (when the below-freezing spray temperature causes some patients to stop inhaling) and reduce the amount of drug deposited in the oropharynx by decreasing the particle size distribution and slowing the aerosol's velocity. The aerosol from a MDI and holding chamber is finer than that with the MDI alone, with an approximate 25% decrease in the MMAD compared with the original aerosol. This finer aerosol is more uniformly distributed in the normal lung, with increased delivery to the peripheral airway. However, in patients with airway obstructions, the addition of a holding chamber to the MDI does not change the distribution of the aerosol.

Breath-actuated MDIs have also been developed to eliminate co-ordination difficulties by firing in response to the patient's inspiratory flow. In patients with poor MDI technique, the breath-actuated pressurized inhaler i.e. Autohaler™ (3M Pharmaceuticals, Minnesota, USA) increased lung deposition from 7.2% (with a conventional MDI) to 20.8% of the dose. However, breath-actuated MDIs do not help patients who stop inhaling at the moment of actuation, nor do they improve lung deposition in those patients with good MDI technique. In addition, the oropharyngeal dose remains the same as for the MDI device. Patients preferred using the Autohaler™ to the MDI even though clinical outcomes were the same.

HFA-BDP solution (QVAR; 3M Pharmaceuticals, St Paul, MN, USA) is an extra-fine solution aerosol with a MMAD of 1.1 μm with 97% of the dose deposited below the larynx within the respirable range (<4.7 μm). In comparison, CFC-BDP suspension (Beclovent; GlaxoSmithKline, Middlesex, UK) has a MMAD of 3.2 μm with 77% of the dose deposited below the larynx in the respirable range. *In vivo* deposition studies in subjects with asthma found 53% of the emitted dose of QVAR deposited in the lung, 3.2-fold greater than the 16.2% of the emitted dose of Beclovent. Efficacy and safety studies have demonstrated that the small particle size of the HFA-BDP solution should be prescribed at half the dose of the CFC-BDP formulation to achieve a similar therapeutic effect without additional side-effects.

Examples of some commercial products

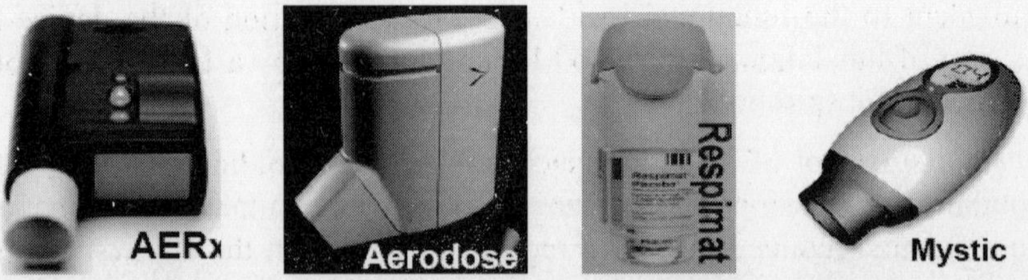

(Metered-dose inhalers)

4.2.3 Dry Powder Inhalers (DPI)

DPIs are an alternative to the aerosol based inhalers commonly MDIs, that deliver a powder dosage form to the lungs in the form of a dry powder. DPIs are commonly used to treat respiratory diseases such as asthma, bronchitis, emphysema and COPD. Although DPIs have also been used in the treatment of diabetes mellitus. Most DPIs include an active ingredient and one or more excipient to aid powder dispersion and flow. DPIs were designed to eliminate the co-ordination difficulties associated with the MDI. Renewed interest in this delivery system has emanated from the urgency to eliminate CFC-containing MDIs. There is a wide range of DPI devices on the market, from single-dose devices (e.g. Aerolizer, Rotahaler) to multiunit dose devices provided in a blister pack (e.g. Diskhaler), multiple unit doses sealed in blisters on a strip which moves through the inhaler (e.g. Diskus) or reservoir-type (bulk powder) systems (e.g. Turbuhaler).

Lung deposition varies among the different DPIs. Approximately 12–40% of the emitted dose is delivered to the lungs with 20–25% of the drug being retained within the device.

Limitations

Poor drug deposition with DPIs can be attributed to inefficient deaggregation of the fine drug particles from coarser carrier lactose particles or drug pellets. Slow IFR, high humidity and rapid, large changes in temperature are known to effect drug deaggregation and hence the efficiency of pulmonary drug delivery with DPIs. With most DPIs, drug delivery to the lungs is augmented by fast inhalation. For example, Borgstrom and colleagues demonstrated that increasing the IFR from 35 l min^{-1} to 60 l min^{-1} through the Turbuhaler™ increased the total lung dose of terbutaline from 14.8% of nominal dose to 27.7%. This is in contrast to the MDI, which requires slow inhalation and breath holding to enhance lung deposition of the drug. The Spiros DPI, however, deposits significantly more in the lung when IFRs <30 l min^{-1} are used.

With DPIs, the drug aerosol is created by directing air through loose powder. Most particles from DPIs are too large to penetrate into the lungs due to large powder agglomerates or the presence of large carrier particles (e.g. lactose). Thus, dispersion of the powder into respirable particles depends on the creation of turbulent air flow in the powder container. The turbulent airstream causes the aggregates to break up into particles small enough to be carried into the lower airways and also to separate carrier from drug. Each DPI has a different air flow resistance that governs the required inspiratory effort. The higher the resistance of the device, the more difficult it is to generate an inspiratory flow great enough to achieve the maximum dose from the inhaler. However, deposition in the lung tends to be increased when using high-resistance inhaler.

Active DPIs are being investigated that reduce the importance of a patient's inspiratory effort. By adding either a battery-driven propeller that aids in the dispersion of the powder (Spiros; Dura Pharmaceuticals, San Diego, CA, USA) or using compressed air to aerosolize the powder and converting it into a standing cloud in a holding chamber, the generation of a respirable aerosol becomes independent of a patient's inspiratory effort (Enhance Pulmonary Delivery System, Inhale Therapeutic Systems, San Carlos, CA, USA). However, DPIs in use today are breath actuated and are dependent on a patient's IFR of 30–130 l min^{-1} to achieve an aerosol within the respirable range.

Examples of some commercial products

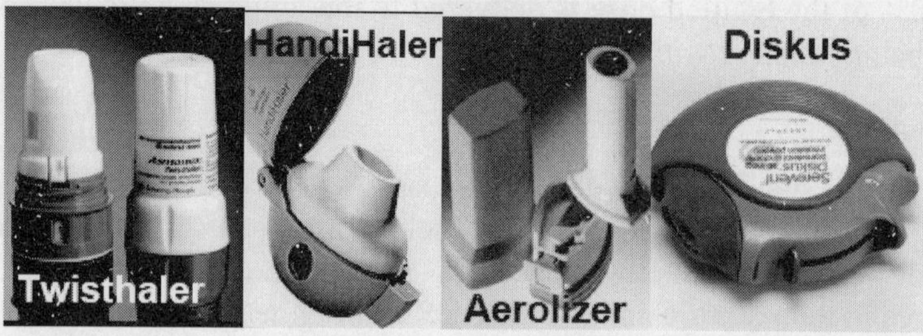

(Passive Dry Powder Inhaler)

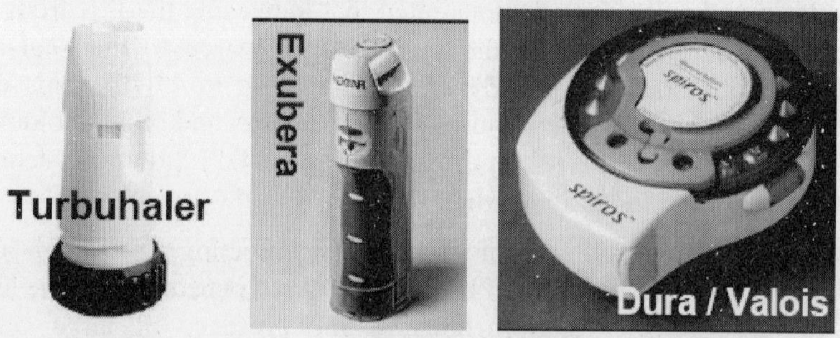

(PDPI) (Active Dry Powder Inhaler)

4.2.4 Strength and Weakness of Inhalers

(MDI)

Strength

- Convenient
- Rugged
- Standardized
- Inexpensive

Weakness

- Poor actuation-inhalation coordination
- Propellant induced gagging
- No dose counter
- Wrong inhaled flow rate
- Non-trivial to replace CFCs using HFAs
- Questionable dose uniformity & Fine Particle Fraction
- Limited dose

(DPI)

Strength
- Propellant-free
- High dose
- Drug stability
- Design flexibility

Weakness
- Inhalation dependent performance
- Non-standard operation
- Difficult to integrate device, formulation and processing,
- Questionable dose uniformity and FPF

(Nebulizer)

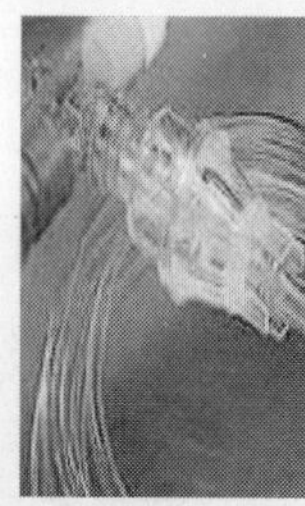

Strength
- Drug-independent performance
- Ease of formulation

Weakness
- Inconvenient
- Bulky
- Require routine Cleaning
- Long dosing time
- Separate regulation of drug & device
- Microbiological contamination
- Drug instability

(Nasal Spray)

Strength
- Ease of formulation
- Flexibility of excipients selection
- Availability

Weakness
- Dripping
- Need for repriming
- Frequent dosing
- No targeting ability
- Limited understanding of optimal mode of use
- Limited range of Drug

4.3 Biomedical Application of Aerosols

4.3.1 Pulmonary Drugs and Targets

Upper Respiratory Tract

Locally acting

- Bronchodilators
- Steroids
- Anti-infectives

Lower Respiratory Tract

Local and systemically acting

- Macromolecules
- Surfactant replacement
- Pain management

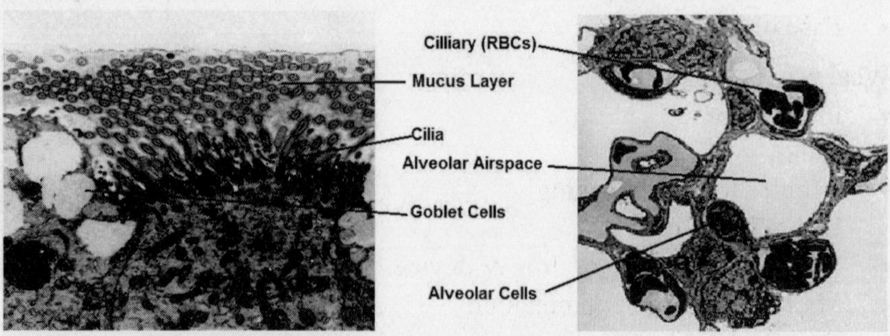

4.3.2 Nasal Drugs and Target

- Decongestants
- Steroids
- Estradiol
- Calcitonin
- Sumatriptan
- Butorphanol
- Zomitriptan
- Nicotine
- Desmopressin
- Metoclopramide

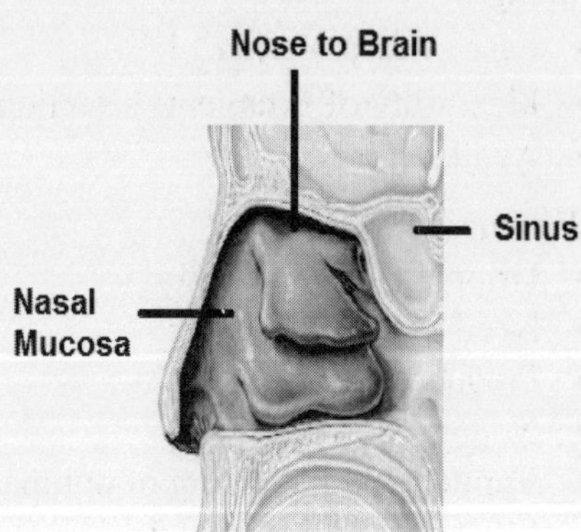

Pulmonary bioavailability of drugs could be improved by including various permeation enhancers such as surfactants, fatty acids, and saccharides, chelating agents and enzyme inhibitors such as protease inhibitors. Some reports suggest that pulmonary absorption of insulin was significantly enhanced in the presence of several adjuvants such as glycocholate, surfactant, span 85, and nafamostat. Calcitonin was delivered with various fatty acids, surfactants, and protease inhibitors and effect of these were studied for enhancement of absorption in the lungs to evaluate the pharmacological response and plasma calcium reduction.

Table 4.1 List of CFC-free inhalers

Name of Product	Active Ingredient of Drug	Manufacturer
Airomir	Salbutamol	3M Drug Delivery Systems
Asmol	Salbutamol	
Epaq	Salbutamol	3M Pharmaceutcials
Ventolin	Albuterol Sulfate	GlaxoSmithKline
Intal Forte CFC-free	Sodium cromoglycate	Aventis Pharmaceutcals
Flixotide	Fluticasone propionate	GlaxoSmithKline
Qvar	Beclomethasone dipropionate	3M Pharmaceuticals
Tilade CFC–free	Nedocromil Sodium	Aventis Pharmaceutcials
Seretide	Salmeterol xinafoate & fluticasone propionate	GlaxoSmithKline

4.3.3 Aerosols in Systemic Drug Delivery

In contrast to oral therapy the Aerosolized medications found to be effective in the fields of systemic drug delivery, gene therapy, and vaccination. This route of administration eliminates the potential for poor absorption and/or high metabolism in the gastrointestinal tract and it eliminates first-pass losses in the liver. Inhalation therapy is not associated with pain and provides patient comfort and compliance, leading to improved treatment outcome.

Inhaled Insulin to Treat Diabetes

Aerosol therapy in terms of systemic drug delivery is the development of insulin as an aerosol to treat diabetes. Devices and formulations that are in the development of inhaled insulin include the Nektar, Aradigm, and Aerogen products (Fig. 4.5).

Each of these products delivers aerosol containing a high percentage of 1–3 μm particles, which are considered optimal for targeting the alveolar lung region, and incorporate methods for controlling breathing parameters that are known to influence aerosol deposition in the lung (eg, inspiratory flow rate and lung volume at the time of inhalation). The Nektar device (Nektar Therapeutics, San Carlos, California) uses compressed air to disperse dry powder insulin into a spacer before inhalation. The patient then inhales insulin from the spacer during a slow, deep breath. The Aradigm device (Aradigm, Hayward, California) is a breath-actuated, aqueous mist inhaler. Liquid insulin aerosol is delivered electronically by means of mechanical extrusion when the patient's inspiratory flow rate and lung volume are appropriate. The Aerogen device (Aerogen Inc, Mountain View, California) is a breath-actuated, liquid aerosol inhaler that delivers insulin to the patient during inspiration by means of vibrating mesh technology

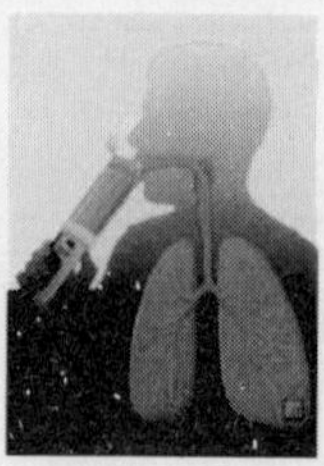

Dry, nonporous particles: Nektar Therapeutics

Liquid particles: Aradigm

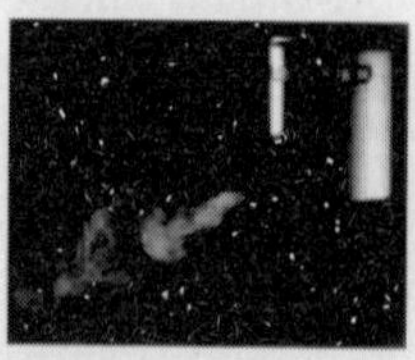

Liquid particles: Aerogen

Fig. 4.5 First-generation devices developed for the delivery of dry powder and liquid aerosol formulations of insulin.

Gene Therapy

It has been found that the role of aerosol therapy is expanded in the treatment of lung diseases with aerosolized gene therapy. There are a number of advantages to this form of therapy. First, aerosolized gene therapy provides a direct, noninvasive means for targeted delivery to different regions of the lung. Second, this route of administration delivers a high dose to the target site. Third, aerosolized gene therapy causes fewer adverse effects than intravenous administration.

Inhaled Complementary DNA to Treat Cystic Fibrosis

CF is an autosomal recessive disease and is the most common lethal genetic disease among whites. It is caused by mutations in the CF transmembrane conductance regulator (CFTR) gene located on chromosome-7 and is associated with defective chloride transport in airway epithelial cells. Lung pathology in CF includes abnormal chloride transport, increased mucus viscosity, decreased mucociliary clearance, recurrent infection, chronic inflammation, and airway destruction. The goal of aerosolized gene therapy in treating CF is to reconstitute CFTR function and normal chloride channel function in the lungs and the development of 3 gene-transfer agents: adenovirus, adeno-associated virus 2, and cationic liposomes. All 3 vectors have demonstrated the proof of principle for gene transfer in the airway. Recently, more efficient liquid aerosol delivery systems (i.e., soft mist inhalers) have been developed and are becoming commercially available

Inhaled DNA to Treat Lung Cancer

In-vivo and *in-vitro* studies have demonstrated that binding of DNA with cationic polypeptides such as polylysine, polyethyleneimine, protamine, and histones may be useful for gene delivery. These are nonviral vectors. Among these polypeptides, polyethyleneimine has received the most attention as a carrier for gene delivery because of its stability during nebulization. Studies have demonstrated that aerosol delivery of polyethyleneimine DNA complexes results in substantial gene expression in the lungs of mice, and aerosol polyethylenimine-p53 therapy and aerosol polyethyleneimine interleukin-12 therapy significantly reduce the number and size of osteosarcoma lung metastases in mice.

Vaccination

Vaccination via inhalation avoids the need for disposal strategies for the large number of needles that would be used in mass vaccination. It prevents the spread of blood-borne diseases such as human immunodeficiency virus (HIV), which can be transmitted by improper use and handling of used sharps. Several diseases are candidates for vaccination via inhalation, including measles, influenza, and rubella plus measles with a combination vaccine. MedImmune Vaccines developed and received Food and Drug Administration approval for FluMist, a live, attenuated influenza vaccine that is a liquid and is administered via nasal spray. It contains attenuated strains of influenza A (H1N1), A (H3N2), and influenza B viruses. Clinical trials with children and adults have demonstrated that intranasal administration of FluMist reduces the incidence of influenza and is well-tolerated. For measles vaccine, nebulizer delivery is efficacious.

Aerosolization of Lipoplexes using AERx® Pulmonary Delivery System

The AERx Pulmonary Delivery System has unique features that guide the patient to breathe in an optimal manner each time a dose is taken, which enhances reproducibility and efficiency of delivery. The AERx System delivers aerosolized medication from a dosage form comprising a blister containing 50 μL of liquid drug formulation and a micromachined nozzle array. The aerosol is generated by extruding the formulation under pressure through the array of holes. The AERx System can also incorporate a temperature controller to minimize the effects of ambient air conditions and enhance the generation of aerosol droplets optimal for pulmonary targeting. The rate limiting factor for gene therapy is the delivery of genetic material to target cell populations. Effective means to deliver genes via the pulmonary route will likely evolve through a combination of an efficient delivery device and an efficient delivery vector. The performance characteristics of the prototype lipoplex and other nonviral formulations with the AERx System suggest that this system can potentially be used for administering DNA-based drug products to the airways for the treatment of respiratory disorders.

References

Newhouse M. T, Managing reversible airflow obstruction: role of MDI accessory devices, powder inhalers and nebulisers. *Respirology*, 1990, June: 1-15.

Lipworth B. J, New perspectives on inhaled drug delivery and systemic bioactivity. *Thorax*, 1995 (50): 105-110.

Ariyananda P. L, S. W. Clarke, and J. E. Agnew, Aerosol delivery systems for bronchial asthma. *Postgrad. Med. J*, 1996 (72): 151-156.

Mortonen T and Y. Yang, Deposition mechanics of pharmaceutical particles in human airways. *In* A. J. Hickey, editor. Inhalational Aerosols: Physical and Biological Basis for Therapy. Marcel Dekker, New York, 1996, 1-21.

Ferron G. A, Aerosol properties and lung deposition. *Eur. Respir. J,* 1994:1392-1394.

Leon Lachman, H. A. Lieberman. The Theory and Practice of Industrial Pharmacy, Special Indian Edition, 2009:589-618.

Martin J Telko and Anthony J Hickey. Dry Powder Inhalar Formulation. Respiratory Care, September, 2005, 50(9): 1209-1227.

Mahavir B. Chougule, Bijay K. Padhi, Kaustubh A. Jinturkar and A. Mishra. Development of Dry Powder Inhalers. *Recent Patents on Drug Delivery and Formulation*, 2007, 1(1):11-21.

Beth L. Laube. The Expanding Role of Aerosols in Systemic Drug Delivery, Gene Delivery, and Vaccination. *Respiratory Care*, September, 2005, 50(9): 1161-1176.

Philip J. Thompson. Drug Delivery to the Small Airways. *Am J Respir Care Med*, 1998 (157): S199-S202.

Vasu V. Sethuraman and Anthony J. Hickey. Powder Properties and their Influence on Dry Powder Inhaler Delivery of an Antitubercular Drug. *AAPS Pharma Sci Tech*, 2002, 3(4): 1-10.

B. J. Lipworth, D. J. Clark. Lung delivery of salbutamol by dry powder inhaler(Turbuhalar) and Small volume antistatic metal spacer (Airomir CFC-free MDI plus NebuChamber). *Eur Respir J*, 1997 (10): 1820-1823.

R. Lizio, D. Marx, T. Nolte, C. M. Lehr, A. Werner S., G. Borchard, W. Jahn and T. Klenner. Development of a new aerosol delivery system for systemic pulmonary delivery in anaesthetized and orotracheal intubated rats. Laboratory Animals, 2001 (35): 261-270.

D. Despande, J Blanchard, S. Srinivasan, D. Fairbanks, J Fujimato, T. Sawa, J. W. Kronish, H. Schreier and I Gonda. Aerosolization of Lipoplexes Using AERx Pulmonary Delivery System. *AAPS Pharma Sci*, 2002, 4(3):1-10.

N R Labiris and M B Dolovich. Pulmonary drug delivery. Part II: The role of inhalant delivery devices and drug formulations in therapeutic effectiveness of aerosolized medications. *Br J Clin Pharmacol.* 2003, December, 56(6): 600–612.

Clark AR. Medical aerosol inhalers. Past, present and future. *Aerosol Sci Technol.* 1995 (22):374–391.

Grossman J. The evolution of inhaler technology. *J Asthma.* 1994 (31): 55–64.

Newman SP, Clarke SW. Inhalation devices and techniques. In: Clark TJH, Godfrey S, Lee TH, editors. Asthma. 3. London: Chapman & Hall; 1992:469–505.

Pedersen S. Inhalers and nebulizers: which to choose and why. *Resp Med.* 1996 (90):69-77.

Ganderton D. Targeted delivery of inhaled drugs: current challenges and future goals. *J Aerosol Med.* 1999, 12(1):s3–s8.

Dolovich M. New propellant-free technologies under investigation. *J Aerosol Med.* 1999, 12(1):s9–s17.

O'Callaghan C, Barry PW. The science of nebulized drug delivery. *Thorax.* 1997, 52(2):s31–s44.

Weber A, Morlin GL, Cohen M, Williams-Warren J, Ramsey BW, Smith AL. Effect of nebulizer type and antibiotic concentration on device performance. *Pediatric Pulmonol.* 1997 (23): 249–260.

Eschenbacher WL, Boushey HA, Sheppard D. Alterations in osmolarity of inhaled aerosols cause bronchoconstriction and cough, but absence of a permeant anion causes cough alone. *Am Rev Respir Dis.* 1984, (129):211–215.

Denyer J, Dyche T, Nikander K, et al. Halolite a novel liquid drug aerosol delivery system. *Thorax.* 1997, 52(6):208.

Smaldone GC, Agosti J, Castillo R, et al. Deposition of radiolabelled protein from AERx in patients with asthma. *J Aerosol Med.* 1999 (12):98.

Newman SP, Pavia D, Moren F, Sheahan NF, Clarke SW. Deposition of pressurized aerosols in the human respiratory tract. *Thorax*. 1981 (36):52–55.

Crompton GK. Problems patients have using pressurized aerosol inhalers. *Eur J Respir Dis*. 1982, 119(Suppl):101–104.

Bennett WD, Smaldone GC. Human ventilation in the peripheral air-space deposition of inhaled particles. *Am J Physiol*. 1987 (62):1603–1610.

Dolovich M, Ryan G, Newhouse MT. Aerosol penetration into the lung. Influence on airway responses. *Chest*. 1981, 80(Suppl 6):834–836.

Newman SP, Pavia D, Garland N, Clarke SW. Effects of various inhalation modes on the deposition of radioactive pressurized aerosols. Eur J Respir Dis. 1982, 63(119):57–65.

Pavia D, Thomson ML, Clarke SW, Shannon HS. Effect of lung function and mode of inhalation on penetration of aerosol into the human lung. *Thorax*. 1977 (32):194–197.

Dolovich M, Ruffin RE, Roberts R, Newhouse MT. Optimal delivery of aerosols from metered dose inhalers. *Chest*. 1981, 80(6):911–915.

Dolovich MB. Characterization of medical aerosols: physical and clinical requirements for new inhalers. *Aerosol Sci Technol*. 1995 (22):39 2–399.

Dolovich M, Chambers C, Mazza M, Newhouse MT. Relative efficiency of four metered dose inhaler (MDI) holding chambers (HC) compared to albuterol MDI. *J Aerosol Med*. 1992 (5):307.

Dolovich M, Ruffin R, Corr D, Newhouse MT. Clinical evaluation of the Aerochamber: a simple demand inhalation MDI delivery device. *Chest*. 1983, (84):36–41.

Dolovich M, Chambers C, Girard L, Newhouse MT. Aerosol delivery through an open tube spacer: importance of inhalation technique. *Am Rev Respir Dis*. 1989 (139):A144.

Byron PR. Aerosol Formulation, Generation, and Delivery Using Metered Systems. Proceedings of Respiratory Drug Delivery. Boca Raton, FL: CRC Press; 1990:167-205.

Davis SS. Merging Protein/Peptide Pharmaceuticals and Drug Delivery Systems. Symposium presented at IIR Drug Delivery Partnerships Conference in Los Angeles, CA; January 2002.

June DS, Schultz RK, Miller NC. A conceptual model for the development of pressurized metered-dose hydrofluoro-alkane-based inhalation aerosols. *Pharmaceut Technol*. 1994, (17):40-52.

Leach CL, Davidson PJ, Boudreau RJ. Improved airway targeting with the CFC-free HFA-beclomethasone metered-dose inhaler compared with CFC-beclomethasone. *Eur Respir J*.1998, 12(6):1346-1353.

5

Biotechnology in Drug Delivery

Introduction

Biotechnology is the use of biological processes, living organisms in engineering, technology, medicine and other fields requiring bio-products, or systems to manufacture products intended to improve the quality of human life. The concept encompasses a wide range of procedures for modifying living organisms according to human purposes-going back to domestication of animals, cultivation of plants, and improvements to these through breeding programs that employ artificial selection and hybridization.

The term "biotechnology" was coined in 1919 by Karl Ereky, a Hungarian engineer. Biotechnology has been described as "Janus faced", which implies that there are two sides. On one hand the techniques that allow DNA to be manipulated, i.e. to move genes from one organism to another. On the other hand, it involves relatively new technologies whose consequences are to be met with caution for the development of stem cells, gene therapy, and genetically modified organisms.

The science of biotechnology can be broken down into sub-disciplines called bioinformatics, red, white, green, and blue biotechnology. **Bioinformatics** is an interdisciplinary field which addresses biological problems using computational techniques, and makes the rapid organization and analysis of biological data. The field may also be referred as *computational biology*, and can be defined as, "conceptualizing biology in terms of molecules and then applying informatics techniques to understand and organize the information associated with these molecules, on a large scale. Bioinformatics plays a key role in various areas, such as **functional genomics**, **structural genomics**. **Red biotechnology** involves medical processes such as getting organisms to produce new drugs, or using stem cells to regenerate damaged human tissues and perhaps re-grow entire organs. **White** (also called gray) **biotechnology** involves industrial processes such as the production of new chemicals or the development of new fuels for vehicles. **Green biotechnology** applies to agriculture and involves such processes as the development of pest-resistant grains or the accelerated evolution of disease-resistant animals. **Blue biotechnology**, rarely mentioned, encompasses processes in marine and aquatic environments, such as controlling the proliferation of noxious water-borne organisms.

Biotechnologies also produced genetically alter bacteria used for cleanup of oils spills. One area of biotechnology uses organisms to manufacture organic products such as beer and milk products. Biotechnology is also used to recycle, treat waste, and clean up sites contaminated by industrial activities (bioremediation). Another area of biotechnology doesn't use living organisms at all, which include DNA micro-arrays used in genetics and radioactive tracers used in medicine. **Modern biotechnology** is often associated with the use of genetically altered microorganisms such as E. coli or yeast for the production of substances like insulin and antibiotics. Another promising new biotechnology application is the development of plant-made pharmaceuticals. Biotechnology is also commonly associated with landmark breakthroughs in new medical therapies to treat hepatitis B, hepatitis-C, cancers, arthritis, hemophilia, bone fractures, multiple sclerosis, and cardiovascular disorders. Modern biotechnology has evolved, making it possible to produce more easily and relatively cheaply human growth hormone, clotting factors for hemophiliacs, fertility drugs, erythropoietin and other drugs.

Biotechnology, like other advanced technologies, has the potential for misuse. Concern led to banning certain processes or programs, such as human cloning and embryonic stem-cell research. There is also concern that if biotechnological processes are used by groups with nefarious intent, the end result could be biological warfare.

5.1 Stem Cell Technology

Stem cells are special kind of cells that self-renew and have the ability to give rise to specialized cell types, which indicate when they divide; each daughter cell has a choice: it can either remain as a stem cell or take a course leading to terminal differentiation, i.e., give rise to a somatic cell body (Fig. 5.1).

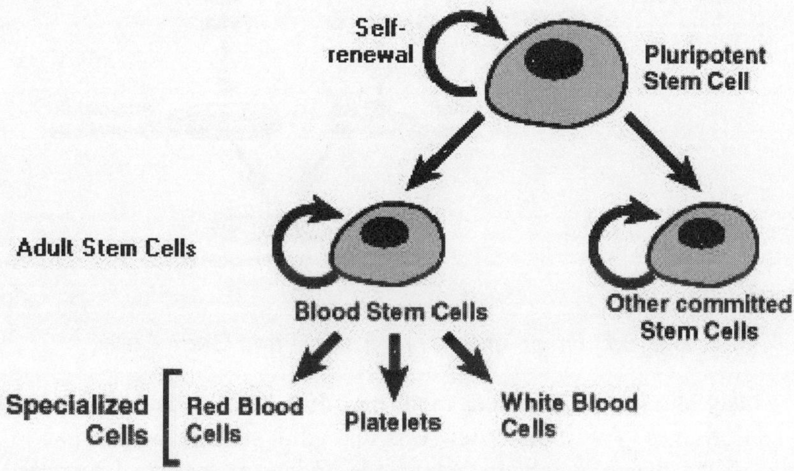

Fig. 5.1 Characteristics of Stem Cells.

In the 1960s, it was recognized that certain mouse cells had the capacity to form multiple tissue types, and the discovery of actual stem cells from mice occurred in 1971. **Zygote** and certain cells of the embryo were identified as the candidates exhibiting these characteristics. Since the zygote has the ability to give rise to the whole organism, it is regarded as ***totipotent*** (total potential). As the development proceeds, the zygote begins to divide giving rise to a hollow sphere of cells called the **blastocyst**. The blastocyst has a cluster of cells inside the hollow sphere, called the inner cell mass. These cells of embryonic origin are the **embryonic stem cells** (ESCs). Further in the developmental pathway, **embryonic germ cells** (EGCs) are derived from the primordial germ cells of the developing embryo (fig. 5.2). These embryonic cells are ***pluripotent*,** having the ability to give rise to cells of all the three germ layers (ectoderm, mesoderm and endoderm) but not the whole organism. In 1998, Dr. Thomson and Dr. Gearhart isolated human embryonic stem cells from excess of IVF clinic embryos and embryonic germ cells from aborted fetuses, respectively.

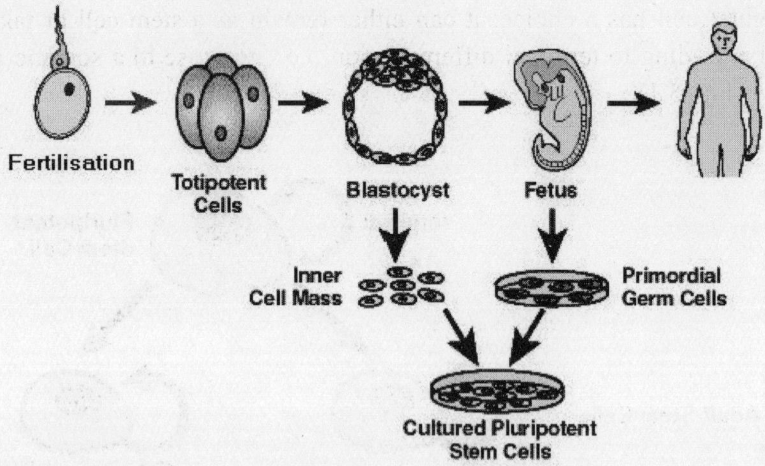

Fig. 5.2 Origin of Embryonic Stem and Germ Cells.

Today studies are focusing more on **adult stem cells** (ASCs) present in adult tissues. Traditional view was that adult stem cells give rise to the cells of the tissue in which they reside and were therefore regarded as *multi-potent*. But recent studies show their ability to give rise to cells of tissues other than the tissue of origin, exhibiting *plasticity*. Thus modern view regards them as *pluripotent*. These cells are responsible for the normal turnover and maintaining **homeostasis**. Example of adult stem cells is **hematopoietic stem cells** present in bone marrow, fetal liver and spleen and in placental and cord blood. There are three reasons for recent emphasis on adult stem cells. First, ethical issues associated with the use of *in vitro* fertilized eggs and aborted fetuses for derivation of ESCs and EGCs; Second, exhibition of plasticity by ASCs; Third, ASCs can be isolated from the patient and injected back after culturing, thereby avoiding the problem of immune rejection.

5.1.1 Development of Human Embryonic Stem Cells (hES)

The murine models of isolation, derivation, culture and characterization in ES cells provide valuable information to the generation of hES cell lines derived from the embryos at the pre-implantation stage, which involves culturing embryos to the morula or blastocyst stage (Fig. 5.3). Thomson et al. isolated the ES from the inner cell mass (ICM) of human blastocyst, placed on inactivated murine feeder cells, and successfully performed initial derivations of hES cell lines.

The inductive differentiation protocols have generated many cell types to further enrich *in vitro* differentiated populations by the use of selective methods that are based on the expression of specific marker protein. Investigators have used the forced gene expression to influence *in vitro* differentiation of ES cells.

For example, constitutive over expression of murine *Pax 4* in ES cells combined with an inductive protocol has resulted in an enrichment of nestin-positive progenitors and insulin-producing cells among other cells found in pancreatic islets.

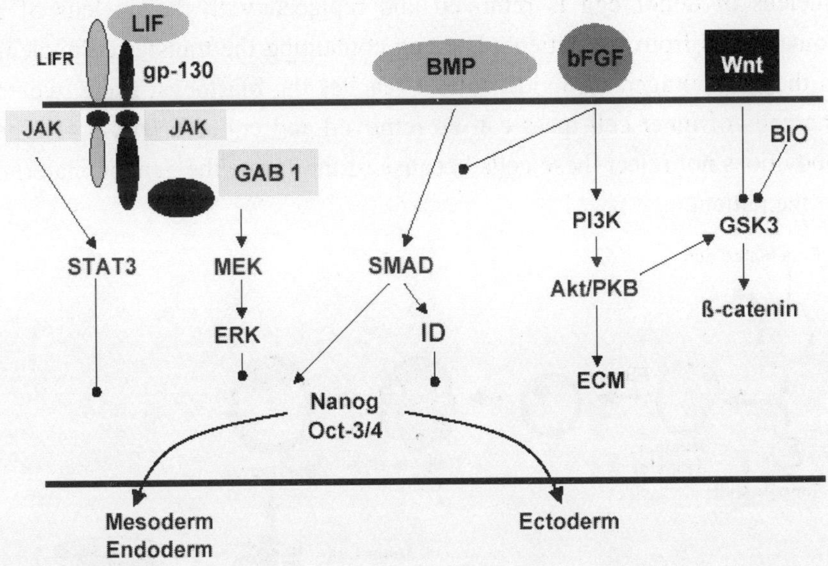

Fig. 5.3 Regulation pathways of self-renewal in the undifferentiated stages of ES cells.

[The undifferentiation of Es cells is regulated by Nanog, Oct-3/4, and interaction between LIF-dependent JAK/ATAT3 pathways principally in mouse EScells. In human Es cells, the mechanism involved is the BMP-dependent activation of ID target genes; the role of bFGF is to activate the (PI3K)/Akt/PKB pathway, which subsequently down regulates the expression of ECM molecules; and finally the Wnt pathway activation by specific pharmacological inhibitor BIO of GSK-3 maintains the undifferentiated phenotype in both mouse and human ES cells. The balanced expression level of Oct-3/4 determines the fate of ES cells. **Abbreviations:** LIF (Leukenia Inhibitor factor), BMP (bone morphogenic protein), bFGF (basic fibroblast growth factor), ECM (extra cellular matrix), BIO (6-bromoinduribin-3'-oxime), GSK3 (glycogen synthase kinage-3), Oct-3/4 (Octomer binding protein-3/4), STAT3 (signal transduction and activation of transcription factor), Nanog (one of the homeodomain protein), P13K/Akt (phosphatidylinositol 3-kinase)

(Figure ref.: ftp://ftb.wiley.com/public/sci_tech_med/pharmaceutical_biotech/.)]

5.1.2 Somatic Cell Nuclear Transfer (SCNT)

Another way of obtaining embryonic stem cells is by the technique known as Somatic Cell Nuclear Transfer (SCNT). In this technique, the nucleus of donor egg is removed and replaced with the nucleus of a somatic cell from the patient. The egg containing the transferred nucleus is then encouraged to divide until it reaches the blastocyst stage, where the cells of inner cell mass can be removed and cultured (fig. 5.4). The body does not reject these cells because of they have the genetic material of the patient.

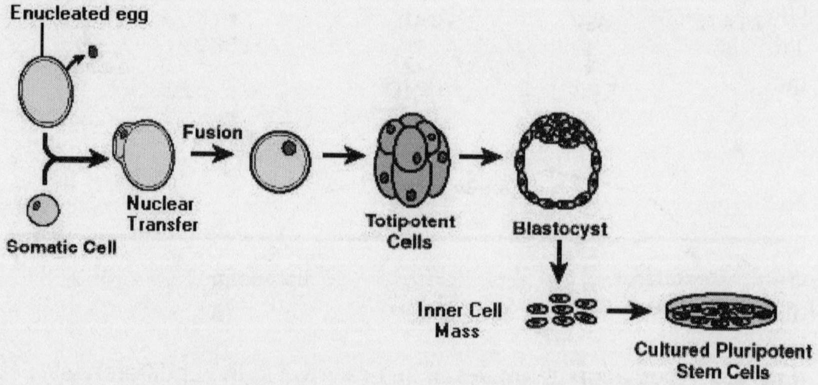

Fig. 5.4 Somatic Cell Nuclear Transfer.

Because of adult somatic stem cells consists of a different population of cells that share common characteristics it may be classified as: a) Hematopoietic stem cells (**HSCs**); b) Mesenchymal stem cells (**MSCs**); c) Multipotent adult progenitor cells (**MAPCs**) isolated by fluorescent-activated cell sorting (**FACs**) from the bone marrow; d) Side-population phenotype cells (**SPs**); e) Tissue-specific cell progenitors (**TSCPs**); and f) Umbelical cord blood-derived stem cells (**UCBDSs**).

5.1.3 Applications

Tissue Engineering: It is an emerging interdisciplinary field that applies the principles of biology and engineering for the development of viable substitutes that restore, maintain, or improve the function of human tissues or organs. This form of therapy differs from standard therapies in that the engineered tissue becomes integrated within the patient, affording a potentially permanent and specific cure of the disease state. This approach involves isolating cells from the body, using such techniques as stem cell therapy, placing them on or within structural matrices, and implanting the new system inside the body or using the system outside the body. Examples of this approach include repairment of bones, muscles, tendons, and cartilages, as well as cell-lined vascular grafts and artificial liver.

Therapeutic Applications

Type 1 Diabetes: Type 1 diabetes is an autoimmune disease characterized by destruction of insulin producing cells in the pancreas. Treatment involves human islet transplantation, which is limited by small number of donated pancreas and problem of graft rejection. Pluripotent stem cells instructed to differentiate into a particular pancreatic cell called β-cell, could overcome the shortage of transplant.

Nervous System Diseases: In Parkinson's disease, neurons that make the neurotransmitter dopamine depletion. In Alzheimer's patients, certain enzymes destroy the proteins produced by neurons and this leads to death of neuron. Similar loss of neurons occurs in spinal cord injury, brain trauma and even stroke. Researchers have isolated neural stem cells, which can replace the lost neurons, and also they have been able to form neuronal cells from hematopoietic stem cells in the laboratory.

Other Diseases: Bone marrow stem cells are used to rescue cancer patients following high dose chemotherapy or radiation therapy. Diseases of the bone and cartilage such as osteogenesis imperfecta and chondrodysplasias, immuno-deficiency diseases such as severe combined immuno-deficiency and many more can be cured using stem cell-based therapy. Apart from these, it also find potential applications in research for studying such as: how the complex developmental events take place; development of animal models of human diseases; use of pluripotent stem cell lines for drug development and safety testing; and use in gene therapy, in order to achieve long-term expression and therapeutic effect.

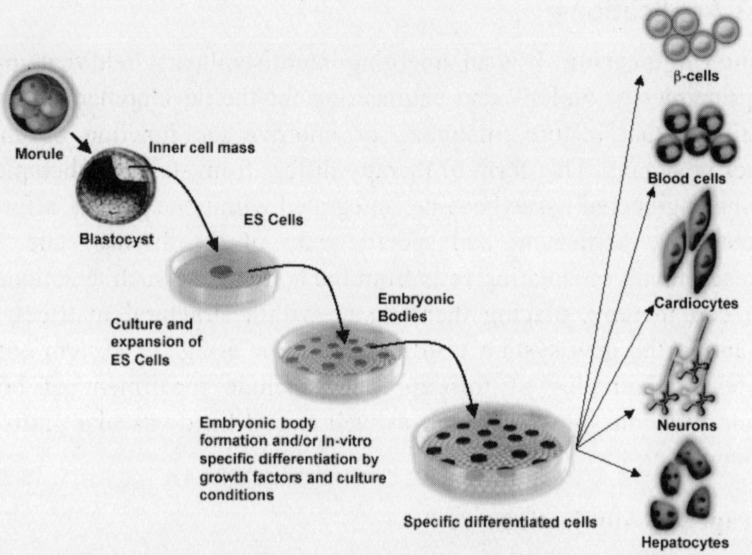

Fig. 5.5 Differentiation potential of embryonic stem cells showing possible applications of ES cells for therapy

(Fig. Ref.: ftp://ftb.wiley.com/public/sci_tech_med/ pharmaceutical_ biotech/)

5.2 Genetics

Genetics derived from the Greek word *genetikos* means origin, a discipline of biology, deals with the science of genes, heredity, and variation in living organisms. The fact that living things inherit traits from their parents has been used since prehistoric times to improve crop plants and animals through selective breeding. However, the modern science of genetics, which seeks to understand the process of inheritance, only began with the work of **Gregor Mendel** in the mid-19th century, who observed that organisms inherit traits through **discrete** units of inheritance, which are now called as **genes.**

Genes correspond to regions within **DNA,** a molecule composed of a chain of four different types of **nucleotides**, and the sequence of these nucleotides inherits the genetic information of the organisms. DNA naturally occurs in a double stranded form, with nucleotides on each strand, which are complementary to each other. Each strand can act as a template for creating a new partner strand and is the physical method for making copies of genes.

The sequence of nucleotides in a gene is **translated** by **cells** to produce a chain of amino acids, which create proteins. The order of amino acids in a protein corresponds to the order of nucleotides in the gene. This relationship between nucleotide sequence and amino acid sequence is known as the genetic code. The amino acids in a protein determine how it folds into a three-dimensional shape; this structure is in turn responsible for the protein's function. Proteins carry out almost all the functions need for the cells to alive. A change to the DNA in a gene can change a protein's amino acids, changing its shape and function: this can have a dramatic effect in the cell and on the organism as a whole.

5.2.1 Molecular Basis for Inheritance

DNA and Chromosomes

The molecular basis for genes is deoxyribonucleic acid (DNA). DNA is composed of a chain of nucleotides, of which there are four types: adenine (A), cytosine (C), guanine (G), and thymine (T). Genetic information exists in the sequence of these nucleotides, and genes exist as stretches of sequence along the DNA chain. Viruses are the only exception to this rule—sometimes viruses use the very similar molecule RNA instead of DNA as their genetic material.

Genes are arranged linearly along long chains of DNA sequence, called **chromosomes.** In bacteria, each cell usually contains a single circular chromosome, while eukaryotic organisms (including plants and animals) have their DNA arranged in multiple linear chromosomes. These DNA strands are often extremely long; the largest human chromosome, for example, is about 247 million base pairs in length. The DNA of a chromosome is associated with structural proteins that organize, compact, and control the access to the DNA, forming a material called chromatin; in eukaryotes, chromatin is usually composed of **nucleosomes**, segments of DNA wound around cores of **histone** proteins. The full set of hereditary material in an organism (usually the combined DNA sequences of all chromosomes) is called the genome.

5.2.2 Deoxyribonucleic Acid (DNA)

DNA is the molecular basis for inheritance. DNA is a nucleic acid that contains the genetic instructions used in the development and functioning of all known living organisms with the exception of some viruses. The main role of DNA molecules is the long-term storage of information. DNA is often compared to a set of blueprints, like a recipe or a code, since it contains the instructions need to construct the other components of cells, such as proteins and RNA molecules. The DNA segments that carry this genetic information are called genes, but other DNA sequences have structural purposes, or are involved in regulating the use of this genetic information.

DNA consists of two long polymers of simple units called nucleotides, with backbones made of sugars and phosphate groups joined each other by ester bonds. These two strands run in opposite directions to each other and are therefore anti-parallel. Attached to each sugar is one of four types of molecules called bases. It is the sequence of these four bases along the backbone that encodes information. This information is read using the genetic code, which specifies the sequence of the amino acids within proteins. The code is read by copying stretches of DNA into the related nucleic acid RNA, in a process called transcription. Within cells, DNA is organized into long structures called **chromosomes**. These chromosomes are duplicated before cells divide, in a process called DNA replication. Eukaryotic organisms (animals, plants, fungi, and protists) store most of their DNA inside the cell nucleus and some of their DNA in organelles, such as mitochondria or chloroplasts. In contrast, prokaryotes (bacteria and archaea) store their DNA only in the cytoplasm. Within the chromosomes, chromatin proteins such as histones compact and organize DNA. These compact structures guide the interactions between DNA and other proteins and help to control, which parts of the DNA are to be transcribed.

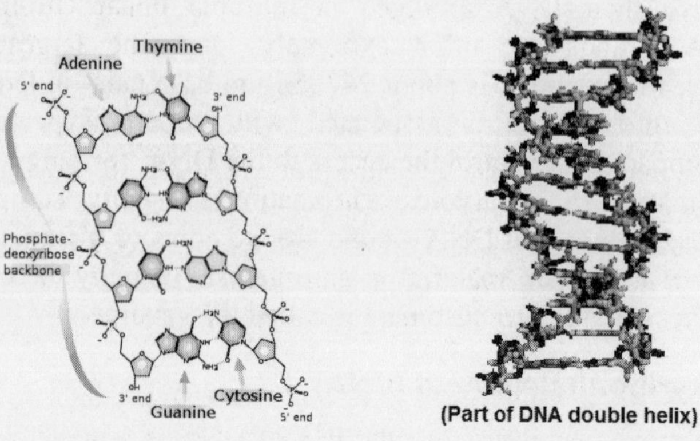

(a) (b)

Fig. 5.6 (a) Chemical structure of DNA (Hydrogen bonds shown as dotted lines) and (b) The structure of part of a DNA double helix.

(Ref: http://en.wikipedia.org/wiki/File:deoxyribonucleicacid...)

DNA is a long polymer made from repeating units called nucleotides. The DNA chain is 22 to 26 Ångströms wide (2.2 to 2.6 nanometres), and one nucleotide unit is 3.3 Å (0.33 nm) long. Although each individual repeating unit is very small, DNA polymers can be very large molecules containing millions of nucleotides. For instance, the largest human chromosome, chromosome number 1, is approximately 220 million base pairs long.

In living organisms, DNA does not usually exist as a single molecule, but instead as a pair of molecules that are held tightly together. These two long strands entwine like vines, in the shape of a **double helix**. The nucleotide repeats contain both the segment of the backbone of the molecule, which holds the chain together, and a base, which interacts with the other DNA strand in the helix (Fig 5.6). A base linked to a sugar is called a **nucleoside** and a base linked to a sugar and one or more phosphate groups is called a **nucleotide**. If multiple nucleotides are linked together, as in DNA, this polymer is called a **polynucleotide**.

The backbone of the DNA strand is made from alternating phosphate and sugar residues. The sugar in DNA is 2-deoxyribose, which is a pentose (five-carbon) sugar. The sugars are joined together by phosphate groups that form **phosphodiester** bonds between the third and fifth carbon atoms of adjacent sugar rings. These asymmetric bonds mean a strand of DNA has a direction. In a double helix the direction of the nucleotides in one strand is opposite to their direction in the other strand: the strands are *antiparallel*. The asymmetric ends of DNA strands are called the 5′ (*five prime*) and 3′ (*three prime*) ends, with the 5' end having a terminal phosphate group and the 3' end a terminal hydroxyl group. One major difference between DNA and RNA is the sugar, with the 2-deoxyribose in DNA being replaced by the alternative pentose sugar ribose in RNA.

The DNA double helix is stabilized by hydrogen bonds between the bases attached to the two strands. The four bases found in DNA are adenine (abbreviated A), cytosine (C), guanine (G) and thymine (T). These four bases are attached to the sugar/phosphate to form the complete nucleotide, as shown for adenosine monophosphate.

These bases are classified into two types; adenine and guanine, which are fused five- and six-membered heterocyclic compounds called purines, while cytosine and thymine are six-membered rings called pyrimidines. A fifth pyrimidine base, called uracil (U), usually takes the place of thymine in RNA and differs from thymine by lacking a methyl group on its ring. Uracil is not usually found in DNA, occurring only as a breakdown product of cytosine. In addition to RNA and DNA, a large number of artificial nucleic acid analogues have also been created to study the properties of nucleic acids, or for use in biotechnology.

5.2.2.1 Base Pairing

Each type of base on one strand forms a bond with the other strand. This is called complementary base pairing, where, purines form hydrogen bonds to pyrimidines, i.e. A bonding only to T, and C bonding only to G. This arrangement of two nucleotides binding together across the double helix is called a base pair.

As hydrogen bonds are not covalent, they can be broken and rejoined relatively easily. The two strands of DNA in a double helix can therefore be pulled apart like a zipper, either by a mechanical force or high temperature. As a result of this complementarities, all the information in the double-stranded sequence of a DNA helix is duplicated on each strand, which is vital in DNA replication. Indeed, this reversible and specific interaction between complementary base pairs is critical for all the functions of DNA in living organisms.

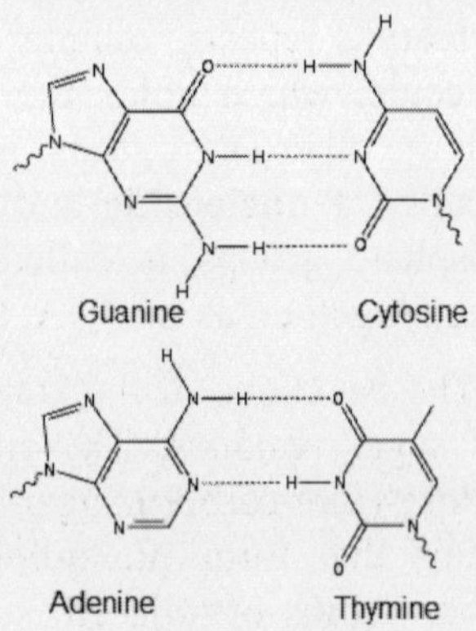

(Base Pair of DNA)

5.2.2.2 Sense and Antisense

A DNA sequence is called "sense", if its sequence is same as that of a messenger RNA. The sequence on the opposite strand is called the "antisense" sequence. Both sense and antisense sequences can exist on different parts of the same strand of DNA (i.e. both strands contain both sense and antisense sequences). In both prokaryotes and eukaryotes, antisense RNA sequences are produced, but the functions of these RNAs are not entirely clear. One proposal is that antisense RNAs are involved in regulating gene expression through RNA-RNA base pairing.

A few DNA sequences in prokaryotes and eukaryotes blur the distinction between sense and antisense strands by overlapping genes. In these cases, some DNA sequences encode one protein, when read along one strand, and a second protein when read in the opposite direction along the other strand. In bacteria, this overlap may be involved in the regulation of gene transcription, while in viruses, overlapping genes increase the amount of information that can be encoded within the small viral genome.

Applications: Fomivirsen® (marketed as Vitravene), was approved by the US FDA in Aug 1998 as a treatment for cytomegalovirus retinitis. The drugs, AVI-6002 and AVI-6003 are novel analogs based on AVI's PMO antisense chemistry in which **anti-viral** potency is enhanced by the addition of positively-charged components to the morpholino oligomer chain. Preclinical results of AVI-6002 and AVI-6003 demonstrated reproducible and high rates of survival in non-human primates challenged with a lethal infection of the **Ebola and Marburg viruses**, respectively. Also in 2006, German physicians reported on a dose-escalation study for the compound phosphorothioate antisense oligodeoxynucleotide specific for the mRNA of human transforming growth factor TGF-beta2) in patients with high grade **gliomas** and reported that the median overall survival had not been obtained and the authors hinted at a potential cure. In February 2010 researchers reported success in reducing **HIV** viral load using patient T-cells which had been harvested, modified with an RNA antisense strand to the HIV viral envelope protein, and re-infused into the patient during a planned lapse in retroviral drug therapy.

5.2.2.3 Base Modifications

The base modifications can be involved in packaging with regions that have low or no gene expression usually containing high levels of methylation of cytosine bases. For example, cytosine methylation, produces 5-methylcytosine, which is important for X-chromosome inactivation. The average level of methylation varies between organisms. The worm *Caenorhabditis elegans* lacks cytosine methylation, while vertebrates have higher levels, with up to 1% of t expression of genes, which is influenced by the chromosomes, in a structure called chromatin, where DNA containing 5-methylcytosine. Despite the importance of 5-methylcytosine, it can deaminate to leave a thymine base, so that methylated cytosines will be particularly prone to mutations. Other base modifications include adenine methylation in bacteria, the presence of 5-hydroxymethyl-cytosine in the brain, and the glycosylation of uracil to produce the "J-base" in kinetoplastids.

5.2.2.4 Damage

DNA can be damaged by many sorts of mutagens, which change the DNA sequence. Mutagens include oxidizing agents, alkylating agents and also high-energy electromagnetic radiation such as ultraviolet light and

X-rays. Many mutagens fit into the space between two adjacent base pairs, this is called *intercalation*. An intercalator can fit between base pairs by distorting the DNA strands and unwinding of the double helix, which inhibits both transcription and DNA replication, causing toxicity and mutations, for which DNA intercalators are carcinogenic.

5.2.3 Biological Functions

DNA usually occurs as linear chromosomes in eukaryotes, and circular chromosomes in prokaryotes. A set of base pairs of DNA consists of 46 chromosomes. The information carried by DNA is held in the sequence of DNA called genes. Transmission of genetic information in genes is achieved via complementary base pairing. For example, in transcription, when a cell uses the information in a gene, the DNA sequence is copied into complementary chromosomes in a cell making a genome. The human genome has approximately billions of RNA sequence, which are used to make a matching protein sequence in a process called translation, which depends on the same interaction between RNA nucleotides.

5.3 Recombinant DNA

Recombinant DNA (rDNA) is a form of artificial DNA, created by combining two or more sequences. In terms of genetic modification, it is created through the introduction of relevant DNA into an existing organism DNA, such as the plasmids of bacteria, to code for or alter different traits for a specific purpose, such as antibiotic resistance. It differs from genetic recombination in that it does not occur through natural processes within the cell, but is engineered. A **recombinant protein** is a protein that is derived from recombinant DNA.

5.3.1 Recombinant DNA Technology

Recombinant DNA technology was made possible by the discovery, isolation and application of restriction endonucleases by Werner Arber, Daniel Nathans, and Hamilton Smith, for which they received Nobel Prize in Medicine in1978. Cohen and Boyer applied for a patent on the Process for producing biologically functional molecular chimeras which could not exist in nature in 1974. The patent was granted in 1980. With the selection of appropriate organisms, strain, and expression systems, a high quality of pharmaceutical products of defined pharmacological properties is possible by recombinant NNA technology.

5.3.1.1 Cloning and Relation to Plasmid

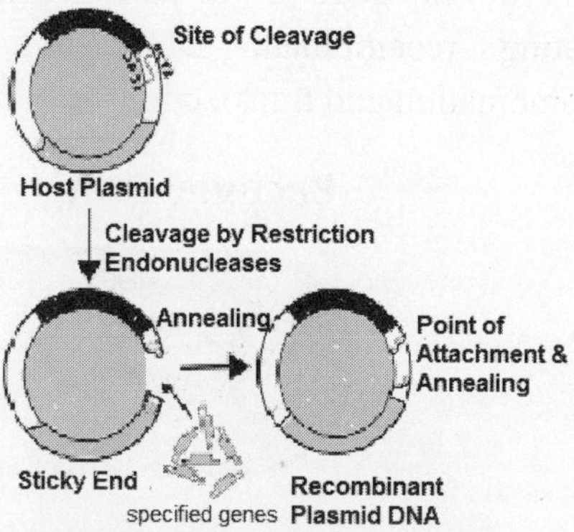

(Insertion of Desired Genes to the Plasmid)

Cloning in biology is the process of producing similar populations of genetically identical individuals that occurs in nature when organisms such as bacteria, insects or plants reproduce asexually. Cloning in biotechnology refers to processes used to create copies of DNA fragments (molecular cloning), cells (cell cloning), or organisms. The term also refers to the production of multiple copies of a product such as digital media or software with bacteria remaining a prime example due to the use of viral vectors in medicine that contain recombinant DNA inserted into a structure known as a plasmid.

5.3.1.2 Plasmids

Plasmids are extrachromosomal self-replicating circular forms of DNA present in most bacteria, such as *Escherichia coli* (E. coli), containing genes related to catabolism and metabolic activity, and allowing the carrier bacterium to survive and reproduce in conditions present within other species and environments. These genes represent characteristics of resistance to bacteriophages and antibiotics and some heavy metals, but can also be fairly easily removed or separated from the plasmid by an enzyme i.e. restriction endonucleases, which regularly produce "sticky ends" and allow the attachment of a selected segment of DNA, which codes for more "reparative" substances, such as peptide hormone medications including insulin, growth hormone, and oxytocin.

Plasmids act as **replicator**, **selectable marker** and a **cloning site**. The replication originates at a specific location. The marker refers to a particular gene that usually confers luminescence to allow identification of successfully recombined DNA. The cloning site is a sequence of nucleotides representing one or more positions, where cleavage occurs by **restriction endonucleases**. Most eukaryotes do not maintain canonical plasmids; yeast is a notable exception. In addition, the Ti plasmid of the bacterium *Agrobacterium tumefaciens* can be used to integrate foreign DNA into the genomes of many plants. Other methods of introducing or creating recombinant DNA in eukaryotes include homologous recombination and transfection with modified viruses.

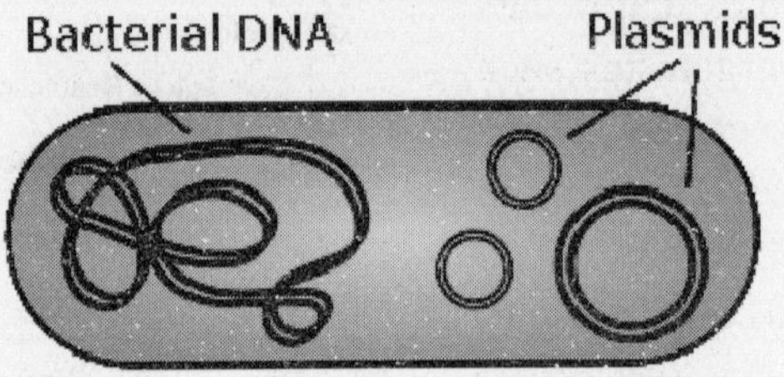

Fig. 5.7 Illustration of a bacterium with plasmid enclosed showing chromosomal DNA and plasmids.

Plasmids used in genetic engineering are called **vectors**. Plasmids serve as important tools in genetics and biotechnology labs, where they are commonly used to multiply (make many copies of) or *express* particular genes. Many plasmids are commercially available for such uses. The gene to be replicated is inserted into copies of a plasmid containing genes that make cells resistant to particular antibiotics and a multiple cloning site (MCS, or polylinker), which is a short region containing several commonly used restriction sites allowing the easy insertion of DNA fragments at this location. Next, the plasmids are inserted into bacteria by a process called *transformation*. Then, the bacteria are exposed to the particular antibiotics. Only bacteria which take up copies of the plasmid survive, since the plasmid makes them resistant.

In particular, the protecting genes are expressed (used to make a protein) and the expressed protein breaks down the antibiotics. In this way the antibiotics act as a filter to select only the modified bacteria. Now these bacteria can be grown in large amounts, harvested and lyses (often using the alkaline lysis method) to isolate the plasmid of interest.

5.3.1.3 Simulation of Plasmids

The use of plasmids as a technique in molecular biology is supported by bioinformatics software. These programmes record the DNA sequence of plasmid vectors; help to predict cut sites of restriction enzymes, and to plan manipulations.

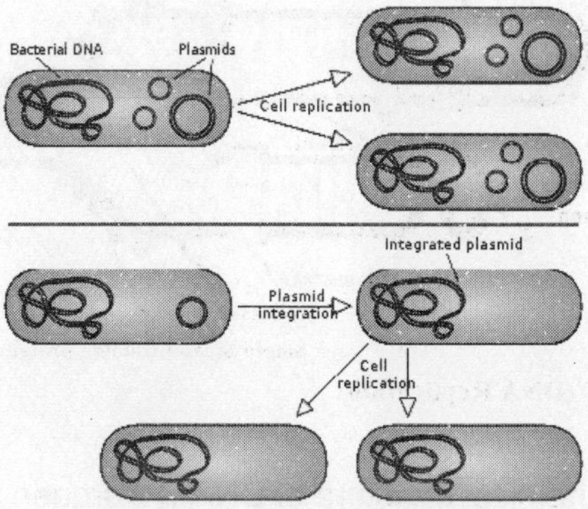

Fig. 5.8 Inserted DNA genes to plasmid vector showing cell replication. (Ref: http://en.wikipedia.org/wiki/File:Plasmid_replication)

5.3.2 Transcription and Translation

A gene is a sequence of DNA that contains genetic information and can influence the phenotype of an organism. Within a gene, the sequence of bases along a DNA strand defines a messenger RNA sequence, which then defines one or more protein sequences. The relationship between the nucleotide sequences of genes and the amino-acid sequences of proteins is determined by the rules of translation, collectively known as genetic code. The genetic code consists of *codons* formed from a sequence of three nucleotides (e.g. ACT, CAG, TTT).

In transcription, the codons of a gene are copied into messenger RNA by **RNA polymerase**. This RNA copy is then decoded by a ribosome that reads the RNA sequence by base-pairing the messenger RNA to transfer RNA, which carries amino acids. Because of there are 4 bases in 3-letter combinations, which results in 64 possible *codons* (4^3 combinations). These encode the twenty standard amino acids, giving most amino acids of more than one possible codon. There are also three 'stop' or 'non-sense' codons signifying the end of the coding region; these are the TAA, TGA and TAG codons.

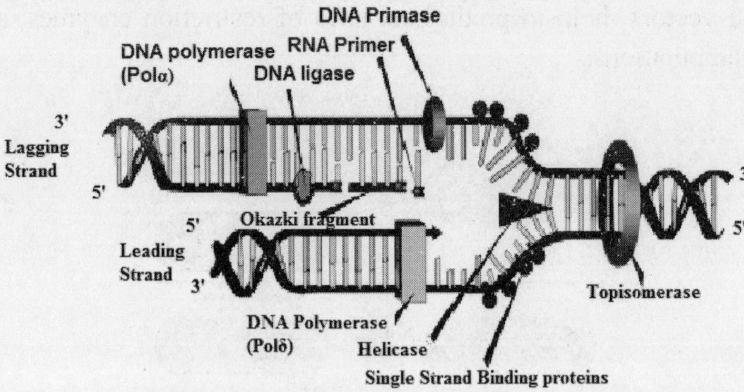

Fig. 5.9 (a) Chemical structure of DNA (Hydrogen bonds shown in dotted lines) and (b) The structure of part of a DNA double helix. (Ref: http://en.wikipedia.org/wiki/File:deoxyribonucleicacid...)

5.3.3 Replication

Cell division is essential for an organism to grow, but, when a cell divides, it must replicate the DNA in its genome so that the two daughter cells have the same genetic information as their parent. The double-stranded structure of DNA provides a simple mechanism for DNA replication. Here, the two strands are separated and then each strand's complementary DNA sequence is recreated by an enzyme called DNA polymerase. This enzyme makes the complementary strand by finding the correct base through complementary base pairing, and bonding it to the original strand. As DNA polymerases can only extend a DNA strand in a 5′ to 3′ direction, different mechanisms are used to copy the anti-parallel strands of the double helix. In this way, the base on the old strand dictates the appearance of the new strand, and the cell ends up with a perfect copy of its DNA.

5.3.3.1 Interactions with Proteins

All the functions of DNA depend on interactions with proteins. These protein interactions can be non-specific, or specific to a single DNA sequence. Enzymes can also bind to DNA and of these, the polymerases that copy the DNA base sequence in transcription and in DNA replication.

5.3.3.2 DNA-Modifying Enzymes

Nucleases and ligases

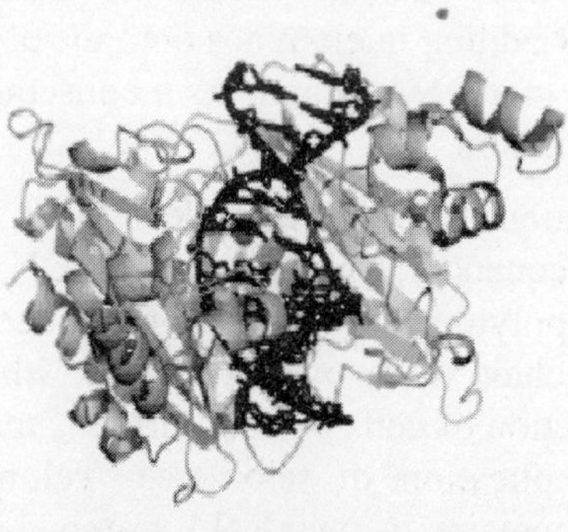

[Restriction enzyme EcoRV (brown) in a complex with its substrate DNA]

Nucleases are enzymes that cut DNA strands by catalyzing the hydrolysis of the phosphodiester bonds. Nucleases that hydrolyze nucleotides from the ends of DNA strands are called exonucleases, while endonucleases cut within strands. The most frequently used nucleases in molecular biology are the restriction endonucleases, which cut DNA at specific sequences. For instance, the EcoRV (Figure Ref: http://en.wikipedia.org/wiki/File:ecorv...) enzyme shown in the figure recognizes the 6-base sequence 5′-GAT|ATC-3′ and makes a cut at the vertical line. In technology, these sequence-specific nucleases are used in molecular cloning and DNA finger printing.

Enzymes called **DNA ligases** can rejoin cut or broken DNA strands. Ligases are particularly important in lagging strand DNA replication, as they join together the short segments of DNA produced at the replication fork into a complete copy of the DNA template. They are also used in DNA repair and genetic recombination.

Topoisomerases are enzymes with both nuclease and ligase activity. These proteins change the amount of supercoiling in DNA. Some of these enzymes work by cutting the DNA helix and allowing one section to rotate, thereby reducing its level of supercoiling; the enzyme then seals the DNA break. **Helicases** are proteins that act as molecular motor. They use the chemical energy in nucleoside triphosphates, predominantly ATP, to break hydrogen bonds between bases and unwind the DNA double helix into single strands.

Polymerases are enzymes that synthesize polynucleotide chains from nucleoside triphosphates. The sequences of their products are copies of existing polynucleotide chains, which are called *templates*. These enzymes function by adding nucleotides onto the 3′ hydroxyl group of the previous nucleotide in a DNA strand. As a consequence, all polymerases work in a 5′ to 3′ direction.

In DNA replication, a DNA-dependent DNA polymerase makes a copy of a DNA sequence. RNA-dependent DNA polymerases are a specialized class of polymerases that copy the sequence of an RNA strand into DNA. They include reverse transcriptase, which is a viral enzyme involved in the infection of cells by retroviruses, and telomerase, which is required for the replication of telomeres. Telomerase is an unusual polymerase, which contains its own RNA template as part of its structure.

5.3.3.3 Transcription

Transcription is carried out by a DNA-dependent RNA polymerase that copies the sequence of a DNA strand into RNA. To begin transcribing a gene, the RNA polymerase binds to a sequence of DNA called a promoter and separates the DNA strands. Then it copies the gene sequence into a messenger RNA and transcript to the region of DNA terminator, where it halts and detaches from the DNA.

5.3.4 Genetic Recombination

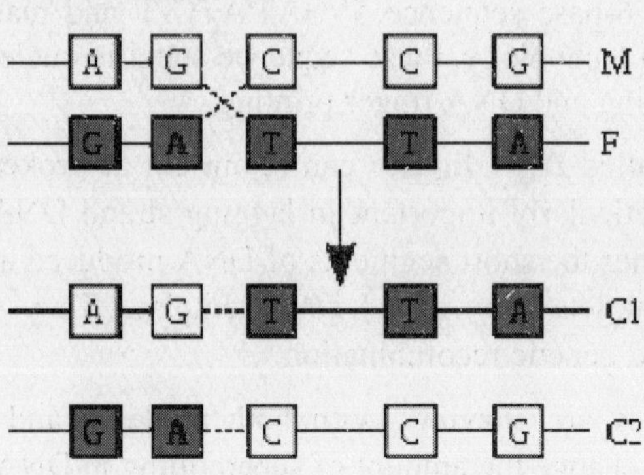

Fig. 5.10 Schematics of genetic recombination.
(Ref: http://en.wikipedia.org/wiki/File:chromosomal_recombination)

Recombination involves the breakage and rejoining of two chromosomes (M and F) to produce two re-arranged chromosomes (C1 and C2). Chromosomal crossover results, when two DNA helices break, swap a section and then rejoin. Recombination allows chromosomes to exchange genetic information and produces new combinations of genes, which increases the efficiency of natural selection and can be important in the rapid evolution of new proteins. Genetic recombination can also be involved in DNA repair, particularly in the cell's response to double-strand breaks.

5.4 Genetic Engineering

Genetic engineering, also called **genetic modification** or **gene splicing**, is the human manipulation of an organism's genetic material in a way that does not occur under natural conditions. It involves the use of recombinant DNA techniques, but does not include traditional animal and plant breeding or mutagenesis. Any organism that is generated using these techniques is considered to be a genetically modified organism. The first organisms genetically engineered were bacteria in 1973 and then mice in 1974. Insulin producing bacteria were commercialized in 1982 and genetically modified food has been sold since 1994.

Production of genetically modified organisms is a multi-step process, which involves the isolating and copying the genetic material of interest, constructing genetic elements for correct **expression** and insertion of the construct into the host organism, either by using a **vector** or directly through injection, in a process called **transformation**. Successfully transformed organisms are then grown and the presence of the new genetic material is tested for.

Genetic engineering alters the genetic makeup of an organism using techniques that introduce heritable material prepared outside the organism either directly into the host or into a cell and then fused or hybridized with the host. This involves using recombinant nucleic acid (DNA or RNA) techniques to form new combinations of heritable genetic material followed by the incorporation of that material either indirectly through a vector system or directly through micro-injection, macro-injection and micro-encapsulation techniques.

(Genetic engineering does not include traditional animal and plant breeding, in vitro fertilization, induction of polyploidy, mutagenesis and cell fusion techniques. Cloning and stem cell research, although not considered genetic engineering, are closely related and genetic engineering can be used within them. Synthetic biology is an emerging discipline that takes genetic engineering a step further by introducing artificially synthesized genetic material from raw materials into an organism).

5.4.1 Process of Gene Expression

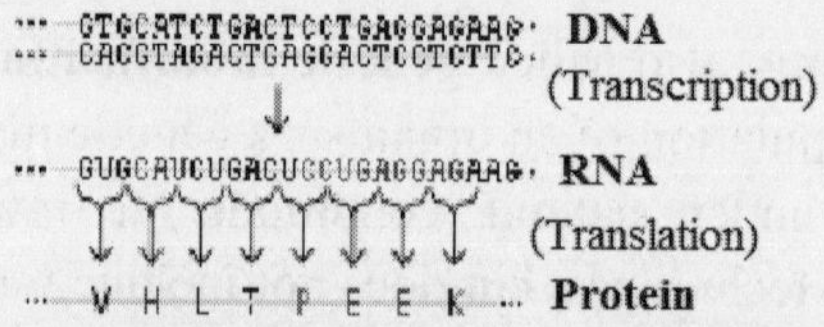

(Gene Expression)

Genes generally express their functional effect through the production of proteins, which are complex molecules responsible for most functions in the cell. Proteins are chains of amino acids, and the DNA sequence of a gene is used to produce a specific protein sequence. This process begins with the production of an RNA molecule with a sequence matching the gene's DNA sequence, a process called transcription. This messenger RNA molecule is then used to produce a corresponding amino acid sequence through translation. Each group of three nucleotides in the sequence able to produce twenty possible amino acids in protein, called as genetic code. The flow of information is unidirectional: information transferred from nucleotide sequences into the amino acid sequence of proteins, but never transfers back from protein to the sequence of DNA.

Isolating the Gene

This typically involves multiplying the gene using polymerase chain reaction (PCR). If the chosen gene or the donor organism's genome has been well studied it may be present in a genetic library. If the DNA sequence is known, but no copies of the gene are available, it can be artificially synthesized. Once isolated, the gene is inserted into a bacterial plasmid.

Constructs

The gene to be inserted into the genetically modified organism must be combined with other genetic elements in order to work properly. The gene can also be modified at this stage for better expression or effectiveness. As well as the gene to be inserted most constructs contain a promoter and terminator region as well as a selectable marker gene. The promoter region initiates transcription of the gene and can be used to control the location and level of gene expression, while the terminator region ends transcription. The constructs are made using recombinant DNA techniques, such as restriction digests, ligations and molecular cloning.

Transformation

About 1% of bacteria are naturally able to take up foreign DNA but it can also be induced in other bacteria. Stressing the bacteria for example, with a heat shock or an electric shock, can make the cell membrane permeable to DNA that may then incorporate into their genome. DNA is generally inserted into animal cells using microinjection, where it can be injected through the cells nuclear envelope directly into the nucleus or through the use of viral vectors. In plants the DNA is generally inserted using *Agrobacterium*-mediated recombination or biolistics. Examples of gene delivery **vectors** include: Retrovirus, Adenovirus, Naked DNA, lipofection, Poxvirus, Vaccinia virus, Herper simplex virus, Adeno-associated virus, RNA transfer, which deliver and express genes at appropriate site.

Regeneration

Because of only a single cell is transformed with genetic material, the organism must be grown from that single cell. As bacteria consist of a single cell can reproduce clonally. In plants this is accomplished through the use of tissue culture. In animals it is necessary to ensure that the inserted DNA is present in the embryonic stem cells. When the offspring is produced they can be screened for the presence of the gene. All offspring from the first generation will be heterozygous for the inserted gene and must meet together to produce a homozygous animal.

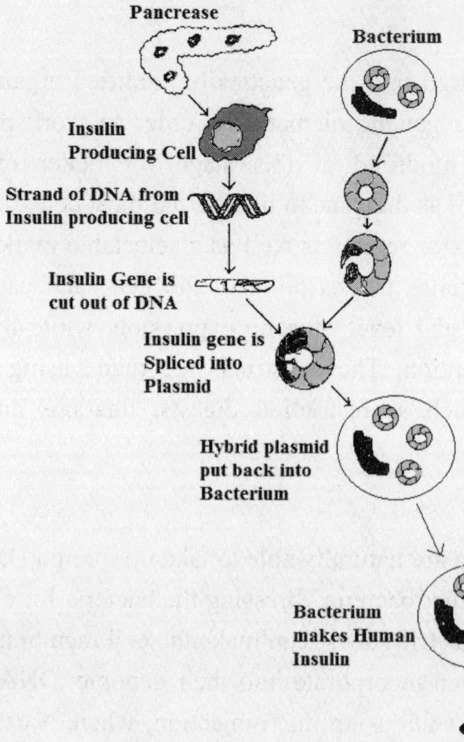

(Production of Insulin through Genetic Engineering)

Confirmation

Further tests using **PCR**, **Southern Blots** and **Bioassays** are needed to confirm that the gene is expressed and functions correctly.

5.4.2 Applications

Genetic engineering has applications in medicine, research, industry and agriculture and can be used on a wide range of plants, animals and micro organism.

Therapeutic Applications

In medicine genetic engineering has been used for mass production of insulin, human growth hormones, follistim (for treating infertility), human albumin, monoclonal antibodies, antihemophilic factors, vaccines and many other drugs. Vaccination generally involves injecting weak live, killed or inactivated forms of viruses or their toxins into the person being immunized. Genetically engineered viruses are being developed that can still confer immunity, but lack the infectious sequences. Mouse hybridoma cells fused together to create monoclonal antibodies have been humanised through genetic engineering to create human monoclonal antibodies.

Genetic engineering is used to create animal models of human diseases. Genetically modified mice are the most common genetically engineered animal model. They have been used to study model cancer (the onco-mouse), obesity, heart disease, diabetes, arthritis, substance abuse, anxiety, aging and Parkinson disease. Potential cures can be tested against these mouse models. Also genetically modified pigs have been bred with the aim of increasing the success of pig to human organ transplantation.

Gene therapy is the genetic engineering of humans by replacing defective human genes with functional copies. This can occur in somatic tissue or germline tissue. If the gene is inserted into the germline tissue it can be passed down to those persons descendants. Gene therapy has been used to treat patients suffering from immune deficiencies (notably Severe combined immunodeficiency) and trials have been carried out on other genetic disorders.

By engineering genes into bacterial plasmids, it is possible to create a biological library that can produce proteins and enzymes. Some genes do not work well in bacteria, so yeast, which is an eukaryote, can also be used. Bacteria and yeast libraries have been used to produce medicines such as insulin, human growth hormone and vaccines, supplements such as tryptophan, aid in the production of food (chymotripsin in cheese making) and fuels. Other applications involving genetically engineered bacteria to perform the tasks such as cleaning up oil spills, carbon and other toxic waste.

Other Fields of Genetics

Medical Genetics encompasses many different areas, including clinical practice of physicians, genetic counselors, and nutritionists, clinical diagnostic laboratory activities, and research into the causes and inheritance of genetic disorders. Examples of conditions that fall within the scope of medical genetics include birth defects and dysmorphology, mental retardation, autism, metabolic and mitochondrial disorders, skeletal dysplasia, connective tissue disorders, cancer genetics, teratogens and prenatal diagnosis. The emerging areas include neurologic, endocrine, cardiovascular, pulmonary, ophthalmologic, renal, psychiatric, and dermatologic conditions. **Clinical genetics** is the practice of clinical medicine with particular attention to hereditary disorders. Referrals are made to genetics clinics for a variety of reasons, including birth defects, developmental delay, autism, epilepsy, short stature, and many others **Metabolic/biochemical genetics** involves the diagnosis and management of inborn errors of metabolism in which patients have enzymatic deficiencies that perturb biochemical pathways involved in metabolism of carbohydrates, amino acids, and lipids. Examples of metabolic disorders include galactosemia, glycogen storage disease, lysosomal storage disorders, metabolic acidosis, peroxisomal disorders, phenylketonuria, and urea cycle disorders. **Cytogenetics** is the study of chromosomes and chromosome abnormalities. **Molecular genetics** involves the discovery of the laboratory testing for DNA mutations that underlie many single gene disorders, which include achondroplasia, cystic fibrosis, Duchenne muscular dystrophy, hereditary breast cancer, Huntington disease, Marfan syndrome, Noonan syndrome, and Rett syndrome, etc

5.5 Genomics

Genomics is the study of the genomes of organisms. The field includes intensive efforts to determine the entire DNA sequence of organisms and fine-scale genetic mapping. The field also includes studies of intragenomic phenomena such as heterosis, epistasis, pleiotropy and other interactions between loci and alleles within the genome.

A genome is the sum total of all individual genes of an organism. Thus, genomics is the study of all the genes of a cell, or tissue, at the DNA (genotype), mRNA (transcriptome), or protein (proteome) levels.

Genomics was established by Fred Sanger when he first sequenced the complete genomes of a virus and a mitochondrion. His group established techniques of sequencing, genome mapping, data storage, and bioinformatics analyses in the 1970-1980s. A major branch of genomics is still concerned with sequencing the genomes of various organisms, but the knowledge of full genomes has created the possibility for the field of functional genomics, mainly concerned with patterns of gene expression during various conditions.

5.5.1 Full Genome Sequencing

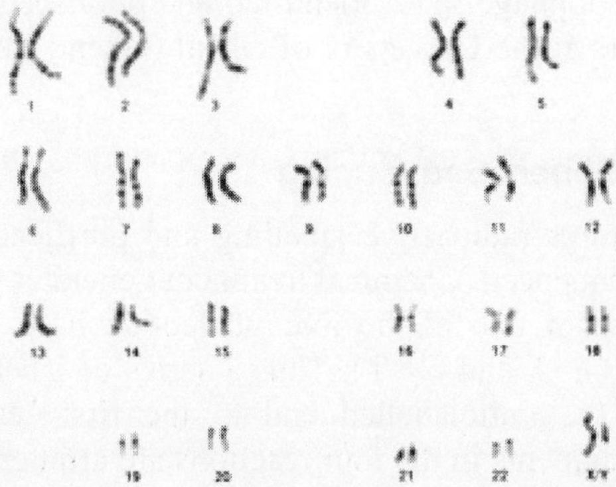

Fig. 5.11 Image of 46 chromosomes, making up the diploid genome of human male.

(Ref: http://en.wikipedia.org/wiki/File:NHGRI_human_male_karyotype.png)

Full genome sequencing (FGS), also known as **whole genome sequencing**, **complete genome sequencing**, or **entire genome sequencing**, is a laboratory process that determines the complete DNA sequence of an organism's genome at a single time. This entails sequencing all of an organism's chromosomal DNA as well as DNA contained in the mitochondria and for plants the chloroplast as well. Because the sequence data that is produced can be quite large (for example, there are approximately six billion base pairs in each human diploid genome), genomic data is stored electronically and requires a large amount of computing power and storage capacity. Full genome sequencing would have been nearly impossible before the advent of the microprocessor, computers, and the Information Age.

The term **DNA sequencing** refers to sequencing methods for determining the order of the nucleotide bases i.e. adenine, guanine, cytosine, and thymine in a molecule of DNA. The rapid speed of sequencing attained with modern DNA sequencing technology, which is instrumental in the sequencing of the human genome, in the **Human Genome Project**.

Methods: The first DNA sequences were obtained in the early 1970s by academic researchers using laborious methods based on two-dimensional chromatography, following the development of dye-based sequencing methods with automated analysis. RNA sequencing was one of the earliest forms of nucleotide sequencing. The major landmark of RNA sequencing is the sequence of the first complete gene and the complete genome of Bacteriophage MS2, identified and published by Walter Fiers and his coworkers at the University of Ghent (Ghent, Belgium), between 1972 and 1976.

5.5.2 Maxam–Gilbert Sequencing

The method requires radioactive labelling and purification of the DNA fragment to be sequenced. Chemical treatment generates breaks at a small proportion of one or two of the four nucleotide bases in each of four reactions (G, A+G, C, and C+ T). Thus a series of labelled fragments is generated from the radiolabelled end to the first "cut" site in each molecule. The fragments in the four reactions are arranged side by side in gel electrophoresis for size separation. To visualize the fragments, the gel is exposed to X-ray film for autoradiography, yielding a series of dark bands each corresponding to a radiolabelled DNA fragment, from which the sequence may be inferred. This method is also sometimes known as "chemical sequencing", and originated in the study of DNA-protein interactions (footprinting), nucleic acid structure and epigenetic modifications to DNA.

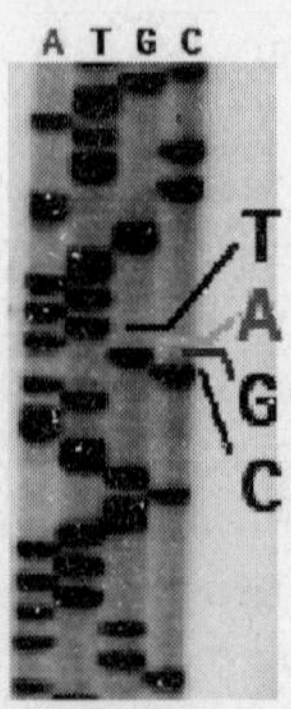

(Part of a radioactively labelled sequencing gel)

5.5.3 Chain-Termination Methods

Chain-terminator method also called as Sanger method after its developer Frederick Sanger, is more efficient and uses fewer toxic chemicals and lower amounts of radioactivity than the method of Maxam and Gilbert, it rapidly became the method of choice. The key principle of the Sanger method was the use of dideoxynucleotide triphosphates (ddNTPs) as DNA chain terminators.

The classical chain-termination method requires a single-stranded DNA template, a DNA primer, a DNA polymerase, radioactively or fluorescently labeled nucleotides, and modified nucleotides that terminate DNA strand elongation. The DNA sample is divided into four separate sequencing reactions, containing all four of the standard deoxynucleotides (dATP, dGTP, dCTP and dTTP) and the DNA polymerase. To each reaction is added only one of the four dideoxynucleotides (ddATP, ddGTP, ddCTP, or ddTTP) which are the chain-terminating nucleotides, lacking a 3'-OH group required for the formation of a phosphodiester bond between two nucleotides, thus terminating DNA strand extension and resulting in DNA fragments of varying length.

The newly synthesized and labeled DNA fragments are heat denatured, and separated by size (with a resolution of just one nucleotide) by **gel electrophoresis** on a denaturing **polyacrylamide-urea gel** with each of the four reactions run in one of four individual lanes (lanes A, T, G, C); the DNA bands are then visualized by autoradiography or UV light, and the DNA sequence can be directly read off the X-ray film or gel image.

5.5.4 Dye-Terminator Sequencing

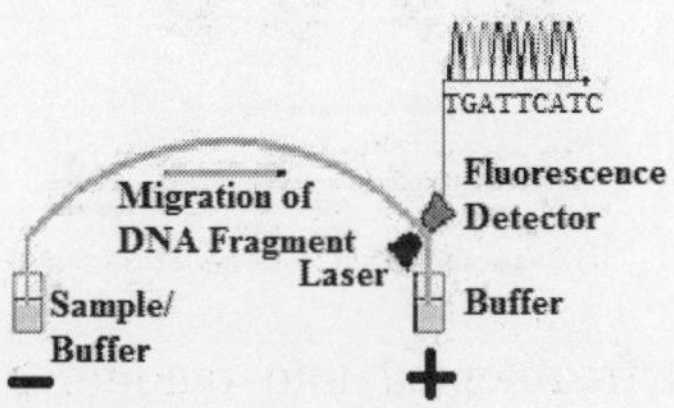

(Capillary electrophoresis)

Dye-terminator sequencing utilizes labelling of the chain terminator ddNTPs, which permits sequencing in a single reaction, rather than four reactions as in the labelled-primer method. In dye-terminator sequencing, each of the four dideoxynucleotide chain terminators is labelled with fluorescent dyes, each of which emits light at different wavelengths. Looking to its greater expediency and speed, dye-terminator sequencing is now the mainstay in automated sequencing. Its limitations include dye effects due to differences in the incorporation of the dye-labelled chain terminators into the DNA fragment, resulting in unequal peak heights and shapes in the electronic DNA sequence trace chromatogram after capillary electrophoresis (see figure to the left, Ref: http://en.wikipedia.org/wiki/File:CE_Basic.jpg). This problem has been addressed with the use of modified DNA polymerase enzyme systems and dyes that minimize incorporation variability, as well as methods for eliminating "dye blobs". The dye-terminator sequencing method, along with automated high-throughput DNA sequence analyzers, is now being used for the vast majority of sequencing projects.

5.5.5 Amplification and Clone Selection

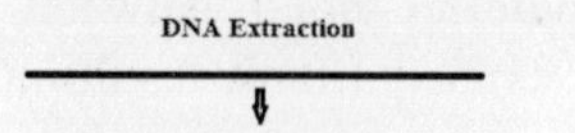

(The fragmented DNA is cloned into a DNA vector)

Genomic DNA is fragmented into random pieces and cloned as a bacterial library. DNA from individual bacterial clones is sequenced and the sequence is assembled by using overlapping DNA regions.

Large-scale sequencing aims at sequencing very long DNA pieces, such as whole chromosomes. Common approaches consist of cutting (with restriction enzymes) or shearing (with mechanical forces) large DNA fragments into shorter DNA fragments. The fragmented DNA is cloned into a DNA vector, and amplified in *Escherichia coli*. Short DNA fragments purified from individual bacterial colonies, individually sequenced and assembled electronically into one long, contiguous sequence. This method does not require any pre-existing information about the sequence of the DNA and is referred to as *de novo* sequencing. Gaps in the assembled sequence may be filled by primer walking.

Most sequencing approaches use an *in vitro* cloning step to amplify individual DNA molecules, because their molecular detection methods are not sensitive enough for single molecule sequencing. Emulsion PCR isolates individual DNA molecules along with primer-coated beads in aqueous droplets within an oil phase. Polymerase chain reaction (PCR) then coats each bead with cloned copies of the DNA molecule followed by immobilization for later sequencing.

5.5.6 Application

Synthetic insulin production using recombinant DNA breaks through in recombinant DNA technology. The specific gene sequence, or oligonucleotide, those codes for insulin production in humans was introduced to a sample colony of *E. coli* (the bacteria found in the human intestine). Only about 1 out of 10^6 bacteria picks up the sequence. However, because the lifecycle is only about 30 minutes for *E. coli*, this limitation is not problematic, and in a 24-hour period, there may be billions of *E. coli* that are coded with the DNA sequences needed to induce insulin production.

5.5.7 Genetic Disorder

A **genetic disorder** is an illness caused by abnormalities in genes or chromosomes. While some diseases, such as cancer, are due to partly genetic disorders, and partly caused by environmental factors.

Single Gene Disorder: is the result of a single mutated gene. There are estimated to be over 4000 human diseases caused by single gene defects. Single gene disorders can be passed on to subsequent generations in several ways. Genomic imprinting and uniparentaldisomy, however, may affect inheritance patterns. The divisions between recessive and dominant types are not "hard and fast" although the divisions between autosomal and X-linked types are based on the chromosomal location of the gene. For example, achondroplasia is typically considered a dominant disorder, but children with two genes for achondroplasia have a severe skeletal disorder that achondroplasics could be viewed as carriers. Sickle-cell anemia is also considered a recessive condition.

Autosomal Dominant Disorder is the disorder in which only one mutated copy of the gene is affected. Each affected person usually has one affected parent. Examples of this type of disorder are Huntington's disease, neurofibromatosis, Marfan syndrome, hereditary nonpolyposis colorectal cancer, and hereditary multiple exostoses, which is a highly penetrant autosomal dominant disorder. Birth defects are also called congenital anomalies.

X-linked dominant disorders are caused by mutations in genes on the X chromosome. Only a few disorders have this inheritance pattern, with a prime example being X-linked hypophosphatemic rickets.

X-linked recessive conditions are also caused by mutations in genes on the X chromosome. Males are more frequently affected than females. X-linked recessive conditions include the serious diseases such as Hemophilia A, Duchenne muscular dystrophy, and Lesch-Nyhan syndrome as well as common and less serious conditions such as male pattern baldness and red-green color blindness.

Y-linked disorders are caused by mutations on the Y chromosome. Because males inherit a Y chromosome from their fathers, *every* son of an affected father will be affected. Because females inherit an X chromosome from their fathers, female offspring of affected fathers are *never* affected. The symptoms include infertility.

Mitochondrial Disorder, also known as maternal inheritance, applies to genes in mitochondrial DNA. Because only egg cells contribute mitochondria to the developing embryo, only mothers can pass on mitochondrial conditions to their children. An example of this type of disorder is Leber's hereditary optic neuropathy.

Multifactorial and Polygenic (Complex Disorders) are genetic disorders which associated with the effects of multiple genes in combination with lifestyle and environmental factors. Multifactorial disorders include heart disease and diabetes.

5.6 Gene Therapy

Human Gene therapy(HGT) involves the transfer of therapeutic genes (DNA or RNA) to somatic cells of patients, resulting in a therapeutic effect, by either a) correcting genetic defects, b) over expressing proteins that are therapeutically useful, or c) inhibiting the production of harmful proteins.

5.6.1 Basic Principles (from Nucleic Acid Delivery to Therapeutic Effect)

Gene therapy is initiated with the introduction of an appropriate vector (viral or non-viral) either into the body locally (direct tissue injection), into body cavities (e.g., peritoneum or cerebrospinal fluid), or into the bloodstream (systemic delivery). The vector needs to find its target tissue, after which it enters the target cell membrane and traffics through the cytoplasm to reach and enter the nucleus. Then the therapeutic (trans) gene needs to be transcribed and the formed mRNA needs to be appropriately translated into the therapeutic protein. The protein then acts on its receptor(s) either on the cell that produced it (intracrine or autocrine mechanism), on neighboring cells (paracrine mechanism), or at distant sited after entering the blood circulation (endocrine mechanism, e.g., erythropoietin, coagulation factors, and growth hormone). Finally, after interacting with its receptor, the protein needs to induce an appropriate biological effect that results in therapeutic benefits.

For gene correction or gene knock-down approaches, the steps are similar except that the last step is modification of genome (i.e., gene correction) or blockade of mRNA transcription (siRNA/shRNA) of dangerous genes, respectively.

5.6.2 Gene Delivery Targeting

Physical targeting can be achieved by catheter-mediated gene transfer to various regions of the body (e.g., feed arteries of the organs, such as leg muscles, heart, and liver, or retrograde injection via veins. Intramuscular injection of plasmid DNA or viral vectors encoding angiogenic growth factors has been used in ischemic myocardium and peripheral vascular disease. **Biological targeting** uses modification of viral coat protein (for viral vectors) or surface properties of synthetic vectors (e.g. liposomes) which may be **passive targeting** possible by shielding the vectors to protect from binding to blood cells, plasma proteins, immunoglobulins and unwanted tissues, which allow them to circulate for longer periods of time in the blood and accumulate in specific tissues with leaky blood vessels(such as tumor) or **active targeting** aims at directing vector binding and uptake to specific surface of the vector by a variety of technologies(e.g., Chemical and genetic). **Transcription targeting** is the receptor targeting possible by the use of cell-, tissue-, and disease-specific promoters

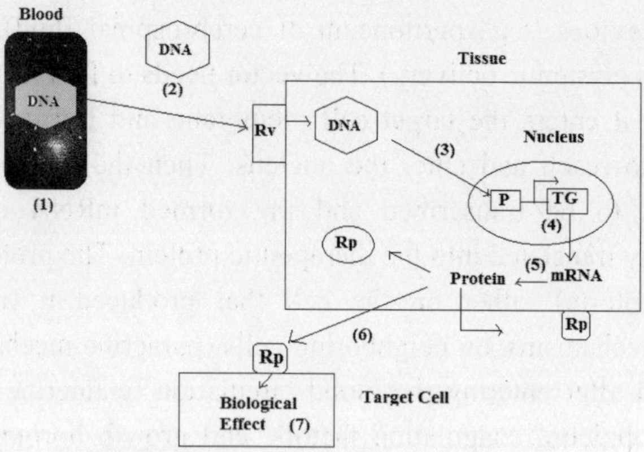

(Multiple biological process of in vivo gene transfer)

Fig. 5.12 Schemetic of gene transfer.

[1) Vector (viral, nonviral, cell-based0 delivery (localized tissue delivery or systemic delivery via blood circulation) of a gene (DNA); 2) vector recognition by specific receptors (Rv) on cells in target tissue; 3) uptake of vector by cells, trafficking to the nucleus; 4) transcription (expression) of therapeutic (trans) gene in the nucleus; 5)Translation of mRNA into the therapeutic protein in the cytoplasm; 6) interaction of therapeutic protein with its receiptor (Rp) within the (autocrine mechanism) or on neighboring target cell (paracrine mechanism)]

Applications

Some examples of applications of gene therapy targeted for monogenic disease mentioned in table 5.1.

The gene therapy targets to multigenic disease include **cancers** such as: breast, ovary, cervix, colon, colorectal, liver, prostate, renal cancer, glioblastoma, leptomeningial carcinoma, adenocarcinoma, melanoma, myeloma, sarcoma; **Infections such** as: HIV/AIDS, Titanus, CMV infection, Adenivirus infection; **Vascular diseases** such as: Peripheral arterial disease, coronary heart disease, Venous ulcer, Vascular complications of diabetes; **Other diseases** such as: Inflammatory disease, Rheumatoid arthritis, Chronic renal disease, Alzhemers disease, Fracture, Diabetes neuropathy, erectile disfunction, Retinitis pigmentosa, glaucoma, etc.

Table 5.1 Disease targets for gene therapy (Monogenic Disease)

Disease	Gene(s)
Cystic fibrosis	CFTR, α-1-anti-trypsin
Severe combined immunodeficiencied (SCID)	ADA
Gaucher disease	Glucocerebrosidase
Canavan Disease	Aspartoacylase
Haemophilia A	Factor VIII
Haemophilia B	Factor IX
Chronic granulomatous disease	Gp91 phox
ALS	CNTF
Familial hypercholesterolemia	LDL-R
Hunter disease	Idurinate-2-sulfatase
Leukocyte adherence deficiency	CD 18
Muscular Distrophy	Sarcoglycan, dystrophin, utrophin
Fanconi anemia	Grou A gene
Purine nucleoside phosphorylase deficiency	PNP
Ornithin transcarbamylase deficiency	OTC

Examples of some therapeutic genes are: Ad5FGF-4 gene thrapy used for chronic myocardiac ischemia; Mda-7/IL-24(DNA) for cancer, HLA-A2 (DNA) for cancer, HLA-B13 (DNA) for cancer, HER-2/neu (DNA) for cancer, Il-12 (DNA) for cancer, HLA-B7 (DNA) for cancer, INF-β (DNA) for cancer therapy; CFTR (DNA) for cystic fibrosis; α_1-Antitrypsin (siRNA) for Antiinflammatory; Neucleoprotein acidic polymerase (siRNA) for viral infection; BMP-4/hPTH1-34(DNA) for bone formation; hPTH1-34(DNA) for bone formation; FGF-4((siRNA) for increase platelet number; Endothelial locus-1(DNA) for small blood vessel formation; FGF-4 (DNA) for angiogenesis, etc. are some of the important areas of gene therapy.

5.7 Proteomics

Proteomics is the large-scale study of proteins, particularly their structures and functions. Proteins are vital parts of living organisms, as they are the main components of the physiological metabolic pathways of cells. The term "proteomics" was first coined in 1997 to make an analogy with genomics, the study of the genes. The word "proteome" is a blend of "**prote**in" and "gen**ome**", and was coined by Marc Wilkins in 1994 while working on the concept as a PhD student. The proteome is the entire complement of proteins, including the modifications made to a particular set of proteins, produced by an organism or system. This will vary with time and distinct requirements, or stresses, that a cell or organism undergoes.

After genomics, proteomics is considered the next step in the study of biological systems. Proteomics confirms the presence of the protein and provides a direct measure of the quantity present.

5.7.1 Proteomic Chemistry

Proteomic chemistry is the systematic study of the interaction of small chemicals obtained as target proteins from humans and other organisms with biological macromolecules. It is a scientific discipline, which is playing major role in current pharmaceutical drug discovery. Proteomic chemistry is working at the interface of the chemical and biological universe, integrating molecular biology, cell biology, structural biology, biochemistry, medicinal chemistry, bioinformatics and cheminformatics. Drug targets are studied as pure components *in-vitro* or as part of their physiological environment in a cell. Proteomic chemistry includes both the study of a very large number of low molecular weight compounds against a single biological drug target (high-throughput screening; lead finding) or the study of a single compound against many different targets within or across different protein families (selectivity or safety profiling). Screening as well as profiling requires a high degree of automation and a substantial infrastructure of assay and information technologies.

5.7.2 Methods of Studying Proteins

Determining proteins which are post-translationally modified (PTM)

One way in which a particular protein can be studied is to develop an antibody which is specific to that modification. For example, there are antibodies which only recognize certain proteins when they are tyrosine-phosphorylated, known as phospho-specific antibodies; also, there are antibodies specific to other modifications. These can be used to determine the set of proteins that have undergone the modification of interest.

For sugar modifications, such as glycosylation of proteins, certain lectins have been discovered which bind sugars, which can be used. A more common way to determine post-translational modification of interest is to subject a complex mixture of proteins to electrophoresis in "two-dimensions", which simply means that the proteins are electrophoresed first in one direction, and then in another. This allows small differences in a protein to be visualized by separating a modified protein from its unmodified form. This methodology is known as "two-dimensional gel electrophoresis". Recently, another approach has been developed called PROTOMAP which combines SDS-PAGE with shotgun proteomics to enable detection of changes in gel-migration such as those caused by proteolysis or post translational modification.

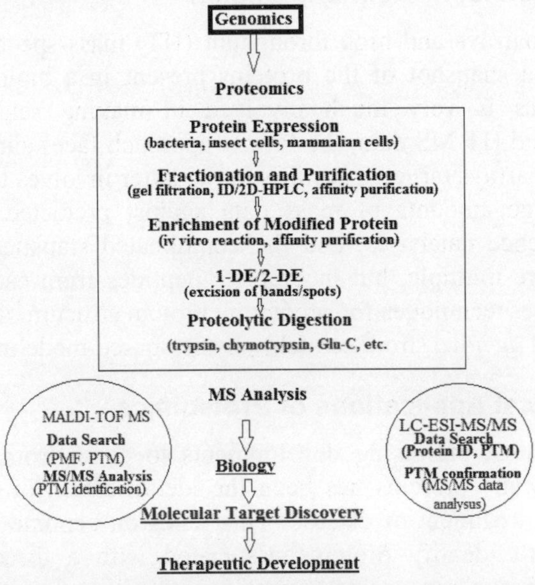

(Strategies and Goal of Proteom Analysis)

5.7.3 Determining the Existence of Proteins in Complex Mixtures

Classically, antibodies to particular proteins or to their modified forms have been used in biochemistry and cell biology studies. These are among the most common tools used by practicing biologists today. For more quantitative determinations of protein amounts, techniques such as ELISAs can be used. For proteomic study, more recent techniques such as matrix-assisted laser desorption/ionization (MALDI) have been employed for rapid determination of proteins in particular mixtures and increasingly electro-spray ionization (ESI)

5.7.4 Establishing Protein-Protein Interactions

Most proteins function in collaboration with other proteins, and one goal of proteomics is to identify which proteins interact. This is especially useful in determining potential partners in cell signaling cascades. Several methods are available to probe protein-protein interactions. The traditional method is yeast two-hybrid analysis. New methods include **protein microarrays**, **immunoaffinity chromatography** followed by

mass spectrometry, dual polarisation interferometry and experimental methods such as phage display and computational methods.

5.7.5 Analysis of Protein Expression

Protein microarrays and high throughput (HT) mass spectrometry (MS) can provide a snapshot of the proteins present in a biological sample. Bioinformatics is very much involved in making sense of protein microarray and HT MS data; the former approach faces similar problems as with microarrays targeted at mRNA, the latter involves the problem of matching large amounts of mass data against predicted masses from protein sequence databases, and the complicated statistical analysis of samples where multiple, but incomplete peptides from each protein are detected. Other techniques for predicting protein structure include protein threading and *de novo* (from scratch) physics-based modeling.

5.7.6 Practical Applications of Proteomics

One of the most promising developments to come from the study of human genes and proteins has been the identification of potential new drugs for the treatment of disease. This relies on genome and proteome information to identify proteins associated with a disease and with computer software can use as targets for new drugs. For example, if a certain protein is implicated in a disease, its 3D structure provides the information to design drugs to interfere with the action of the protein. A molecule that fits the active site of an enzyme, but cannot be released by the enzyme, will inactivate the enzyme. This is the basis of new drug-discovery tools, which aim to find new drugs to inactivate proteins involved in disease. A computer technique which attempts to fit millions of small molecules to the three-dimensional structure of a protein is called "virtual ligand screening". The computer rates the quality of the fit to various sites in the protein, with the goal of either enhancing or disabling the function of the protein, depending on its function in the cell. A good example of this is the identification of new drugs to target and inactivate the HIV-1 protease. The HIV-1 protease is an enzyme that cleaves a very large HIV protein into smaller, functional proteins. The virus cannot survive without this enzyme; therefore, it is one of the most effective protein targets for killing HIV.

Classes of Therapeutic Proteins and Peptides: Whether naturally occurring, genetically engineered, or semi-synthetic, there is a broad spectrum of protein and peptide drugs, including: (1) hormones and growth factors, (2) clotting factors and anticoagulants, (3) bacterial or plant toxins, (4) drug-activating enzymes, and (5) antibody-based drugs.

Most of the above classes of protein and peptide drugs have been used one way or another in various forms of cancer therapy. It is highly desirable that these therapeutic proteins and peptides possess an "active targeting" capability to reach intended target cells and leave the normal cells unharmed. The "self-homing" hormones (LHRH, luteinizing hormone-releasing hormone also known as GRH, gonadotropin-releasing hormone; somatostatin), growth factors (VEGF, vascular endothelial growth factor; EGF, epidermal growth factor; ILs, interleukins), and their agonists or antagonists by and large fall into this category. They have demonstrated some clinical success either by themselves or as delivering ligands (eg, denileukin diftitox). However, it is the monoclonal antibodies that gained the most attention in the past decades as the successful example of targeted protein therapeutics in oncology, either free (rituximab, trastuzumab, cetuximab) or in forms of immunoconjugates (gemtuzumab ozogamicin, ^{131}I-tositumomab, ^{90}Y-ibritumomab).

For proteins and peptides that lack tumor selectivity, coupling to a tumor-specific ligand can significantly modify its pharmacological properties and enhance its tumor specificity. For example, bacterial (diphtheria toxin, *Pseudomonas* exotoxin) and plant (gelonin, ricin) toxins are among the first proteins that have been explored in cancer therapy in a ligand-targeted fashion. To achieve tumor selectivity, protein toxins are structurally altered to remove their normal tissue-binding function before genetically or biochemically linking to a tumor-specific ligand. Although limited by their toxicities, protein toxins are therapeutically beneficial for advanced hematologic malignancies that have become resistant to chemotherapy or radiation. Prodrug-activating enzymes (eg, carboxypeptidase G2, β-glucuronidase) constitute a unique class of protein drugs that have been selected for site-specific drug delivery. In this approach, enzymes of nonhuman origin are coupled to a tumor-selective antibody, a growth factor, or a small molecular ligand (folic acid, carbohydrates). After allowing the enzyme conjugate or fusion protein to localize in tumor and clear from the circulation, a specially designed nontoxic substrate (pro-drug) is administered and converted to an active drug capable of rapid diffusion into the target tissue (a bystander effect). This 2-step process is designed to improve the efficiency of drug delivery and make the cytotoxic agent more tolerable in humans. Finally, immuno-modulators such as cytokines (eg, IL-2, GM-CSF [granulocyte-macrophage colony-stimulating factor], TNF (tumor necrosis factor)-α) and co-stimulatory molecules (B7) have been coupled to antibodies or other ligands directed at tumor cells as a method of activating an immune response on the cell surface. A list of examples of tumor ligand-targeted protein therapeutics is provided in Table-2.

5.8 Biomarkers

The FDA defines a biomarker as, "A characteristic that is objectively measured and evaluated as an indicator of normal biologic processes, pathogenic processes, or pharmacologic responses to a therapeutic intervention".

Understanding the proteome, the structure and function of each protein and the complexities of protein-protein interactions will be critical for developing the most effective diagnostic techniques and disease treatments in the future.

An interesting use of proteomics is using specific protein biomarkers to diagnose disease. A number of techniques allow testing for proteins produced during a particular disease, which helps to diagnose the disease quickly. Techniques include western blot, immunohistochemical staining, enzyme linked immunosorbent assay (ELISA) or mass spectrometry.

5.9 Bioinformatics

is the application of statistics and computer science to the field of molecular biology. The term *bioinformatics* was coined by Paulien Hogeweg in 1979 for the study of informatics processes in biotic systems. Its primary use since at least the late 1980s has been in genomics and genetics, particularly in those areas of genomics involving large-scale DNA sequencing.

Bioinformatics now entails the creation and advancement of databases, algorithms, computational and statistical techniques and theory to solve formal and practical problems arising from the management and analysis of biological data.

Common activities in bioinformatics include mapping and analyzing DNA and protein sequences, aligning different DNA and protein sequences to compare them and creating and viewing 3-D models of protein structures.

The primary goal of bioinformatics is to increase the understanding of biological processes. What sets it apart from other approaches, however, is its focus on developing and applying computationally intensive techniques (e.g., pattern recognition, data mining, machine learning algorithms, and visualization) to achieve this goal. Major research efforts in the field include sequence alignment, gene finding, genome assembly, drug design, drug discovery, protein structure alignment, protein structure prediction, prediction of gene expression and protein-protein interactions, genome-wide association studies and the modeling of evolution.

Table 5.2 Classes of Therapeutic proteins and its Application

Protein Drugs	**Targeting Ligands**	**Disease Indications**
Diphtheria toxin	IL-2	Cutaneous T-cell lymphoma
	Transferrin	Malignant glioma
	GM-CSF	Acute myeloid leukemia
Pseudomonas exotoxin	anti-CD25	CD25-positive hematologic malignancy
	anti-CD22	Hairy cell leukemia
	TGF-α	Malignant brain tumors
	Folic acid	FR-expressing cancer
Momordin	Folic acid	FR-expressing cancer
Gelonin	Folic acid	FR-expressing cancer
Carboxypeptidase G2 /mustard prodrug	anti-CEA	CEA-expressing cancer
β-glucuronidase/ doxorubicin prodrug	anti-EpCAM scFv	EpCAM-expressing cancer
Carboxypeptidase G2 /CMDA prodrug	VEGF	VEGF-expressing cancer
Penicillin-V-amidase/ doxorubicin prodrug	Folic acid	FR-expressing cancer
α-rhamnosidase/ doxorubicin prodrug	Gal	Hepatocarcinomas
IL-2	anti-GD2	Metastatic melanoma
	anti-EpCAM	Epithelial ovarian cancer
	anti-CD20	CD20-positive lymphoma
	MOv19 scFv	FR-expressing cancer
GM-CSF	anti-GD2	Neuroblastoma
TNF-α	anti-HER-2/ neu scFv	HER-2/neu-expressing cancer
B7	anti-CD64	Acute myeloid leukemia blasts
Abbreviations: IL-2 indicates interleukin 2; GM-CSF, granulocyte-macrophage colony-stimulating factor; TGF-α, transforming growth factor α; FR, folate receptor; CEA, carcinoembryonic antigen; EpCAM, epithelial cell adhesion molecule; VEGF, vascular endothelial growth factor; Gal, galactose alpha 1,3 galactose; GD2, disialoganglioside; TNF-α, tumor necrosis factor α; HER-2, human epidermal growth factor receptor 2.		

Table *contd...*

Protein Drugs	Targeting Ligands	Disease Indications
Antibody-Targeted Antimitotic Peptides		
Monomethylauristatin E	Anti-CD20	B-cell malignancies
Auristatin E Monomethylauristatin E	Anti-CD30	Hematological malignancies
Auristatin E Monomethylauristatin E	Anti-Lewis Y	Carcinomas
Monomethylauristatin E	Anti-E-selectin	Prostate cancer
Auristatin E	Anti-TMEFF2	Prostate cancer
Targeting tumor vasculature		
TNF-α	CNGRC (linear), GNGRG (cyclic)	Targeting tumor vasculature
Doxorubicin	CDCRGDCFC CNGRC	Targeting tumor vasculature
Tachyplesin (antimicrobial peptide)	CRGDCGG	Targeting tumor vasculature
$(KLAKLAK)_2$ (pro-apoptotic peptide)	CNGRC	Targeting tumor vasculature
Targeting cell-surface hormone receptors		
^{111}In, ^{90}Y, ^{177}Lu	Somatostain	Targeting cell-surface hormone receptors
^{99m}Tc	Bombesin	Targeting cell-surface hormone receptors
^{99m}Tc	Gastrin-releasing peptide (GRP)	Targeting cell-surface hormone receptors
Doxorubicin 2-pyrrolino-doxorubicin	Somatostain, Bombesin, LHRH	Targeting cell-surface hormone receptors
Hecate (membrane-lytic peptide)	LHRH, LH β-chain (AAs 81–95), HCG fragment	Targeting cell-surface hormone receptors
Targeting tumor vasculature		
ATWLPPR	VEGFR	VEGF antagonist
Thrombospondin (TSP)-1-mimetics	CD36	Apoptosis
ACDCRGDCFCG (cyclic), SCH 221153 (RGD peptidomimetic)	Integrins $\alpha_v\beta_3$ and $\alpha_v\beta_5$	Ligand mimics
CNGRC (cyclic)	Aminopeptidase N	Inhibitor
CTTHWGFTLC	MMP-2 and MMP-9	Inhibitor
CGNKRTRGC (LyP-1)	Lymphatic vessels	Apoptosis
Abbreviations: TNF-α, tumor necrosis factor α; LHRH, luteinizing hormone-releasing hormone; HCG, human chorionic gonadotropin		

References

Abbott, A. "Gene therapy. Italians first to use stem cells". *Nature,* 1992, 356 (6369): 465–199.

Adams MD, Kelley JM, Gocayne JD, *et al.* "Complementary DNA sequencing: expressed sequence tags and human genome project". *Science,* 1991, 252 (5013): 1651–1656.

Lu Y, Yang J, Sega E. Issues Related to Targeted Delivery of Proteins and Peptides. *AAPS Journal.*, 2006, 8(3): E466-E478. DOI: 10.1208 /aapsj080355

Adams, M. D. et al. The genome sequence of Drosophila melanogaster. *Science,* 2000, 287 (5461): 2185–95.

Ahn SM, Kim TH, Lee S, Kim D, Ghang H, Kim D, Kim BC, Kim SY, Kim WY, Kim C, Park D, Lee YS, Kim S, Reja R, Jho S, Kim CG, Cha JY, Kim KH, Lee B, Bhak J, Kim SJ . "The first Korean genome sequence and analysis: Full genome sequencing for a socio-ethnic group". *Genome Research,* 2009, 19 (9): 1622–1629.

Albà M. "Replicative DNA polymerases". *Genome Biol,* 2001, 2 (1): REVIEWS3002. doi:10.1186/gb-2001-2-1-reviews3002

Anderson N L, Anderson N G. "Proteome and proteomics: new technologies, new concepts, and new words". *Electrophoresis, 1998,* 19 (11): 1853–1861.

Avery O, MacLeod C, McCarty M. "Studies on the chemical nature of the substance inducing transformation of pneumococcal types. Inductions of transformation by a desoxyribonucleic acid fraction isolated from pneumococcus type III". *J Exp Med,* 1944, 79 (2): 137–158.

www. Becomehealthynow.com/category/bodyimmune/….

Basu H, Feuerstein B, Zarling D, Shafer R, Marton L. "Recognition of Z-RNA and Z-DNA determinants by polyamines in solution: experimental and theoretical studies". *J Biomol Struct Dyn,* 1988, 6 (2): 299–309.

Baum C, Düllmann J, Li Z, *et al.* "Side effects of retroviral gene transfer into hematopoietic stem cells". *Blood,* 2003, 101 (6): 2099–114.

Bentley D. R. "Whole-genome re-sequencing". *Curr. Opin. Genet. Dev,* 2006, 16 (6): 545–52. doi:10.1016/j.gde.2006.10.009

Berg, P.; Mertz, J. Personal reflections on the origins and emergence of recombinant DNA technology. *Genetics, 2010,* 184 (1): 9–17.

Berger AB, et al. Activity-based protein profiling: applications to biomarker discovery, *in vivo* imaging and drug discovery. *American Journal of Pharmacogenomics,* 2004.

Bickle T, Krüger D. "Biology of DNA restriction". *Microbiol Rev,*1993, 57 (2): 434–50

Blackstock WP, Weir MP. "Proteomics: quantitative and physical mapping of cellular proteins". *Trends Biotechnol.* 1999, 17 (3): 121–7.doi:10.1016/S0167-7799(98)01245-1

Braña M, Cacho M, Gradillas A, de Pascual-Teresa B, Ramos A. "Intercalators as anticancer drugs". *Curr Pharm Des,* 2001, **7** (17): 1745–80.

Braslavsky I, Hebert B, Kartalov E, Quake SR. "Sequence information can be obtained from single DNA molecules". *Proc. Natl. Acad. Sci.,* 2003, 100 (7): 3960–4. doi:10.1073/pnas.0230489100

Brenner, Sidney; Johnson, M; Bridgham, J; Golda, G; Lloyd, DH; Johnson, D; Luo, S; McCurdy, S *et al.* "Gene expression analysis by massively parallel signature sequencing (MPSS) on microbead arrays". *Nature Biotechnology*, 2000, 18 (6): 630–634. doi:10.1038/76469.

Brown BD, Venneri MA, Zingale A, Sergi Sergi L, Naldini L. "Endogenous microRNA regulation suppresses transgene expression in hematopoietic lineages and enables stable gene transfer". *Nat Med.* 2006, 12 (5): 585–91. doi:10.1038/nm1398.

Burge S, Parkinson G, Hazel P, Todd A, Neidle S. "Quadruplex DNA: sequence, topology and structure". *Nucleic Acids Res,* 2006, 34 (19): 5402–15.

Butler, John M. Forensic DNA Typing. Elsevier. ISBN 978-0-12-147951-0. OCLC 45406517 223032110 45406517. 2001: 14–15.

Cadet J, Delatour T, Douki T, Gasparutto D, Pouget J, Ravanat J, Sauvaigo S. "Hydroxyl radicals and DNA base damage". *Mutat Res*, 1999, 424 (1–2): 9–21.

Chalikian T, Völker J, Plum G, Breslauer K. "A more unified picture for the thermodynamics of nucleic acid duplex melting: a characterization by calorimetric and volumetric techniques". *Proc Natl Acad Sci,* 1999, 96 (14): 7853–8. doi:10.1073/pnas.96.14.7853.

Champoux J. "DNA topoisomerases: structure, function, and mechanism". *Annu Rev Biochem,* 2001 (70): 369–413. doi:10.1146/annurev.biochem.70.1.369

Chen I, Dubnau D. "DNA uptake during bacterial transformation". *Nat. Rev. Microbiol.* 2004, **2** (3): 241–9. doi:10.1038/nrmicro844

Church GM. "Genomes for all". *Sci. Am.* 2006, 294 (1): 46–54. doi:10.1038/scientificamerican0106-46

Clausen-Schaumann H, Rief M, Tolksdorf C, Gaub H (2000). "Mechanical stability of single DNA molecules". *Biophys J,* 2000, 78 (4): 1997–2007. doi:10.1016/S0006-3495(00)76747-6

Cremer T, Cremer C. "Chromosome territories, nuclear architecture and gene regulation in mammalian cells". *Nat Rev Genet*, 2001, 2 (4): 292–301. doi:10.1038/35066075

Cuello JC, Engineering to biology and biology to engineering, The bi-directional connection between engineering and biology in biological engineering design, *Int J Engng Ed*, 2005 (21):1-7

Dahm R. "Discovering DNA: Friedrich Miescher and the early years of nucleic acid research". *Hum. Genet.* 2008, 122 (6): 565–81.

Dame RT. "The role of nucleoid-associated proteins in the organization and compaction of bacterial chromatin". *Mol. Microbiol.* 2005, 56 (4): 858–70.

Davenport R. "Ribozymes. Making copies in the RNA world". *Science,* 2001, 292 (5520):1278. doi:10.1126/science.292. 5520. 1278a

David M. Suter, Michel Dubois-Dauphin, Karl-Heinz Krause. "Genetic engineering of embryonic stem cells". *Swiss Med Wkly,* 2006, 136 (27-28): 413–415.

De Grouchy J. "Chromosome phylogenies of man, great apes, and Old World monkeys". *Genetica,* 1987, 73 (1-2): 37–52.

De Haseth P, Helmann J. "Open complex formation by Escherichia coli RNA polymerase: the mechanism of polymerase-induced strand separation of double helical DNA". *Mol Microbiol,* 1995, 16 (5): 817–24. doi:10.1111/j.1365-2958.1995.tb02309.x.

Doherty A, Suh S. "Structural and mechanistic conservation in DNA ligases". *Nucleic Acids Res,* 2000, 28 (21): 4051–8. doi:10.1093/nar/28.21.4051

Douki T, Reynaud-Angelin A, Cadet J, Sage E. "Bipyrimidine photoproducts rather than oxidative lesions are the main type of DNA damage involved in the genotoxic effect of solar UVA radiation". *Biochemistry,* 2003, 42 (30): 9221–6.

Durai S, Mani M, Kandavelou K, Wu J, Porteus MH, Chandrasegaran S. "Zinc finger nucleases: custom-designed molecular scissors for genome engineering of plant and mammalian cells". *Nucleic Acids Res.* 2005, 33 (18): 5978–90.

Edwards A, Voss H, Rice P, Civitello A, Stegemann J, Schwager C, Zimmermann J, Erfle H, Caskey CT, Ansorge W. "Automated DNA sequencing of the human HPRT locus". *Genomics* 1990, 6 (4): 593–608.

Ehrenfeld, David. "Transgenics and Vertebrate Cloning as Tools for Species Conservation". *Conservation Biology*, 2006, 20 (3): 723–732.

Ernesto Andrianantoandro, Subhayu Basu, David K Kariga & Ron Weiss. "Synthetic biology: new engineering rules for an emerging discipline". *Molecular Systems Biology,* 2006 (2):28.

Ferguson L, Denny W. "The genetic toxicology of acridines". *Mutat Res,* 1991, 258 (2): 123–60.

Fields, S. Song, O. "A novel genetic system to detect protein-protein interactions". *Nature,* 1989, 340 (6230): 245–246. doi:10.1038/340245a0.

Fiers W, Contreras R, Duerinck F, *et al.* "Complete nucleotide sequence of bacteriophage MS2 RNA: primary and secondary structure of the replicase gene". *Nature,* 1976, 260 (5551): 500–7. doi:10.1038/260500a0

Fiers W, Contreras R, Duerinck F, Haegeman G, Iserentant D, Merregaert J, Min Jou W, Molemans F, Raeymaekers A, Van den Berghe A, Volckaert G, Ysebaert M. "Complete nucleotide sequence of bacteriophage MS2 RNA: primary and secondary structure of the replicase gene". *Nature,* 1976, 260 (5551): 500–507. doi:10.1038/260500a0

Finan TM, Weidner S, Wong K, Buhrmester J, Chain P, Vorhölter FJ, Hernandez-Lucas I, Becker A, Cowie A, Gouzy J, Golding B, Pühler A. "The complete sequence of the 1,683-kb pSymB megaplasmid from the N2-fixing endosymbiont". *Proceedings of the National Academy of Sciences,* 2001, 98 (17):9889. doi:10.1073/pnas.161294698

Fisher, Jennifer. "Murine Gene Therapy Corrects Symptoms of Sickle Cell Disease - The Scientist - Magazine of the Life Sciences". The Scientist. http://www.the-scientist.com/article/display/12938.

Fleischmann RD, Adams MD, White O, Clayton RA, Kirkness EF, Kerlavage AR, Bult CJ, Tomb JF, Dougherty BA, Merrick JM. "Whole-genome random sequencing and assembly of Haemophilus influenzae Rd". *Science (journal),* 1995, 269 (5223): 496–512.

Friedmann, T.; Roblin, R. "Gene Therapy for Human Genetic Disease?". *Science,* 1973, 175 (25): 949. doi:10.1126/science.175.4025.949

Gardlík R, Pálffy R, Hodosy J, Lukács J, Turna J, Celec P. "Vectors and delivery systems in gene therapy". *Med Sci Monit.* 2005, 11 (4): RA110–21. PMID 15795707

Gaucher's disease:Treatments and drugs, eMedicine WebMD, 2009-07-11

Gelvin, S. B. "Agrobacterium-Mediated Plant Transformation: the Biology behind the "Gene-Jockeying" Tool". *Microbiology and Molecular Biology Reviews,* 2003, 67 (1): 16.

Abbott A. Gene therapy. Italians first to use stem cells. *Nature.* 1992, 356(6369):465.

Ghosh A, Bansal M. "A glossary of DNA structures from A to Z". *Acta Crystallogr D Biol Crystallogr,* 2003, 59 (4):620–6. doi:10.1107/S0907444903003251.

Gibson, D.; Glass, J.; Lartigue, C.; Noskov, V.; Chuang, R.; Algire, M.; Benders, G.; Montague, M. *et al.* "Creation of a Bacterial Cell Controlled by a Chemically Synthesized Genome". *Science (New York, N.Y.),* 2010, 329 (5987): 52–56. doi:10.1126/science.1190719

Gilbert, D. *Bioinformatics software resources.* Briefings in Bioinformatics, Briefings in Bioinformatics, 2004 5(3):300-304.

Goeddel DV, Heyneker HL, Hozumi T, *et al*. "Direct expression in *Escherichia coli* of a DNA sequence coding for human growth hormone". *Nature*, 1979, 281 (5732): 544–8. doi:10.1038/281544a0. PMID 386136

Goeddel, David; Dennis G. Kleid, Francisco Bolivar, Herbert L. Heyneker, Daniel G. Yansura, Roberto Crea, Tadaaki Hirose, Adam Kraszewski, Keiichi Itakura, AND Arthur D. Riggs. "Expression in Escherichia coli of chemically synthesized genes for human insulin". *PNAS,* 1979, 76 (1): 106–110. doi:10.1073/pnas.76.1.106

Goff SP, Berg P. "Construction of hybrid viruses containing SV40 and lambda phage DNA segments and their propagation in cultured monkey cells". *Cell,* 1976, 9 (4 PT 2): 695–705

Gregory S; Barlow, KF; McLay, KE; Kaul, R; Swarbreck, D; Dunham, A; Scott, CE; Howe, KL *et al*. "The DNA sequence and biological annotation of human chromosome 1". *Nature,* 2006, 441 (7091): 315–21. doi:10.1038/nature04727.

Greider C, Blackburn E. "Identification of a specific telomere terminal transferase activity in Tetrahymena extracts". *Cell,* 1985, 43 (2 Pt 1): 405–13.

Griffith J, Comeau L, Rosenfield S, Stansel R, Bianchi A, Moss H, de Lange T (1999). "Mammalian telomeres end in a large duplex loop". *Cell,* 1999, 97 (4): 503–14.

Grosschedl R, Giese K, Pagel J. "HMG domain proteins: architectural elements in the assembly of nucleoprotein structures". *Trends Genet,* 1994, 10 (3): 94–100. doi:10.1016/0168-9525(94)90232-1.

Hall N. "Advanced sequencing technologies and their wider impact in microbiology". *J. Exp. Biol.* 2007, 210 (9): 1518–25. doi:10.1242/jeb.001370.

Harrison P, Gerstein M. "Studying genomes through the aeons: protein families, pseudogenes and proteome evolution". *J Mol Biol,* 2002, 318 (5): 1155–74.

Hershey A, Chase M. "Independent functions of viral protein and nucleic acid in growth of bacteriophage" (PDF). *J Gen Physiol*, 1952, 36 (1): 39–56.

Hinnebusch J, Tilly K. "Linear plasmids and chromosomes in bacteria". *Mol. Microbiol,* 1993, 10 (5): 917–22. doi:10.1111/j.1365-2958.1993.tb00963.x

Horn PA, Morris JC, Neff T, Kiem HP. "Stem cell gene transfer—efficacy and safety in large animal studies". *Mol. Ther.* 2004, 10 (3): 417–31. doi:10.1016/j.ymthe.2004.05.017.

Houdebine L. "Transgenic animal models in biomedical research". *Methods Mol Biol,* 2007 (360): 163–202. doi:10.1385/1-59745-165-7:163

http://en.wikipedia.org/wiki/Use_of_biotechnology_in_pharmaceutical_manufacturing"

http://www.ncbi.nlm.nih.gov/genome/seq/ for more information on the Human Genome Project.

Hubscher U, Maga G, Spadari S. "Eukaryotic DNA polymerases". *Annu Rev Biochem, 2002* (71): 133–63.

Human Insulin: Seizing the Golden Plasmid". *Science News* 1978, 114 (12): 195. doi:10.2307/3963132

Hüttenhofer A, Schattner P, Polacek N. "Non-coding RNAs: hope or hype?". *Trends Genet,* 2005, 21 (5): 289–97. doi:10.1016/j.tig.2005.03.007

Illmensee K, Levanduski M, Vidali A, Husami N, Goudas VT. "Human embryo twinning with applications in reproductive medicine". *Fertil. Steril.* 2009, 93 (2): 423–7. doi:10.1016/j.fertnstert.2008.12.098. PMID 19217091

Inoue, Noboru; Takeuchi, Hideya; Ohashi, Makoto; Suzuki, Takamoto; Makoto Takeuchi. "The production of recombinant human erythropoietin". *Biotechnology Annual Review*, 1995 (1): 297–300. doi:10.1016/S1387-2656(08)70055-3

Isaksson J, Acharya S, Barman J, Cheruku P, Chattopadhyaya J. "Single-stranded adenine-rich DNA and RNA retain structural characteristics of their respective double-stranded conformations and show directional differences in stacking pattern". *Biochemistry,* 2004, 43 (51): 15996–6010. doi:10.1021/bi048221v.

Ishitsuka Y, Ha T. "DNA nanotechnology: a nanomachine goes live". *Nat Nanotechnol,* 2009, 4 (5): 281–2.

Jackson, DA; Symons, RH; Berg, P. "Biochemical Method for Inserting New Genetic Information into DNA of Simian Virus 40: Circular SV40 DNA Molecules Containing Lambda Phage Genes and the Galactose Operon of Escherichia coli". *PNAS,* 1972, 69 (10): 2904–2909.

Jaenisch, R. and Mintz, B. "Simian virus 40 DNA sequences in DNA of healthy adult mice derived from preimplantation blastocysts injected with viral DNA.". *Proc. Natl. Acad. Sci*. 1974 (71): 1250–1254

Jeffrey A. "DNA modification by chemical carcinogens". *Pharmacol Ther*, 1985, 28 (2): 237–72.

Jenuwein T, Allis C. "Translating the histone code". *Science,* 2001, 293 (5532): 1074–80. doi:10.1126/science.1063127.

Johnson Z, Chisholm S. "Properties of overlapping genes are conserved across microbial genomes". *Genome Res,* 2004, 14 (11): 2268–72. doi:10.1101/gr.2433104

Joyce C, Steitz T. "Polymerase structures and function: variations on a theme?". *J Bacteriol,* 1995, 177 (22): 6321–9.

Kandavelou K; Chandrasegaran S. "Plasmids for Gene Therapy". *Plasmids: Current Research and Future Trends*. Caister Academic Press. 2008, ISBN 978-1-904455-35-6.

Kaufman RJ, Wasley LC, Furie BC, Furie B, Shoemaker CB. "Expression, purification, and characterization of recombinant gamma-carboxylated factor IX synthesized in Chinese hamster ovary cells". *J. Biol. Chem.* 1986, 261 (21): 9622–8. PMID 3733688.

Klose R, Bird A. "Genomic DNA methylation: the mark and its mediators". *Trends Biochem Sci,* 2006, 31 (2): 89–97

Klose R, Bird A. "Genomic DNA methylation: the mark and its mediators". *Trends Biochem Sci,* 2006, 31 (2): 89–97. doi:10.1016/j.tibs.2005.12.008.

Komáromy, A.; Alexander, J.; Rowlan, J.; Garcia, M.; Chiodo, V.; Kaya, A.; Tanaka, J.; Acland, G. *et al*. "Gene therapy rescues cone function in congenital achromatopsia". *Human molecular genetics,* 2010, 19 (13): 2581–2593. doi:10.1093/hmg/ddq136.

L. P. Gianessi, C. S. Silvers, S. Sankula and J. E. Carpenter. Plant Biotechnology: Current and Potential Impact for Improving Pest management in US Agriculture, An Analysis of 40 Case Studies (Washington, D.C.: National Center for Food and Agricultural Policy, 2002: 5–6

Lamb R, Horvath C. "Diversity of coding strategies in influenza viruses". *Trends Genet,* 1991, 7 (8): 261–6.

Leslie AG, Arnott S, Chandrasekaran R, Ratliff RL. "Polymorphism of DNA double helices". *J. Mol. Biol.* 1980, 143 (1):49–72. doi:10.1016/0022-2836(80)90124-2

Levene P. "The structure of yeast nucleic acid". *J Biol Chem,* 1919, 40 (2): 415–24

Levine BL, Humeau LM, Boyer J, *et al.* "Gene transfer in humans using a conditionally replicating lentiviral vector". *Proc Natl Acad Sci,* 2006, 103 (46): 17372–7. doi:10.1073/pnas.0608138103.

Levy S, Sutton G, Ng PC, Feuk L, Halpern AL, Walenz BP, Axelrod N, Huang J, Kirkness EF, Denisov G, Lin Y, MacDonald JR, Pang AW, Shago M, Stockwell TB, Tsiamouri A, Bafna V, Bansal V, Kravitz SA, Busam DA, Beeson KY, McIntosh TC, Remington KA, Abril JF, Gill J, Borman J, Rogers YH, Frazier ME, Scherer SW, Strausberg RL, Venter JC. "The diploid genome sequence of an individual human". *PLoS Biol.,* 2007, 5 (10): e254. doi:10.1371/journal.pbio.0050254.

Ley TJ, Mardis ER, Ding L, Fulton B, McLellan MD, Chen K, Dooling D, Dunford-Shore BH, McGrath S, Hickenbotham M, Cook L, Abbott R, Larson DE, Koboldt DC, Pohl C, Smith S, Hawkins A, Abbott S, Locke D, Hillier LW, Miner T, Fulton L, Magrini V, Wylie T, Glasscock J, Conyers J, Sander N, Shi X, Osborne JR, Minx P, Gordon D, Chinwalla A, Zhao Y, Ries RE, Payton JE, Westervelt P, Tomasson MH, Watson M, Baty J, Ivanovich J, Heath S, Shannon WD, Nagarajan R, Walter MJ, Link DC, Graubert TA, DiPersio JF, Wilson RK. "DNA sequencing of a cytogenetically normal acute myeloid leukaemia genome". *Nature,* 2008, 456 (7218): 66–72. doi:10.1038/nature07485

Lu XJ, Shakked Z, Olson WK. "A-form conformational motifs in ligand-bound DNA structures". *J. Mol. Biol.* 2000, 300 (4): 819–40. doi:10.1006/jmbi.2000.3690

Luger K, Mäder A, Richmond R, Sargent D, Richmond T. "Crystal structure of the nucleosome core particle at 2.8 A resolution". *Nature,* 1997, 389 (6648): 251–60. doi:10.1038/38444.

Maguire AM, Simonelli F, Pierce EA, *et al*. "Safety and efficacy of gene transfer for Leber's congenital amaurosis". *N Engl J Med.* 2008, 358 (21): 2240–8. doi:10.1056/NEJMoa0802315

Makalowska I, Lin C, Makalowski W. "Overlapping genes in vertebrate genomes". *Comput Biol Chem,* 2008, 29 (1): 1–12. doi:10.1016/j.compbiolchem.2004.12.006

Mandelkern M, Elias J, Eden D, Crothers D. "The dimensions of DNA in solution". *J Mol Biol,* 1981, 152 (1): 153–61. doi:10.1016/0022-2836(81)90099-1

Marc R. Wilkins, Christian Pasquali, Ron D. Appel, Keli Ou, Olivier Golaz, Jean-Charles Sanchez, Jun X. Yan, Andrew. A. Gooley, Graham Hughes, Ian Humphery-Smith, Keith L. Williams & Denis F. Hochstrasser. "From Proteins to Proteomes: Large Scale Protein Identification by Two-Dimensional Electrophoresis and Arnino Acid Analysis". *Nature Biotechnology,* 1996, 14 (1): 61–65. doi:10.1038/nbt0196-61.

Martinez E. "Multi-protein complexes in eukaryotic gene transcription". *Plant Mol Biol,* 2002, 50 (6): 925–47

Maxam AM, Gilbert W. "A new method for sequencing DNA". *Proc. Natl. Acad. Sci. U.S.A.,* 1997, 74 (2): 560–4. doi:10.1073/pnas.74.2.560

McCabe LL, McCabe ER. "Postgenomic medicine. Presymptomatic testing for prediction and prevention". *Clin Perinatol,* 2001, 28 (2): 425–34. doi:10.1016/S0095-5108(05)70094-4

McFarland, Douglas. "Preparatioii of pure cell cultures by cloning". *Methods in Cell Science,* 2000, 22 (1): 63–66. doi:10.1023/A:1009838416621. PMID 10650336.

Min Jou W, Haegeman G, Ysebaert M, Fiers W. "Nucleotide sequence of the gene coding for the bacteriophage MS2 coat protein". *Nature,* 1972, 237 (5350): 82–88. doi:10.1038/237082a0

Min Jou W, Haegeman G, Ysebaert M, Fiers W. "Nucleotide sequence of the gene coding for the bacteriophage MS2 coat protein". *Nature,* 1972, 237 (5350): 82–8. doi:10.1038/237082a0

Morcos, PA. "Achieving targeted and quantifiable alteration of mRNA splicing with Morpholino oligos". *Biochem Biophys Res Commun,* 2007, 358 (2): 521–7. doi:10.1016/j.bbrc.2007.04.172.

Morgan RA, Dudley ME, Wunderlich JR, *et al.* "Cancer regression in patients after transfer of genetically engineered lymphocytes". *Science,* 2006, 314 (5796): 126–9. doi:10.1126/science.1129003.

Mount, David W. *Bioinformatics: Sequence and Genome Analysis* Spring Harbor Press, 2002. ISBN 0-87969-608-7

Mukhopadhyay R. "DNA sequencers: the next generation". *Anal. Chem.* 2009, 81 (5): 1736–40. doi:10.1021/ac802712u

Munroe S. "Diversity of antisense regulation in eukaryotes: multiple mechanisms, emerging patterns". *J Cell Biochem,* 2004, 93 (4): 664–71. doi:10.1002/jcb.20252

Murphy, K.; Berg, K.; Eshleman, J. "Sequencing of genomic DNA by combined amplification and cycle sequencing reaction". *Clinical chemistry*, 2005, 51 (1): 35–39. doi:10.1373/clinchem.2004.039164

Myers L, Kornberg R. "Mediator of transcriptional regulation". *Annu Rev Biochem,* 2000, (69): 729–49. doi:10.1146/annurev.biochem.69.1.729

Neale MJ, Keeney S. "Clarifying the mechanics of DNA strand exchange in meiotic recombination". *Nature,* 2006, 442 (7099): 153–8.

Nugent C, Lundblad V. "The telomerase reverse transcriptase: components and regulation". *Genes Dev,* 1998, 12 (8): 1073–85.

O'Driscoll M, Jeggo P. "The role of double-strand break repair - insights from human genetics". *Nat Rev Genet,* 2006, 7 (1): 45–54. doi:10.1038/nrg1746

Oh D, Kim Y, Rich A. "Z-DNA-binding proteins can act as potent effectors of gene expression in vivo". *Proc. Natl. Acad. Sci. U.S.A.* 2002, 99 (26): 16666–71. doi:10.1073/pnas.262672699.

Olsvik O, Wahlberg J, Petterson B, *et al.* "Use of automated sequencing of polymerase chain reaction-generated amplicons to identify three types of cholera toxin subunit B in Vibrio cholerae O1 strains". *J. Clin. Microbiol.*, 1993, 31 (1): 22–5.

Ott MG, Schmidt M, Schwarzwaelder K, *et al.* "Correction of X-linked chronic granulomatous disease by gene therapy, augmented by insertional activation of MDS1-EVI1, PRDM16 or SETBP1". *Nat Med.*, 2006, 12 (4): 401–9. doi:10.1038/nm1393.

P. James. "Protein identification in the post-genome era: the rapid rise of proteomics.". *Quarterly reviews of biophysics,* 1997, 30 (4): 279–331. doi:10.1017/S0033583597003399

Pabo C, Sauer R. "Protein-DNA recognition". *Annu Rev Biochem,* 1994 (53): 293–321. doi:10.1146/annurev.bi.53.070184.001453.

Pabo C, Sauer R. "Protein-DNA recognition". *Annu Rev Biochem* 1984 (53): 293–321. doi:10.1146/annurev.bi.53.070184.001453

Painter PC, Mosher LE, Rhoads C. "Low-frequency modes in the Raman spectra of proteins". *Biopolymers,* 1982, 21 (7): 1469–72.

Pál C, Papp B, Lercher M. "An integrated view of protein evolution". *Nat Rev Genet*, 2006, 7 (5): 337–48. doi:10.1038/nrg1838

Parkinson G, Lee M, Neidle S. "Crystal structure of parallel quadruplexes from human telomeric DNA". *Nature,* 2002, 417 (6891): 876–80. doi:10.1038/nature755

Pettersson E, Lundeberg J, Ahmadian A. "Generations of sequencing technologies". *Genomics,* 2009, 93 (2): 105–11.

Pevsner J. *Bioinformatics and functional genomics* (2nd ed.), 2009. Hoboken, NJ: Wiley-Blackwell

Pradella S, Hans A, Spröer C, Reichenbach H, Gerth K, Beyer S. "Characterisation, genome size and genetic manipulation of the myxobacterium Sorangium cellulosum So ce56". *Arch Microbiol,* 2002, 178 (6): 484–92.

Ratel D, Ravanat J, Berger F, Wion D. "N6-methyladenine: the other methylated base of DNA". *Bioessays,* 2006, 28 (3): 309–15. doi:10.1002/bies.20342

Retinitis Pigmentosa: Treatment & Medication, eMedicine WebMD, 2009-09-19, accessed 2010-03-31.

Richard Williams, Sergio G Peisajovich, Oliver J Miller, Shlomo Magdassi, Dan S Tawfik, Andrew D Griffith. "Amplification of complex gene libraries by emulsion PCR". *Nature methods,* 2006, 3 (7): 545–550. doi:10.1038/nmeth896

Roach JC, Boysen C, Wang K, Hood L. "Pairwise end sequencing: a unified approach to genomic mapping and sequencing". *Genomics,* 1995, 26 (2): 345–53. doi:10.1016/0888-7543(95)80219-C

Rodriguez, L.; Grubman, M. "Foot and mouth disease virus vaccines". *Vaccine,* 2009, 27 (Suppl 4): D90–D94. doi:10.1016/j.vaccine. 2009.08.039

Ronaghi M, Karamohamed S, Pettersson B, Uhlén M, Nyrén P. "Real-time DNA sequencing using detection of pyrophosphate release". *Anal. Biochem.,* 1996, 242 (1): 84–9. doi:10.1006/abio.1996.0432

Ronaghi M, Uhlén M, Nyrén P. "A sequencing method based on real-time pyrophosphate". *Science (journal),* 1998, 281 (5375): 363, 365. doi:10.1126/science.281.5375.363.

Roque AC, Lowe CR, Taipa MA. "Antibodies and genetically engineered related molecules: production and purification". *Biotechnol Proress,* 2004, 20 (3): 639–54. doi:10.1021/bp030070k

Rothemund PW. "Folding DNA to create nanoscale shapes and patterns". *Nature,* 2006, 440 (7082): 297–302

Rothenburg S, Koch-Nolte F, Haag F. "DNA methylation and Z-DNA formation as mediators of quantitative differences in the expression of alleles". *Immunol Rev,* 2001 (184): 286–98. doi:10.1034/j.1600-065x.2001.1840125.x

Russell, Peter. *iGenetics*. New York: Benjamin Cummings. 2001, ISBN 0-805-34553-1.

Saenger, Wolfram. *Principles of Nucleic Acid Structure*. New York: Springer-Verlag.1984, ISBN 0387907629

Saiki, RK; Scharf, S; Faloona, F; Mullis, KB; Horn, GT; Erlich, HA; Arnheim, N. "Enzymatic amplification of beta-globin genomic sequences and restriction site analysis for diagnosis of sickle cell anemia.". *Science,* 1995, 230 (4732): 1350–4. doi:10.1126/ science.2999980

Salmons B, Günzburg WH. "Targeting of retroviral vectors for gene therapy". *Hum Gene Ther.,* 1993, 4 (2): 129–41. doi:10.1089/hum. 1993.4.2-129

Sandman K, Pereira S, Reeve J. "Diversity of prokaryotic chromosomal proteins and the origin of the nucleosome". *Cell Mol Life Sci,* 1998, 54 (12): 1350–64. doi:10.1007/s000180050259.

Sandman K, Reeve JN. "Structure and functional relationships of archaeal and eukaryal histones and nucleosomes". *Arch. Microbiol.* 2000, 173 (3): 165–9. doi:10.1007/s002039900122.

Sanger F, Air GM, Barrell BG, Brown NL, Coulson AR, Fiddes CA, Hutchison CA, Slocombe PM, Smith M. "Nucleotide sequence of bacteriophage phi X174 DNA". *Nature,* 1997, 265 (5596): 687–695

Sanger F, Coulson AR. "A rapid method for determining sequences in DNA by primed synthesis with DNA polymerase". *J. Mol. Biol.,* 1975, 94 (3): 441–8. doi:10.1016/0022-2836(75)90213-2

Sanger F, Nicklen S, Coulson AR. "DNA sequencing with chain-terminating inhibitors". *Proc. Natl. Acad. Sci. U.S.A.,* 1977, 74 (12): 5463–7.

Schoeffler A, Berger J. "Recent advances in understanding structure-function relationships in the type II topoisomerase mechanism". *Biochem Soc Trans,* 2005, 33 (6): 1465–70. doi:10.1042/BST20051465

Smith LM, Fung S, Hunkapiller MW, Hunkapiller TJ, Hood LE. "The synthesis of oligonucleotides containing an aliphatic amino group at the 5' terminus: synthesis of fluorescent DNA primers for use in DNA sequence analysis". *Nucleic Acids Res.,* 1985, 13 (7): 2399–412. doi:10.1093/nar/13.7.2399.

Soinov, L. Bioinformatics and Pattern Recognition Come Together Journal of Pattern Recognition Research (JPRR), 2006, 1(1):37-41

Spiegelman B, Heinrich R. "Biological control through regulated transcriptional coactivators". *Cell,* 2004, 119 (2): 157–67. doi:10.1016/j.cell.2004.09.037.

Springham, D.; Springham, G.; Moses, V.; Cape, R.E. "Biotechnology: The Science and the Business." 1999, Taylor & Francis. p. 1.

Tarrago-Litvak L, Andréola M, Nevinsky G, Sarih-Cottin L, Litvak S. "The reverse transcriptase of HIV-1: from enzymology to therapeutic intervention". *FASEB J,* 1994, 8 (8): 497–503.

Thanbichler M, Shapiro L. "Chromosome organization and segregation in bacteria". *J. Struct. Biol.,* 2006, 156 (2): 292–303. doi:10.1016/j.jsb.2006.05.007

Thanbichler M, Wang S, Shapiro L. "The bacterial nucleoid: a highly organized and dynamic structure". *J Cell Biochem,* 2005, 96 (3): 506–21. doi:10.1002/jcb.20519.

The ENCODE Project Consortium. "Identification and analysis of functional elements in 1% of the human genome by the ENCODE pilot project". *Nature,* 2007, 447 (7146): 799–816. doi:10.1038/nature05874

Thomas J . "HMG1 and 2: architectural DNA-binding proteins". *Biochem Soc Trans,* 2001, 29 (4): 395–401. doi:10.1042/BST0290395.

Thrasher AJ, Gaspar HB, Baum C, *et al.* "Gene therapy: X-SCID transgene leukaemogenicity". *Nature,* 2006, 443 (7109): E5–6; discussion E6–7.

Tinkov, S., Bekeredjian, R., Winter, G., Coester, C., Polyplex-conjugated microbubbles for enhanced ultrasound targeted gene therapy. AAPS Annual Meeting and Exposition, 2008:16–20 November, Georgia World Congress Center, Atlanta, GA, USA, http://www.aapsj.org/abstracts/AM_2008/AAPS2008-000838.PDF.

Toole JJ, Knopf JL, Wozney JM, *et al.* "Molecular cloning of a cDNA encoding human antihaemophilic factor". *Nature,* 1984, 312 (5992): 342–7. doi:10.1038/312342a0. PMID 6438528. "page 343".

Tuteja N, Tuteja R. "Unraveling DNA helicases. Motif, structure, mechanism and function". *Eur J Biochem,* 2004, 271 (10): 1849–63.

Valerie K, Povirk L. "Regulation and mechanisms of mammalian double-strand break repair". *Oncogene,* 2003, 22 (37): 5792–812.

Venter JC, Adams MD, Myers EW, *et al.* "The sequence of the human genome". *Science,* 2001, 291 (5507): 1304–51. doi:10.1126/science.1058040

Verma S, Eckstein F. "Modified oligonucleotides: synthesis and strategy for users". *Annu. Rev. Biochem.,* 1998 (67): 99–134. doi:10.1146/annurev.biochem.67.1.99

Vikas Dhingraa, Mukta Gupta, Tracy Andacht and Zhen F. Fu. "New frontiers in proteomics research: A perspective". *International Journal of Pharmaceutics,* 2005, 299 (1–2): 1–18. doi:10.1016/j.ijpharm.2005.04.010.

Vitturi R, Colomba MS, Pirrone AM, Mandrioli M. "rDNA (18S-28S and 5S) colocalization and linkage between ribosomal genes and (TTAGGG)(n) telomeric sequence in the earthworm, Octodrilus complanatus (Annelida: Oligochaeta: Lumbricidae), revealed by single- and double-color FISH". *J. Hered.,* 2002, 93 (4): 279–82. doi:10.1093/jhered/93.4.279

W. Bains, Genetic Engineering For Almost Everybody: What Does It Do? What Will It Do? London: Penguin Books, 1987:99

Wahl M, Sundaralingam M. "Crystal structures of A-DNA duplexes". *Biopolymers,* 1997, 44 (1): 45–63. doi:10.1002/(SICI)1097-0282(1997)44:1<45::AID-BIP4>3.0.CO;2-#.

Walker F O. "Huntington's disease". *Lancet*, 2007, 369 (9557): 221. doi:10.1016/S0140-6736(07)60111-1

Walsh C, Xu G. "Cytosine methylation and DNA repair". *Curr Top Microbiol Immunol,* 2006, (301): 283–315. doi:10.1007/3-540-31390-7_11.

Wang J. "Cellular roles of DNA topoisomerases: a molecular perspective". *Nat Rev Mol Cell Biol,* 2002, 3 (6): 430–40. doi:10.1038/nrm831

Wang, Hongjie; Dmitry M. Shayakhmetov, Tobias Leege, Michael Harkey, Qiliang Li, Thalia Papayannopoulou, George Stamatoyannopolous, and André Lieber. "A capsid-modified helper-dependent adenovirus vector containing the beta-globin locus control region displays a nonrandom integration pattern and allows stable, erythroid-specific gene expression". *Journal of Virology,* 2005, 79 (17): 10999–1013. doi:10.1128/JVI.79.17.10999-11013.2005

Watson, J. D.; Crick, FH. "Genetical Implications of the Structure of Deoxyribonucleic Acid". *Nature,* 1953, 171 (4361): 964. doi:10. 1038/171964b0

Watson, J. D., Crick, F H. "Molecular Structure of Nucleic Acids: A Structure for Deoxyribose Nucleic Acid". *Nature,* 1953, 171 (4356): 737. doi:10.1038/171737a0. http://www.nature.com/nature/dna50/ watsoncrick.pdf.

Weiling, F. "Historical study: Johann Gregor Mendel 1822–1884.". *American journal of medical genetics,* 1991, 40 (1): 1–25; discussion 26. doi:10.1002/ajmg.1320400103.

Weir B, Triggs C, Starling L, Stowell L, Walsh K, Buckleton J. "Interpreting DNA mixtures". *J Forensic Sci,* 1997, 42 (2): 213–22.

Weiss, B. (ed.): Antisense Oligodeoxynucleotides and Antisense RNA : Novel Pharmacological and Therapeutic Agents, CRC Press, Boca Raton, FL, 1997

Weiss, B., Davidkova, G., and Zhou, L-W.: Antisense RNA gene therapy for studying and modulating biological processes. *Cell. Mol. Life Sci.,* 1999 (55):334-358.

Wheeler DA, Srinivasan M, Egholm M, Shen Y, Chen L, McGuire A, He W, Chen YJ, Makhijani V, Roth GT, Gomes X, Tartaro K, Niazi F, Turcotte CL, Irzyk GP, Lupski JR, Chinault C, Song XZ, Liu Y, Yuan Y, Nazareth L, Qin X, Muzny DM, Margulies M, Weinstock GM, Gibbs RA, Rothberg JM. "The complete genome of an individual by massively parallel DNA sequencing". *Nature,* 2008, 452 (7189): 872–876.

Wilkins M.H.F., A.R. Stokes A.R. & Wilson, H.R. "Molecular Structure of Deoxypentose Nucleic Acids". *Nature,* 1953, 171 (4356): 738–740. doi:10.1038/171738a0.

Wing R, Drew H, Takano T, Broka C, Tanaka S, Itakura K, Dickerson R. "Crystal structure analysis of a complete turn of B-DNA". *Nature,* 1980, 287 (5784): 755–758. doi:10.1038/287755a0

Wolfsberg T, McEntyre J, Schuler G. "Guide to the draft human genome". *Nature,* 2001, 409 (6822): 824–826. doi:10.1038/35057000

Woods NB, Bottero V, Schmidt M, von Kalle C, Verma I M. "Gene therapy: therapeutic gene causing lymphoma". *Nature,* 2006, 440 (7088): 1123. doi:10.1038/4401123a

Xu M, Fujita D, Hanagata N. "Perspectives and challenges of emerging single-molecule DNA sequencing technologies". *Small,* 2009, 5 (23): 2638–49. doi:10.1002/smll.200900976

Yakovchuk P, Protozanova E, Frank-Kamenetskii M D. "Base-stacking and base-pairing contributions into thermal stability of the DNA double helix". *Nucleic Acids Res.,* 2006, 34 (2): 564–74. doi:10.1093/nar/gkj454

Frank-Ranier Schmidt. From gene to product: “The advantage of Integrative biotechnology”. Hand book of Pharmaceutical Biotechnology, by Shayne Cox Gad. Willey, 2007:1-52.

Eugene Zabarovsky. Sequencing the Human genome: What it worth it? Hand book of Pharmaceutical Biotechnology, by Shayne Cox Gad. Willey, 2007:53-87.

Mary Jane Cunningham and Mrinal Shah. Toxiconenomics. Hand book of Pharmaceutical Biotechnology, by Shayne Cox Gad. Willey, 2007:229-251.

M D Mostaqul Huq and Li-Na Wei. Protein posttranslational modification target in Pharmaceutical development. Hand book of Pharmaceutical Biotechnology, by Shayne Cox Gad. Willey, 2007:417-441.

Steve Pascolo. Plasmid DNA and Messenger RNA for Therapy. Hand book of Pharmaceutical Biotechnology, by Shayne Cox Gad. Willey, 2007:991-1011.

Tatiana Sengupta. Formulations and Delivery limitations of Nucleic-Aci-Based Therapeutics. Hand book of Pharmaceutical Biotechnology, by Shayne Cox Gad. Willey, 2007:1013-1059.

6

Immunotechnology in Drug Delivery

Introduction

Immunotechnology, is an important arm of biotechnology, constituting the industrial application of immunological procedures to produce vaccines for mass immunization and to prevent diseases and/or producing immunological products on large-scale use for a long time. The current trend is to develop the more specific DNA vaccines using recombinant DNA technology and therapeutic agents to cure the afflicted. Today, immunology is a very complex and sophisticated area of biology, which has become one of the most versatile research tools in biology and medicine, as well as a powerful weapon in the armory of prevention and management of several viral and bacterial diseases. Technology based on applications of cells and molecules of the immune system, a major research interest is created in the field of human recombinant antibodies and antibody fragments in medical and industrial applications, as well as studies of mechanisms underlying somatic mutations in B cells and IgE switch in allergy. The use of synthetic antibodies in proteome analysis, including protein array technology is also pursued as well as gene array analysis of the transcriptome. B-cell malignancies are one focus in antibody and gene therapy projects as well as viral infection in molecular breeding projects.

6.1 Immunity

Immunity is the capacity of the body to resist infection. Natural immunity is the resistance to disease possessed as a part of an individual's constitutional make-up, which results in differences between species, races and individuals. Acquired immunity, when the natural immunity is inadequate to protect the infection additional immunity can be acquired either by actively, due to stimulation of the individual's antibody producing cells (active immunity) may be possible by vaccination or passively, as a result of the introduction of antibodies from another person or animals (Passive immunity) immune to that infection.

6.1.1 Immune System

Immunological defensive response

When an infecting organism enter into the mammalian system for the first time, the immune system of the mammal reacts in response to the proteins of the invading organism, generally called **antigens,** by producing a special classes of protein called **antibodies**. The antibody recognizes the antigen and binds with it forming an antigen-antibody complex, and neutralizes the pathogen's potential to cause the disease. The antigen-antibody complex is scavenged by the body, mainly through the lymphatic system and often seen as **pus**, in dermal eruptions in the form of pustules. Pus, the fluid from pustules, contains serum, antigen-antibody complexes, expended white blood cells, dead and lives pathogens, and debris of tissue.

This first encounter is very important because the system gains a '**memory**' template of the three dimensional structural configuration of the **epitopes** of the antigens. This memory may last for a few hours or days (common colds), a few months (cholera, or **haptens**, which are sourced in the protein coats of viruses, cell walls of tetanus) or the lifetime (smallpox) of the individual.

When the same antigen gains entry a second and the subsequent times, the immune system recognizes the foreign entity and produces antibodies specific to the antigen, basing on the memory, developed on the first encounter. This is **immune response**, which results in a) the production of antibodies, b) antibody-bearing cells or c) cell mediated hypersensitivity reaction (**allergy**). Each re-entry of the exogenous entities triggers enhanced production of the corresponding antibodies (**booster reaction**).

The antibodies recognize their antigens, bind with them, and neutralize them, before they can cause harm to the individual or cause the disease specific to them. This is **immunological defense**. The antigen-antibody recognition is a highly specific phenomenon of **biorecognition** at the molecular level.

Immune response is a selective reaction of a mammalian body to substances that are foreign (**exogenous**) to it or those that the immune system identifies as foreign. The three important aspects are:

(a) **Memory**: the primary response of the formation of the memory template at the first encounter,

(b) **Distinction between self- and non-self**: distinction between the organism's endogenous proteins and those that are foreign (exogenous), and

(c) **Specificity**: the secondary response of production of antibodies very specific to each foreign agent. Mammalian systems produce a highly specific antibody to each of the pathogens and even their different strains.

6.1.2 Components of the Immunological System and Their Role

6.1.2.1 The Fluid Components

Serum: The liquid part of blood, without the cells and the coagulating factors, but containing antigens and antibodies; it is the storehouse and means of transport of immunological components.

Lymphatic system: is parallel to the blood conducting system and is constituted of the lymph, lymphocytes, lymph vessels, lymph nodes and lymph glands. The lymph is a watery, transparent or slightly yellow, fluid conducted through the lymph vessels. Lymph contains only one kind of cells, the lymphocytes, unlike blood that contains several different kinds of cells including lymphocytes. The blood and the lymphatic systems come into a sort of confluence in the lymph nodes and the tissues. The lymph cells that secrete lymph are aggregated into lymphatic tissue in the form of glands or occur in small groups of cells in different parts of the body.

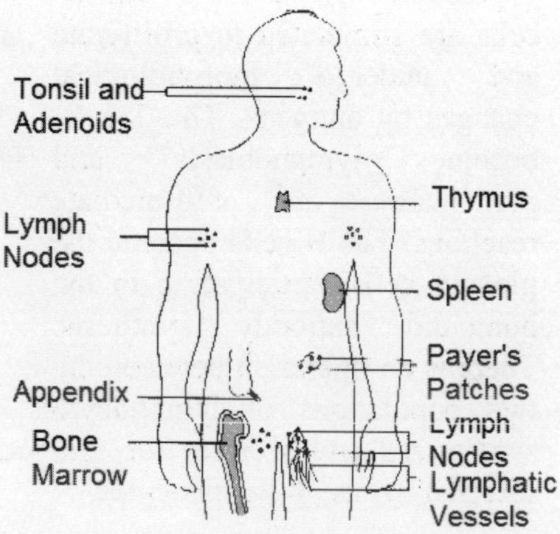

(Lymphatic System)

Lymphocytes: The cells of the lymphatic system (lymphoid group) which play the main role in immune responses

Role of lymphocytes: The lymphocytes have an important role to play both in humoural and cell-mediated immunity. The lymphocytes re-circulate in the blood, lymph nodes, spleen and other tissues and back to blood by the lymphatic vessels.

Kinds of lymphocytes: **T-lymphocytes** and the **B-lymphocytes** (T-cells and B-cells) basing on origin, whether Thymus derived and Bursa derived respectively.

6.1.2.2 The Cell Components

(a) Antibody synthesizing cells called **plasma cells**, or

(b) **Effectors cells** called **lymphoblasts**.

The lymphoblasts, along with blood group antigens, are responsible for immunological tissue rejection reactions in transplantations. The small lymphocytes also carry the memory of the first contact with an antigen. Without this memory mechanism, there can be no secondary response and so no immunological defense.

Modified T- and B-cells: The populations of both the T- and B-cells are stimulated to proliferate and undergo morphological changes by antigens. The T-cells become lymphoblasts and participate in cell-mediate reactions. The B-cells become the **plasma cells** participating in the humoural antibody synthesis. There is co-operation between the two populations of lymphocytes. The mature plasma cell actively synthesizes and secretes the antibody. There are no antibodies in, or secreted by, the T-lymphocytes.

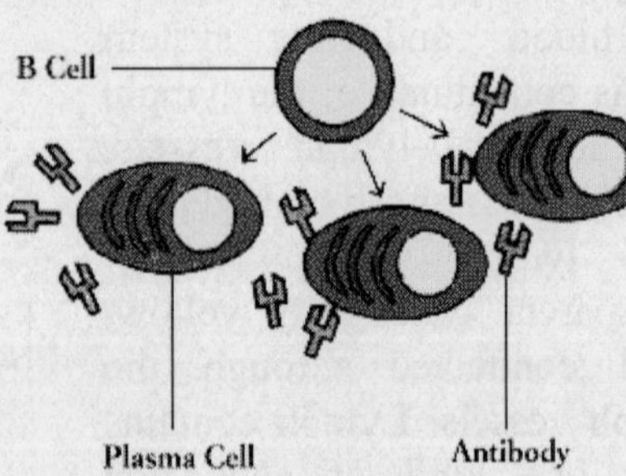

T-cell dependence of B-lymphocytes: Certain B-lymphocytes in mammals are dependent upon the T-lymphocytes for their function (T-cell dependent) while the others are independent of the T-lymphocytes (T-cell independent).

T cells work primarily by secreting substances known as **cytokines** or, more specifically, **lymphokines**. Lymphokines (which are also secreted by B cells) and their relatives, the **monokines** produced by **monocytes** and macrophages, are diverse and potent chemical messengers. Binding to specific receptors on target cells, lymphokines call into play many other cells and substances, including the elements of the inflammatory response. They encourage cell growth, promote cell activation, direct cellular traffic, destroy target cells, and incite macrophages. A single cytokine may have many functions; conversely, several different cytokines may be able to produce the same effect.

Cytokines are also known as interferons, produced by T cells and macrophages. Interferons are a family of proteins with antiviral properties. Interferon from immune cells, known as immune interferon or gamma interferon, activates macrophages. Two other cytokines, closely related to one another are lymphotoxin (from lymphocytes) and tumor necrosis factor.

Many cytokines are initially given descriptive names but, as their basic structure is identified, they are renamed as "**interleukins**", which is the messenger between **leukocytes**, or white cells. Interleukin-1 or IL-1 is a product of macrophages that helps to activate B cells and T cells. IL-2, originally known as T cell growth factor or TCGF, is produced by antigen-activated T cells and promotes the rapid growth or differentiation of mature T cells and B cells. IL-3 is a T-cell derived member of the family of protein mediators known as colony-stimulating factors (CSF); one of its many functions is to nurture the development of immature precursor cells into a variety of mature blood cells. IL-4, IL-5, and IL-6 help B cells growth and differentiation; IL-4 also affects T cells, macrophages, **mast cells**, and **granulocytes**.

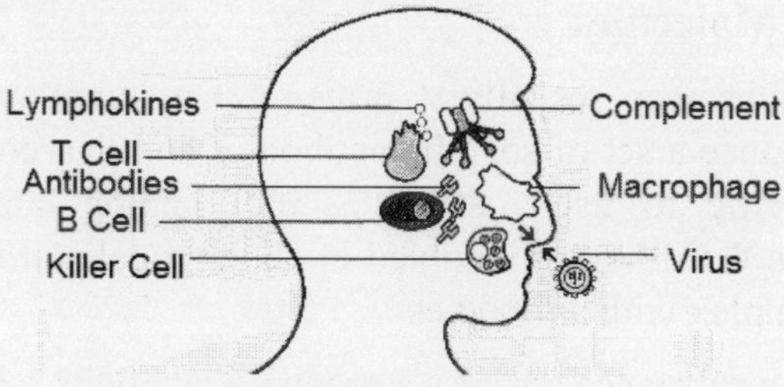

(Immune Response)

Natural Killer (NK) cells are yet another type of lethal lymphocyte. Like cytotoxic T cells, they contain granules filled with potent chemicals. They are called "natural" killers because they, unlike cytotoxic T cells, do not need to recognize a specific antigen before swinging into action. They target tumor cells and protect against a wide variety of infectious microbes. In several immunodeficiency diseases, including AIDS, natural killer cell function is abnormal. Natural killer cells may also contribute to immuno-regulation by secreting high levels of influential lymphokines.

Complement is a system made up of a series of about 25 proteins that work to "complement" the activity of antibodies in destroying bacteria, either by facilitating phagocytosis or by puncturing the bacterial cell membrane. Complement also helps to rid the body of antigen-antibody complexes. In carrying out these tasks, it induces an inflammatory response.

6.1.2.3 Monocytes, Macrophages and Phagocytes

Monocytes, originate from stem cells, have a single nucleus and develop into macrophages, which are the phagocytic cells engulf particulate matter in a non-specific defense mechanism.

Mast cells: Mast cells occur in the skin and epithelial layers. They contain histamine in the form of granules bound to membranes. Explosive de-granulation results in the release of histamine, which increases the permeability of the blood vessels, causing **inflammatory reactions**. Mast cells have a key role in **allergy**.

Eosinophils: These are cells with granules in the cytoplasm (one kind of granulocytes), also known as polymorphonuclear leucocytes, stainable with the reddish biological stain eosin. The mast cells and eosinophils have an important role in allergy.

6.1.2.4 The Molecules

Antigen: a substance, usually a protein, that stimulates the immune system to produce a set of specific antibodies and that combines with an antibody specific to itself, at a specific binding site; differs from immunogen in that it is not involved in eliciting cellular response and in that it can complex with antibodies.

Immunogen: a substance, usually a protein, that elicits a cellular immune response, and/or antibody production; differs from antigen in that it mainly elicits cellular response but does not complex with an antibody.

Hapten: a low-molecular weight non-protein molecule which contains an antigenic determinant but which is not itself antigenic unless it complexes with an antigenic carrier, such as a protein; once an antibody is available, it can readily recognise the hapten, even without the carrier, and bind with it. To be antigenic, the hapten must bind to an exogenous protein carrier.

Epitope: a part of a protein molecule that acts as an immunogenic / antigenic determinant, and so determines specificities; a macromolecule, such as a protein, may contain many different epitopes, each capable of stimulating the production of specific antibodies, each with a correspondingly specific binding site.

Antibodies: Globulins (roughly spherical in shape and extractable in saline solutions) are glycoproteins (proteins with a carbohydrate content ranging from 3 to 13%), produced by the immune system of an organism in response to exposure of a foreign molecule and characterised by its specific binding to a site, related to an epitope of that molecule, which are induced response proteins. The antibodies, like all proteins, are formed of chains of amino acids, which undergo very complex packing, giving the proteins a specific and functionally significant final shape (tertiary configuration), which determines most of the characteristics of the protein. As globulin proteins are involved in immune reactions, antibodies are called immunoglobulins (abbreviated to Ig).

Antiserum: Blood serum containing antibodies arising out of immunization or after an infectious disease.

Vaccine: An agent containing antigens/immunogens produced from killed, attenuated or lives pathogenic microorganisms, synthetic peptides, by recombinant organisms or DNA, used for stimulating the immune system of the recipient to produce specific antibodies providing active immunity and/or passive immunity.

6.1.2.5 The Immunoglobulins

Immunoglobulin (Ig): A protein molecule of the globulin-type, found in the serum or other body fluids and that possess antibody activity; there are five classes of immunoglobulins (IgA, IgD, IgE, IgG and IgM), based on antigenic and structural differences. In addition to these five classes, there are several subclasses (four in IgG) and other variants of Ig molecules.

Classes of antibodies: There are five classes of immunoglobulins in the human system: Immunoglobulin G (the gammaglobulins; IgG), IgA, IgM, IgD and IgE.

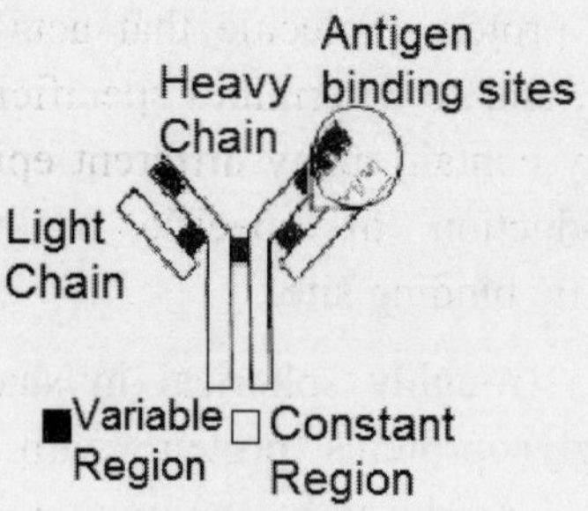

(Antibody)

Molecular structure of the antibodies: The conventional model of the Ig molecules is a 'Y' shaped configuration, with two heavy chains and two light chains, with two open arms containing the antigen combining sites, which occur on both the light and the heavy chains. The two heavy chains are bound together by disulphide bonds. At any point, the molecule has two chain sections, parallel to each other.

The modern view of the structure of the Ig molecules is to look at them as containing series of regions activity called **domains.** Variable light, variable heavy, constant heavy 1, 2, 3 and constant light are the domains recognized on Ig molecules. The constant domains provide for the identity of the molecules and the variable regions are responsible for the diversity in the specificity of the antibodies.

Antibodies can work in several ways, depending on the nature of the antigen. Antibodies that interlock with toxins produced by certain bacteria can disable them directly (and are known as antitoxins). Other antibodies, by coating (or opsonizing) bacteria, make the microbes highly palatable to scavenger cells equipped to engulf and destroy them. More often an antigen-antibody combination unleashes a group of lethal serum enzymes known as complement (Complement). Yet other antibodies block viruses from entering into cells (a quality that is exploited in making vaccines) in a phenomenon known as antibody-dependent cell-mediated cytotoxicity (ADCC), cells coated with antibody become vulnerable to attack by several types of white blood cells.

Immunoglobulin A (IgA): With a molecular weight of about 1,60,000, IgA molecules are only slightly heavier than the IgG molecules but they can form aggregates of higher molecular weights. IgA are about 13% of the total Ig with a concentration of 1.4 to 4 mg/ml in the normal serum. The IgA are the major Ig in the serum and mucous secretions, such as saliva, tears, nasal fluids, sweat, lung and the gastrointestinal tract. They defend the exposed external surfaces of the body against the attack of microorganisms. IgA antibodies seem to inhibit adherence of the microorganisms to the surface of the mucosal cells and thus prevent their entry into the body tissues. IgA molecules differ from the other Ig classes in having three disulphide bonds holding the two heavy chains, instead of two bonds in the others.

Immunoglobulin M (IgM): The IgM molecules are the heaviest of all Ig. They have a molecular weight of 900,000 and are often known as the macroglobulins. They form about 6% of the total Ig and occur in a concentration of 0.5 to 2% of the normal serum. IgM are very efficient agglutinators of bacterial cells and are effective cytolytic agents. They form the most immediate and effective first line defense against bacteraemia. Since they appear in response to infection they are mostly confined to the blood stream. The anti-A and anti-B haemagglutinins and many antimicrobial antibodies as well as typhoid exotoxin antibodies are all IgM. During the course of evolution of Ig, IgM seem to have appeared earliest.

Immunoglobulin G (IgG): IgG molecules are the lightest of all the Ig and have a molecular weight of about 1,50,000 and about 3% carbohydrate content. They form about 80% of the total Ig of the human body. In the normal serum their concentration ranges from 8 to 16 mg/ml. These are the most abundant component of Ig in the body fluids particularly the blood vessels, where they combat microorganisms and their toxins. IgG are the only antibody that can get across the placenta and so provide the major line of defense during the first few weeks of the life of a foetus. IgG also diffuse very readily from the blood vessels into the body spaces. When, IgG molecules attach to microorganisms, the susceptibility of the latter for phagocytosis increases. In a germ free environment, the IgG concentration of the serum is very low and increases with infection. IgG are the major antibody synthesized during the secondary response, their synthesis being entirely governed by the antigenic situation. All the IgG molecules are seemingly identical. The most fascinating thing is that there are an infinitesimal number of antigens, with each pathogenic organism producing several of them. During the course of our lifetime we develop immunity against a very large number of infections, some on a long-term basis and some are short term, but repeated infection renewing our ability to combat the disease. The key to understanding this versatility of the IgG molecule lies in the fact that the IgG molecule has a part that is invariable and this gives the basic characteristics for it to function as an antibody. Another part of the IgG molecule is variable in its amino acid content and sequence and this gives the molecule the ability to be a specific antibody against a particular antigen. This is nothing surprising. Almost all proteins have variable and invariable regions.

Immunoglobulin E (IgE): The molecular weight of IgE is about 200,000 and they form only 0.002% of the total Ig, with a serum concentration of 17 to 450 ng/ml. IgE protect the external mucosal surfaces of the body through plasma factors. Pathogens crossing the IgA line combine with IgE molecules specific to them. This results in the release of amines (eg. histamine) that increase the permeability of the blood vessels causing the symptoms of allergy. The release of amines is due to a degranulation of the mast cells. The level of IgE is raised during parasitic infections but the importance of IgE lies with atopic allergy.

Immunoglobulin D (IgD): IgD have a molecular weight of about 1,85,000 and form only about 1% of the total Ig. They occur at a concentration of 0 to 0.4% of the normal serum. They are present only on the surface of the lymphocytes along with IgM. The IgD are susceptible to enzyme degradation and so have a very short life span (2.8 days) in the plasma. IgD have the highest carbohydrate content (13%) of all Ig. The exact function of IgD is not understood.

6.1.3 Types of Immunological Reactions

Agglutination: an immunological or chemical reaction leading to the aggregation of particulate matter such as bacteria, erythrocytes or other cells, or synthetic particles such as plastic beads coated with antigens or antibodies.

Precipitin reaction: When an antigen and its antibody are brought together in solution, a precipitate is formed due to the binding of the antigen and the antibody. If unrelated antigen and antibody are brought together no binding occurs and hence no precipitate is formed. Antigen-antibody binding occurs when they come together in the blood stream or in the tissues. Precipitation occurs because the antigen-antibody complexes form a three-dimensional lattice. Precipitin reactions are a very useful tool in several areas of biological research. In the case of both antigen-antibody and enzyme-substrate affinity, there is a complementarities of the molecular shape between the antigen/enzyme and the antibody/substrate and the fit is exact like that of a key in its lock.

In semisolid media, such as bacto-agar, the precipitin reaction results in the formation of lines called **precipitin lines**. Such reactions are studied by **Ouchterlony's** double **diffusion method**, where the antigen and the antibody diffuse towards each other from two spaced wells cut in semisolid agar. This method provides only qualitative data. A variant of this method is **single radial diffusion**, which helps to quantify the antigen with reference to the antibody.

Basis of recognition of the antigen by the antibody and their binding: The overall physical configuration of the antigen seems to be more important than its chemical structure which means that the antigen is recognized by the three-dimensional shape of its outer electron cloud. Chemical composition and reactivity are less important.

Binding site: a specific region in a molecular entity, such as an antigen, that is capable of entering into a stabilizing interaction with another molecular entity, such as the corresponding antibodies.

Forces of antigen-antibody binding:

One or more of the following forces appear to be involved in antigen-antibody binding: electrostatic forces, hydrogen bonding, hydrophobic (water repulsion) forces and Van der Waals attractions between molecules. What is surprising is that the very same forces also operate between unrelated proteins or other macromolecules in normal chemical reactions.

6.1.4 Types of Immunity

Humoural immunity: When microorganisms enter the body, antibodies are synthesized and released into the blood and other body fluids. These antibodies circulate throughout the body. The free antibodies coat the cells of the organism and enhance their phagocytosis and also neutralize the toxins released by the organisms. This type of immune response is called the humoural immunity.

Cell-mediated immunity: In response to the presence of antigens, the body produces lymphocytes with antibodies or antibody-like molecules on their surface. This is cell-mediate immunity, which offers protection, particularly against organisms, which live and multiply within the host cells. Tubercle bacteria, small poxvirus, etc., are subject to the action by cell-bound antibodies.

Acquired immunity: Not all antibodies are synthesized in the body. Some are pre-natal acquisitions from the mother through the placenta and some are post-natal through breast-feeding. These constitute acquired immunity. Immunity is also acquired through one's own body's experience gained on encountering a pathogen.

Specific and non-specific defense: Immunological defense is specific to particular pathogens, and even to their strains, which is **specific defense**.

Mammals have also a **non-specific defense** mechanism. For example, the **macrophages,** that are associated with the lumenal side of the walls of the blood vessels and the connective tissue, physically engulf cells of pathogens or complexes of proteins, to remove them from the system.

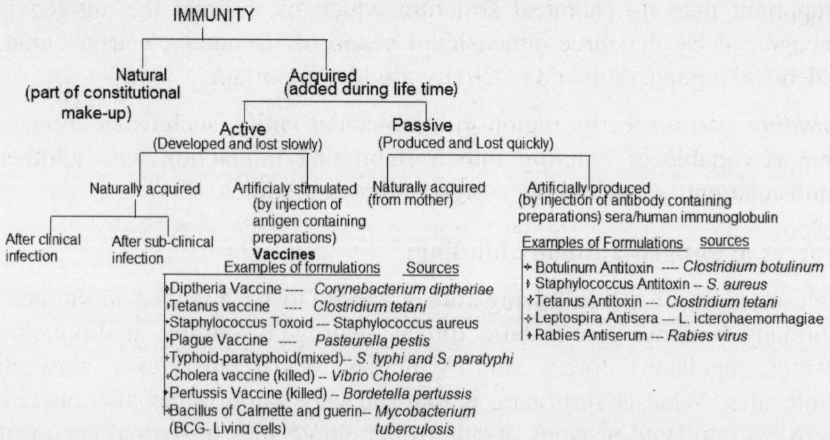

(Classes of Immunity with Examples of Classical Immunizing Agents)

6.1.5 Body's Immunological Reaction

Allergy: A hypersensitivity reaction of the body to antigens. In a sense it is the immunological system that has gone wrong. An allergen is an antigen that stimulates the production of IgE antibodies, although low titres of IgG molecules are also formed. The IgE antibodies bind to **mast cells** resulting in the 'explosion' of the mast cells leading to the release of histamine that triggers an **inflammatory response** in the skin, mucosa or epithelial cells, a syndrome termed allergy.

Allergen: an antigen that can induce an allergic reaction, thorough eliciting IgE antibodies. Some allergens are haptens, as for example parthenin which is a sesquiterpene lactone. Some haptens bind to endogenous proteins in the individual, because of which IgG antibodies cannot be produced against such a hapten-carrier complex, a situation that makes it almost impossible to treat the patient through immunization.

Anaphylaxis: a sudden and severe form of IgE based reaction, occurring on the second encounter with the allergen (antigen) that can be fatal. Penicillin may induce a severe anaphylactic reaction in some individuals sensitive to it. In fact, purified or synthetic penicillin does not cause anaphylaxis, but the impurities in biologically produced penicillin or protein compounds such as procaine added to penicillin injection, are responsible for the reaction.

Inflammatory response: This is the body's reaction to injury or infection/antigens, in the form of a syndrome constituted of swelling, redness (erythematic) and heat (collectively called **inflammation**), in the affected part of the body. Inflammation controls the spread of infection. Uncontrolled inflammation causes tissue damage.

6.1.6 Inhibition of Immune Response

Antigen-antibody competing, phagocytosis, inflammation during immune response, etc., all leads to tissue damage. Uncontrolled immune reactions can be dangerous to the body. There are some factors that inhibit immune response and some situations where immune response fails to materialize.

(a) Antibody suppressor cells and/or factors are present in the serum or tissues. Prostaglandins, the compounds secreted by the organs of the human body into the blood stream to perform various functions, such as muscle contraction, may also inhibit immune reactions. This is the body's way of controlling immune reactions to minimize tissue damage.

(b) While our body prepares for extensive warfare, the pathogens themselves would not be idle. A number of immunosuppressive agents like lipo-polysaccharides, lipoteichoic acid, dextran, Levan, etc., are produced by bacteria. They also produce proteinases that denature some classes of Ig, particularly IgA. Modulation of immune responses both by the host and the pathogen ultimately regulates the host-pathogen interaction and the development of disease.

(c) The antibodies can be defeated in their function by slight changes in the chemical (and consequently physical) structure of the antigen. This happens with the antigens of viruses and bacteria, which grow very rapidly and develop into new strains through mutations and other evolutionary processes depending upon several conditions, particularly environmental. Under these conditions, the host's immune defenses become inadequate. For example, we never seem to acquire immunity from colds. In fact, we do get immunized to a particular strain of cold causing virus of a given time but in no time the virus modifies itself in some minute way and we have no immediate defense against this modified version of the virus. Antibody production against the ever-changing organisms is a race between the host and the pathogen.

6.1.7 Clinical Suppression of Immune Response

At the time of tissue and organ transplantation from one individual to another, the immune system of the recipient's body, especially the lymphoblast component triggers the production of antibodies against the antigens in the tissue/organ of the donor, which results in the rejection of the transplant. This is also because of the presence of blood group antigens in tissues, in addition to principally being on the surface of the erythrocytes. In order to prevent this situation, the immune responses are deliberately suppressed by the use certain drugs such as azathioprine, cyclophosphamide and cyclosporine A, prior to transplantation. Such drugs are also used in the event of autoimmune diseases, like rheumatoid arthritis.

6.1.8 Secretors and Non-Secretors

All of us secrete antigens and antibodies in our body fluids such as sweat, tears, saliva, semen, etc., to some degree or the other. For example, the IgA in saliva serves as the first line of defense of the oral route. However, a certain percentage of human populations secrete antigens and antibodies in high titres and are called secretors, the other group being non-secretors. The status of an individual as a secretor is genetically determined and offers certain advantages to the secretors in terms of body hygiene. Their surface and first line defenses are quite high compared to those of the non-secretors. The frequency of secretor *vs* non-secretor alleles in different human populations is of interest to the population geneticist. The status of an individual as a secretor (or non-secretor) is easily determined by the use of appropriate lectins, a class of proteins that can recognize and bind to cell surface carbohydrates, resulting in agglutination of cells, such as erythrocytes and lymphocytes.

6.2 Frontiers in Immunology

6.2.1 Hybridoma Technology

A hybridoma is created by fusing two cells, a secreting cell from the immune system and a long-lived cancerous immune cell, within a single membrane. The resulting hybrid cell can be cloned, producing many identical offspring. Each of these daughter clones will secrete, over a long period of time, the immune cell product. A B-cell hybridoma secretes a single specific antibody.

The term hybridoma is myeloma cell culture applied to fused cells resulting due to fusion of following two types of cells:

(i) an antibody producing lymphocyte cell (e.g. a spleen cell of mouse immunized with red blood cells from sheep), and (ii) a single myeloma cell (bone marrow tumour cell) which is capable of multiplying indefinitely. These fused hybrid cells or hybridoma have the antibody producing capability inherited from lymphocytes and have the ability to grow continuously (immortal) like malignant cancer cells.

***Monoclonal Antibodies (mAb)*:** Monoclonal antibodies can be produced in specialized cells through a technique now popularly known as hybridoma technology. This technology was discovered in 1975 by two scientists; Georges Kohler of West Germany and Cesal Milstein of Argentina, jointly with Niels Jerne of Denmark were awarded the Nobel Prize for Physiology and Medicine in 1984.

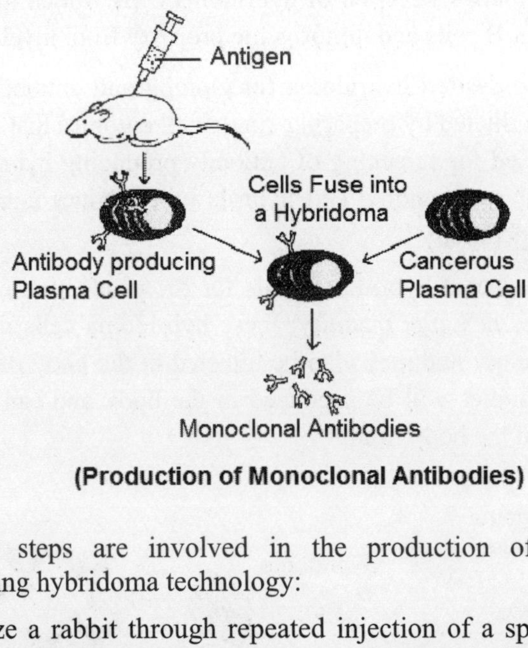

(Production of Monoclonal Antibodies)

Following steps are involved in the production of monoclonal antibodies using hybridoma technology:

(i) Immunize a rabbit through repeated injection of a specific antigen for the production of specific antibody, facilitated due to proliferation of the desired B cells.

(ii) Produce tumours in a mouse or a rabbit.

(iii) From the above two types of animals, culture separately spleen cells (spleen cells are rich in B cells and T cells) that produce specific antibodies, and myeloma cells that produce tumours (the myeloma cell line used, is unusual in two ways; it has stopped synthesizing antibodies and it is a mutant called HGPRT that can not synthesize the enzyme hypoxanthine guanine phosphoribosyl transferase or HGPRT).

(iv) Induce fusion of spleen cells to myeloma cells, using polyethylene glycol (PEG), to produce hybridoma; the hybrid cells are grown in selective hypoxanthine aminopterin thymidine (HAT) medium. HAT medium contains a drug aminopterin, which blocks one pathway for nucleotide synthesis, making the cells dependent on another pathway that needs HGPRT enzyme absent in myeloma cells. Therefore, myeloma cells that do not fuse with B cells will die since they are HGPET-.B cells that do not fuse will also die because they lack tumorigenic property of immortal growth. Therefore HAT medium allows selection of hybridoma cells, which inherit HGPRT gene from B cells and tumorigenic property from myeloma cells.

(v) Select the desired hybridoma for cloning and antibody production; this is facilitated by preparing single cell colonies that will grow and can be used for screening of antibody producing hybridomas ; only one in several hundred cell hybrids will produce antibodies of the desired specificity ;

(vi) Culture selected hybridoma cells for the production of monoclonal antibodies in Large quantity; these hybridoma cells may be frozen for future use and may also be injected in the body of an animal so that antibodies will be produced in the body and can be recovered later from the body fluid.

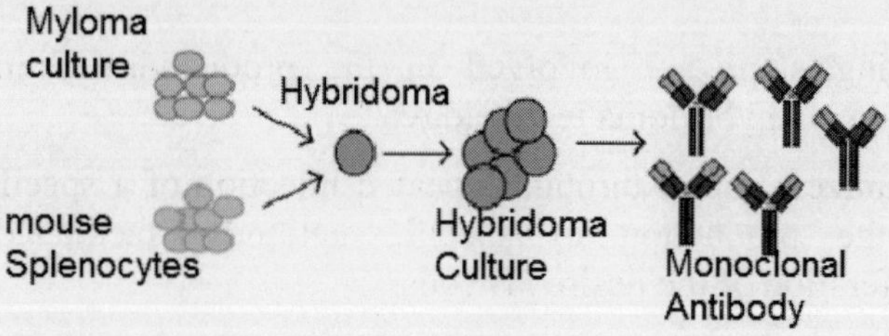

(Schematic description of the synthesis of monoclonal antibodies)

Human Monoclonal Antibodies

The advantages of using human monoclonal antibodies can be summarized as follows.

(i) To reduce the possibility of an immune response: the administration of human immunoglobulins will not lead to the development of strong immune responses, such as those seen against mouse Igs, although antiidiotypic responses could appear.

(ii) The repertoire of antigenic epitopes is recognized in a distinct way among the different species.

(iii) Human immunoglobulins interact better with human effector systems, Fc receptors, opsonization, and so forth, than antibodies of murine origin.

(iv) There are differences associated with the distinct patterns of glycosylation among different species, which can affect the effectiveness of the antibody as a therapeutic agent.

(v) The half-life of these antibodies is longer. Human antibodies have a half-life of 11– 24 days.

Applications

In Diagnosis

The monoclonal antibodies (mAbs) have wide application in the field of clinical diagnosis, for the purification of products (coagulation factors, interferon), for the design of new technologies, and also they have been actively introduced into the treatment of various diseases (Table 6.1). In diagnosis and research, they have helped to increase knowledge of a large variety of molecules, to define the stages of leukemia, to identify tumours and transcription factors, to define species, to quantify hormones, to study blood groups, to characterize infectious agents, besides an enormous number of targets that are now known because of monoclonal antibodies. In order to standardize the names given to the antibodies produced against the cell membrane, especially that of leukocytes, an agreement has been reached to group under the same cluster or numbers all those antibodies that recognize the same membrane molecule. The cluster differentiation (CD) concept has emerged as a consequence of this agreement, in which, to date, more than 300 different molecules have been classified. These are revised and updated every four years at an international CD workshop. As examples, a helper T lymphocyte is known as CD3+ CD4+ for presenting those membrane molecules, which are recognized by specific antibodies; a cytotoxic T lymphocyte is CD3+ CD8+, and a B lymphocyte is CD19+ CD20+. These and many other markers have improved the study of cell subpopulations, have allowed better definition of differentiation states, and have helped in the identification of leukemia, lymphomas, breast tumours, and so forth. In addition to the antibodies directed against molecules of the cell membrane, there is a long list of antibodies that recognize transcription factors, pathogenic agents, hormones which are now essential in the laboratories of Immunology, Microbiology, Physiology, Virology, Pathology, Cell Biology, and many others, and are mainly used in diagnostic techniques, such as ELISA, immunofluorescence, inmunohistochemistry, Western blot, immunochromatography or lateral flow, nephelometry, agglutination and precipitation methods, are some examples.

In Therapy

The enormous therapeutic potential of mAbs was initially undermined by certain "technical" difficulties, which explain the long gap from obtaining mAbs in sufficiently large quantities in 1975 to the beginning of their clinical application. This started in 1986, with the use of the anti-CD3 antibody for avoiding rejection in heart transplantation, and it was not until 1997 that mAbs entered in the field of onco-haematology, when the US Food and Drug Administration (FDA) approved Rituximab (a humanized anti-CD20 monoclonal antibody) as the first antibody for antitumoral therapy (http://www.fda.gov/). In antitumoral therapy, when compared to conventional chemotherapeutic agents, mAbs are high molecular weight proteins with slow distribution kinetics and limited tissue penetrating ability. However, their enormous advantage is that they are specific, thereby minimizing secondary effects, as they would be directed at their tumor targets (antigens) without affecting or with only minimum effect on healthy tissues. The capacity of monoclonal antibodies to penetrate tumours or to access inflammation sites is low. In particular, in antitumoral therapy, antigen expression and blood irrigation are the factors limiting the effectiveness of the mAbs. Therefore, for optimal effectiveness, the target antigen should be tumour-specific, intensely expressed, and its expression should not be susceptible to being reduced. The mAbs requirements for achieving a maximum response would be: a long half-life, the power to penetrate, not to induce an immune response, and a maximum cytotoxic effectiveness.

Table 6.1 Examples of some Monoclonal Antibodies approved for human therapy and their applications

Product	Company	Applications
Muromomab (Orthoclone OKT3)	Ortho Biotech, Inc. (John & Johnson)	Mouse IgG2a anti-CD3. Transplantation Rejection
Nofetumomab(Verluma)	Boehringern Ingelheim Pharma KG	Mouse Fab IgG_{21} directed against the glycoprotein 40 kD (Expressed in several tumors). Conjugated to Tec^{99}
Capromob Pendetide prostaScint	Cytogen Crop	Mouse Ab-$Indio^{111}$. Detection of prostate tumour
Daclizumab(Zenapax)	Hoffman-LA Roche Inc.	Humanized IgG^{I} anti-IL-2α. To avoid transplant rejection
Rituximab(rituxan)	Genentech Inc.	Chimeric Ig anti-CD20 for Non-Hodkin lymphoma
Basiliximab(simulect)	Novartis Pharma Crop.	Chimeric anti-IL-2α. To avoid renal transplant
Trastuzumab(Herceptin)	Genentech Inc.	Humanized IgG_1 anti-HER2; breast cancer
Remicade(Infliximab)	Centocor(Johnson & Johnson)	Chimeric anti TNF- α. Rhumatoid Arthritis, Crohn's disease
Gemtuzumab Ozogamicin(Mylotarg)	Wyeth Averst	Humanized Ig anti CD52; acute myloid leukaemia
Omalizumab (Xolair)	Genetech Inc./Roche	Humanized IgE. Severe Asthma
Bevacizumab(Avastin)	Genetech Inc.	Humanized anti VEG; tumor
Natalizumab(Tysabri)	Biogen Inc.	Humanized anti CD49d, Multiple sclerosis, Chron's disease
Panimumab(Vectibix)	Amgen	Human anti EGFR(epidermal growth factor receptor). Metastatic colorectal carcinoma
Eculizumab(Soliris)	Alexion pharma Inc.	Humanized anti CD59. proximal nocturnal hemoglobinuria

6.2.2 Recombinant Antibody (rAb) Technology

The recombinant DNA technology opens up new horizon in production of antibody, where antibody gene can be amplified and selected through phase display, cell surface display, or cell-free display systems (i.e., ribosomal display. This system directly links the genotype and the phenotype of displayed antibodies during selection and allows for simultaneous co-selection of the desired antibodies and their encoding gene as a result of the binding characteristics of the displayed antibodies. The gene encoding of the desired antibodies can be further manipulated for improvements in affinity and/or specificity, increased expression, post-translation modification or fusion of secondary protein.

Recombinant monoclonal antibodies from various animals can be produced without a myeloma cell line as fusion partner. The animals can be immunized by exposure to a desired antigen, and the antibody genes from these animals are amplified by polymerase chain reaction (PCR) and expressed in different expression systems in various formats, such as Fab, single-chain variable fragment (scFv), and single domain antibodies (V_H, V_{HH} and V_{NAR}).

Fab: Fab fragments are the monovalent antibody fragment, can be produced from an IgG by treatment with papain.

Scfv: This antibody fragment is composed of the variable segments of both the heavy and light antibody chains. The two fragments are typically joined by a flexible peptide linker, most commonly the 15 amino acids of $(Gly_4Ser)_3$, to improve scFv folding and stability. Due to rapid *in vivo* blood clearance of scFv, the half life is in the range of minutes to tens of minutes. The smaller size allows them to penetrate the tissues.

V_H : This single-domain antibodies fragment is composed of a single-variable antibody domain. They rarely retain the affinity of the parent antibody and were poorly soluble and often prone to aggregation. Exposure of the hydrophobic surface of the V_H to V_L in solvents forms a sticky behavior of the isolated V_{HS}. Because of small size, it can target some epitopes, which are inaccessible to larger antibody fragments.

V_{HH}: The member of the *Camelidae* family have a single V-like domains that are mounted on a Fc-equivalent constant domain frame work, which is having the heavy-chain variable domain, referred as V_{HH}. The V_{HH} are in the nano-molar range and are comparatively stable up to a temperature of 90^0C in comparison to Fab and scFv.

V_{NAR}: The immunoglobulin isotype novel antigen receptor (IgNAR) is a homodimeric heavy-chain complex found in the serum of the nurse shark (*Ginglymostroma cirratum*). IgNAR do not contain light chain. Each molecule consists of a single-variable domain (V_{NAR}) and five constant domains (C_{NAR}). The presence of disulfide bonds insists affinities in monomolar range.

Antibody Libraries: Three types of libraries generally can be constructed for selection of rAbs such as:
(a) immune,
(b) non-immune, and
(c) Synthetic/semi-synthetic.

Immune libraries are generated by cloning antibody genes from B-cells of immunized animals. These libraries are having added advantage of antigen specific. Hybridoma cell lines can also be used for the formation of immune libraries. As each hybridoma cell line theoretically produces only a single type of functional antibody, only one set of immunoglobulin genes is available. To select for functional immunoglobulin genes from hybridoma cell lines, intensive screening of cloned genes through planning of phase-display antibodies is necessary.

Non-immune libraries are constructed in the same manner as immune libraries, but B-cell from non-immunized donors are used as sources for antibody genes.

Synthetic/semi-synthetic libraries are artificially constructed by using PCR to randomly assemble the genes encoding the mHV regions from a nonimmune B-cell. By employing this process, a unique set of VH and VL genes are recombined. These are some times referred as single-pot libraries, because each such library can be source of antibodies directed against multiple antigens. This technology can be used to display the immunoglobulin genes of conventional hybridomas for the purpose of rescuing unstable clones or improve their binding specificities and/or affinities through genetic manipulation such as mutagenesis or heavy/light-chain shuffling.

***Planning Antibody Libraries*:** Through planning library, the desired rAbs can be selected from an antibody library. During planning, recombinant antibody libraries are incubated with the target antigen to bind clones bearing specificities of interest. Unbound or weak clones are to be washed away, and rAbs, which remain bound to the antigen, are eluted and amplified for further propagation. After two to four rounds of planning, all binders should be specific to the antigen of interest. Antibody-displaying complexes bearing expressed rAb can be used directly in diagnostic assays or can be genetically modified to express soluble antibody molecules. The clones are stable and capable of self replication. They can be stored as bacterial stocks, phase stocks, or plasmid DNA.

***Phase Display*.** Phase display is often used to screen the recombinant antibodies. It is based on the functional expression of peptides and proteins on the surfaces of bacteriophages.

Bacterial Display. The display of recombinant peptides on the surface of bacteria such as E. *coli* was first described, in which several peptides have been displayed on bacterial surfaces, including antigenic determinants, heterologous enzymes, antibody fragments, and peptide libraries. Expression of scFv on the surfaces of Gram-positive bacteria has been investigated. ScFv antibodies have been displayed on the surfaces of *Staphylococcus xylosus* and *Stphylococcus carnosus* through the fusion to the C-terminus of protein A.

Yeast Display. The eukaryotic system of yeast offers an advantage of posttranslational modification and processing of mammalian proteins and is better suited for expression and secretion of human-derived antibody fragments than a host such as *E. coli*. Yeasts displaying antibodies share many of the same advantages as bacterial display. Yeasts have a rigid, thick cell wall that enables the stable maintenance of surface displayed proteins.

Ribosome Display. This cell-free display system used the same principle as the phase-display systems in bacteria. The phenotype and genotype of a peptide are linked together and, therefore, can be simultaneously selected based on the function of peptides. However, the linkage is a physical one in that the genetic material is covalently linked to its encoded product in the formation of antibody-ribosome-mRNA (ARM) complex through *in vitro* transcription/translation. Mutations can be incorporated in each cycle of selection using PCR-based mutagenesis, thereby allowing for the introduction of diversification during selection. Antibodies with equilibrium dissociation constant improve from the nano-molar to pico-molar range using ribosome display.

Soluble Antibody Production

Several methods are available for the expression of soluble recombinant antibodies. Bacteria, yeasts, plants, insects and mammalian cell lines, and cloned transgenic animals have all been used for rAb expression. Bacteria, particularly *E. Coli*, are favored for expression of small, non-glycosylated Fab, scFv, and V_H fragments and diabodies. ScFv fragments are generally expressed to a higher level over Fab in bacteria. Bacterial expression of rAbs is typically achieved by fusion of N-terminal signal peptides, which targets the protein to the periplasmic space of *E. coli*. This periplasmic expression may render a large fraction of the produced rAb insoluble. Soluble bacterial antibody can also be directed in to the culture medium. The risk is that the production of inclusion bodies often results in insoluble protein aggregates. Secretion of the rAb can be associated with cell lysis and subsequent product loss.

Purification of Recombinant Antibodies. Purification of bacterial expressed rAb is assisted by the presence of fusion polypeptides such as c-Myc, poly-His, and the FLAG epitope, which have been added to allow for affinity purification. Subseqent cleavage of the fusion protein yields the rAb of interest.

Applications

Immunodetection of Pathogens: It has been found that rAbs possess affinities rivaling those of monoclonal antibodies. The efficiencies of immunoassay relies on four components such as: 1) the target antigen to be detected; 2) the antibody used for detection; 3) the method to separate bound antigen and antibody complexes from unbound reactants (if heterogeneous materials involve); and 4) the detection method.

Immunoassays: A common way to detect antibody-antigen binding in an immunoassay is the enzyme-linked immunosorbant assay (ELISA). ELISA is based on the target antigen being captured on to a plastic multiwell or tube by a capture antibody previously bound to the solid matrix. Bound antigen is then detected using a secondary antibody. The detector antibody can be directly labeled with a signal-generating molecule, or it can be detected with another antibody that is labeled with an enzyme. These enzymes catalyze the chemical reaction with a substrate that results a colorimetric change. The colour intensity can be detected by a spectrophotometer.

Immunotherapy: Immunotherapy is a treatment to stimulate or restore the ability of the immune system to fight infection and disease. It can be either specific or non specific. Cytokines, growth factors and leukocytes can enhance the host defense against a variety of pathogens, which are the examples of non-specific immunotherapies. In contrast, the administration of antibodies is a form of specific immunotherapy similar to antimicrobial therapy is directed specifically at the pathogen or toxin. The increase in production of antibodies for therapeutic use may be due to: a) increase the resistance of pathogens to antibiotics; b) the lack of effective chemotherapeutic agent; c) the toxicity of available drugs; d) the emergence of new micro-organisms and viruses; e) the epidemiological evidence of an important role for the humoral immune system in host control of infection; f) the increased incidence of chronic disease in individuals with various antibody deficiencies; and g) the use of antibodies is currently the only means to provide immediate immunological protection against biological agents. Most antibodies currently being used for immunotherapy are monoclonal immunoglobulins (i.e., IgG and IgM). Their multivalency and high specificity and affinity make them attractive for therapies (Table 6.2).

Table 6.2 A list of some recombinant Antibodies that May be Theoretically effective in Immunotherapy

Antibody Target	rAb type	Therapeutic status
Bacterial		
Bacillus anthracis Toxin	Fab	Rat; *In Vivo*
Neisseria meningitidis	scFv	Mouse; *In Vivo*
Staphylococcus aurious-Methicillin Resistant	Aurograb	Human; Clinical Trial
Fungal		
Candida albicans	scFv	Rat; *In Vivo*
Candida spp.	Mycograb	Human; Clinical Trial
Cryptococcus neoformas	scFv	Mouse; *In Vivo*
Protozoan		
Plasmodium falciparum	Recombinant peptide	Cell culture; *In Vivo*
Plasmodium yoelii	scFv	Mouse; *In Vivo*
Plasmodium berghei	scFv	Mosquito; *In Vivo*
Viral		
Cucumber Mosaic Virus	scFv	In Planta
Foot and Mouth Disease	Fab	PRNT; *In Vitro*
Herpes Simplex Virus 1,2	Fab	Cell Culture; *In Vivo*
HIV-1 Reverse Transcriptase	scFv	Cell Culture; *In Vitro*
HIV-1 Vif Protein	scFv	Cell Culture; *In Vitro*
Japanese Encephalitis Virus	Fab	PRNT; *In Vitro*
Measles Virus	Fab	PRNT; *In Vitro*
Rabies Virus	scFv-Fc	Mouse; *In Vivo*
Rotavirus	Fab	Cell Culture; *In Vitro*
Varicella-zoster Virus	scFv	Cell Culture; *In Vitro*
Other		
Prion PcP^{sc}	scFv	Cell Culture; *In Vitro*
Prion PcP^{sc}	Fab	Cell Culture; *In Vitro*

[Abbreviation: PRNT = Plaque Reduction Nutralization Test. PcP^{sc} = The modified form of PrP^{c} (a normal cellular Protein thought to be involved in synaptic function). Which may cause diseases: i.e., the prion is known as PrP^{sc} (for scrapie). Aurograb = commercially available recombinant antibody from Neu Tech Pharma plc. Mycograb = Commercially available recombinant antibody from Neu Tech Pharma plc.]

6.2.3 Vaccine / Active Immunotherapy

The recombinant DNA technology is used for development of vaccines for active immunization. The recombinant antigens are synthesized to represent the most immunogenic antigens from a particular pathogen.

After purification recombinant antigens is administered using traditional immunization protocol and are considered to be safe. List of pathogens being treated by active immunotherapy are as follows (Table 6.3):

Table 6.3 List of pathogens being treated by active immunotherapy

Antibody Target	Type of Vaccine	Therapeutic Status
Chlamydia pneumoniae	DNA/Recomb Virus	Mouse ; *In Vivo*
Cryptosporidium parvum	Recomb antigen	Bovine; *In Vivo*
Entamoeba hystolytica	Recomb antigen	Mouse; *In Vivo*
HIV-1 Envelop Protein	Recomb Virus	Mouse; *In Vivo*
Listeria monocytogenes	DNA Vaccine	Mouse; *In Vivo*
Mycobacterium avium	Recomb antigen	Lamb; *In Vivo*
Plasmdodium falciparum	Recomb Virus	Human; *In Vivo*
Vibrio anguillarum	Recomb peptide	Piscine; *In Vivo*
Virulent infectious Bursal Disease Virus	DNA Vaccine	Chicken; *In Vivo*

6.2.4 Nanoparticle-Based Immunotherapy

Biodegradable polymeric particles have attracted considerable attention as potential drug delivery devices of low molecular weight drugs, as well as bio-macromolecules such as proteins, peptides or DNA. In particular, polymeric nanoparticles with entrapped immuno-stimulatory antigens or antibodies represent an exciting approach to control the release of therapeutics and to optimize the desired immune response via selective targeting of the antigen to antigen-presenting cells (APC).

The antibody-conjugated nanoparticles can be used principally in two biomedical applications such as: therapy and diagnosis. In therapy, the development of targeted drug delivery represents, together with tissue repair, the main applications of antibody-conjugated nanoparticles. In diagnosis, the applications can be divided into those using *in vivo* and those using *in vitro* experimentation and include contrast agents for magnetic resonance imaging (MRI), sensing, cell sorting, bioseparation, enzyme immobilization, immunoassays, transfection (gene delivery), purification, and so forth. The importance of all these applications can be demonstrated by a list of all the companies involved in the synthesis and applications of antibody conjugated nanoparticles (Table 6.1)

Commercial antibodies are already on the market either attached to drugs (Mylortag) or to radioisotopes (i.e., ProstaScint), used in the treatment of acute myeloid leukaemia and prostate cancer, respectively.

Poly-(d,l-lactide-co-glycolide)/montmorillonite nanoparticles (PLGA/MMT NPs) have been decorated with human epidermal growth factor receptor-2 (HER_2) antibody, Trastuzumab ($Herceptin_2$), for targeted breast cancer chemotherapy with paclitaxel as a model anticancer drug. Their *in vitro* studies revealed that the therapeutic effects of the drug formulated in the conjugated nanoparticles could be 12.7 times higher than that of the bare nanoparticles, and 13.1 times higher than Taxol. The same humanised IgG1 monoclonal antibody has been covalently attached to human serum albumin-based nanoparticles, and the resulting carrier showed specific targeting to HER_2-positive breast cancer cells.

Paclitaxel-loaded poly (lactide-co-glycolide) (PLGA) nanoparticles coated with cationic SM5-1 singlechain antibody (scFv), containing a polylysine (SMFv-polylys), were synthesized and effectively tested *in vitro*. The purpose of tagging the antibody with the cationic polypeptide polylysine is to achieve electrostatic attraction with the negatively charged nanoparticles. When compared to nontargeted paclitaxel-loaded PLGA nanoparticles, the antibodyconjugated nanoparticles showed enhanced *in vitro* cytotoxicity against human hepatocellular carcinoma cell lines (Ch-hep-3).

Two different monoclonal antibodies have been covalently coupled to the same poly-(malic acid)-based nanoparticle, one of them, monoclonal anti-TfR, to direct the conjugate across the blood-tumour barrier, and the other, monoclonal 2C5, to target tumourcell surface-bound nucleosomes. The presence of both antibodies on the same nanoparticle and their biological activity were confirmed by ELISA. *In vivo* experiments showed the significantly higher accumulation of the conjugated nanoparticles in human glioma.

Tyner et al. have recently reported the surface functionalization of inorganic nanoparticles (made of magnesium–aluminium layered double hydroxides) with disuccinimidyl carbonate (DSC), which was then loaded with a huA33 antibody and a blood plasma protein (serum albumin). The biological *in vitro* tests showed that LDH-DSC-huA33 nanobiohybrids had an activity against human A33 antigen 30 times higher than that of LDH-DSC-albumin.

Hayes et al. described the use of an antibodylipopolymer (anti-HER2 scFv (F5)-PEG-DSPE) conjugated to a cationic lipid nanoparticle with the DNA encapsulated in its interior, which achieves a high degree of specific transfection activity.

The Recombinant Molecules Derived from Antibodies may be utilized to develop a more heterogeneous group of new proteins, all based on the structure of antibodies, either as autonomous recombinant entities (Fab type fragments, Fv of the simple chain form of scFv, diabodies, triabodies, bispecifics, minibodies, phage antibodies), or as fusion proteins or conjugated antibodies, in which the Fc or the Fab portion is combined with new properties that have a toxin, an enzyme, a cellular receptor, a cytokine, and so forth for a greater solution for human health.

References

A. Aigner, "Tumor-targeting nanosystems for the delivery of siRNA," *Nanomedicine*, 2007, 2 (4):569–572.

A. J. Haes, D. A. Stuart, S. Nie, and R. P. Van Duyne, "Using solution-phase nanoparticles, surface-confined nanoparticle arrays and single nanoparticles as biological sensing platforms," *Journal of Fluorescence*, 2004,14 (40: 355–367.

A. Mayer, E. Tsiompanou, D. O'Malley, et al., "Radioimmunoguided surgery in colorectal cancer using a genetically engineered anti-CEA single-chain Fv antibody," *Clinical Cancer Research*, 2000, 6 (5): 1711–1719.

A. Molina, M. Valladares, S. Magad´an, et al., "The use of transgenic mice for the production of a human monoclonal antibody specific for human CD69 antigen," *Journal of Immunological Methods*, 2003, 282 (1-2):147–158.

A. Pl¨uckthum, A. Krebber, C. Krebber, et al., "Producingg antibodies in Escherichia coli: from PCR to fermentation," in *Antibody Engineering: A Practical Approach*, J. McCafferty, H. R. Hoogenboom, and D. J. Chiswell, Eds., 1996:203–252, Oxford IRL Press, Oxford, UK.

A. Tramontano, E. Bianchi, S. Venturini, F. Martin, A. Pessi, and M. Sollazzo, "The making of the minibody: an engineered beta-protein for the display of conformationally constrained peptides," *Journal of Molecular Recognition*, 1994, 7(1):9–24.

B. D´ıaz, I. Sanjuan, F. Gamb´ on, C. Loureiro, S.Magad´an, and A´. Gonza´lez-Ferna´ndez, "Generation of a human IgM monoclonal antibody directed against HLA class II molecules: a potential agent in the treatment of haematologicalmalignancies," *Cancer Immunology, Immunotherapy*, 2009, 58(3): 351–360.

B. Gupta and V. P. Torchilin, "Monoclonal antibody 2C5- modified doxorubicin-loaded liposomes with significantly enhanced therapeutic activity against intracranial human brain U-87 MG tumor xenografts in nude mice," *Cancer Immunology, Immunotherapy*, 2007, 56(8):1215–1223.

B. Sun, B. Ranganathan, and S.-S. Feng, "Multifunctional poly(d,l-lactide-co-glycolide)/montmorillonite (PLGA/MMT) nanoparticles decorated by Trastuzumab for targeted chemotherapy of breast cancer," *Biomaterials*, 2008, 29 (4):475–486.

C. Rader and C. F. Barbas III, "Phage display of combinatorial antibody libraries," *Current Opinion in Biotechnology*, 1997, 8 (4):503–508.

C. S. Owen and N. L. Sykes, "Magnetic labelling and cell sorting," *Journal of Immunological Methods*, 1984, vol. 73, pp. 41–48.

D. A. Bell, B. Hahn, and G. Harkiss, "Idiotypes, antibodies and immunopathology," *Lupus*, 1992, 1 (5):335–337.

D. R. Burton, L. Gregory and R. Jefferis. "Aspects of the molecular structure of IgG subclasses," *Monographs in Allergy*, 1986 (19): 7–35.

E. A. Kabat, "The structural basis of antibody complementarity," *Advances in Protein Chemistry*, 1978 (32):1-75.

E. A. Kabat, T. T.Wu, and H. Bilofsky, "Unusual distributions of amino acids in complementarity determining (hypervariable) segments of heavy and light chains of immunoglobulins and their possible roles in specificity of antibody-combining sites," *The Journal of Biological Chemistry*, 1977, 252 (19):6609–6616.

E. Aboud-Pirak, B. Lesur, K. S. P. B. Rao, R. Baurain, A. Trouet, and Y.-J. Schneider, "Cytotoxic activity of daunorubicin or vindesin conjugated to a monoclonal antibody on cultured MCF-7 breast carcinoma cells," *Biochemical Pharmacology*, 1989, 38 (4):641–648.

E. Li, A. Pedraza, M. Bestagno, S. Mancardi, R. Sanchez, and O. Burrone, "Mammalian cell expression of dimeric small immune proteins (SIP)," *Protein Engineering*, 1997, 10 (6):731–736.

E. M. Yoo, K. R. Chintalacharuvu, M. L. Penichet, and S. L. Morrison, "Myeloma expression systems," *Journal of Immunological Methods*, 2002, 261(1-2):1–20.

G. Kou, J. Gao, H. Wang, et al., "Preparation and characterization of paclitaxel-loaded PLGA nanoparticles coated with cationic SM5-1 single-chain antibody," *Journal of Biochemistry and Molecular Biology*, 2007, 40 (5):731–739.

G. L. Boulianne, N. Hozumi, and M. J. Shulman, "Production of functional chimaeric mouse/human antibody," *Nature*, 1984, 312(5995):643–646.

G. Winter and W. J. Harris, "Humanized antibodies," *Immunology Today*,1993 (6): 243–246.

H. R. Hoogenboom, J. D.Marks, A. D. Griffiths, and G. Winter, "Building antibodies from their genes," *Immunological Reviews*, 1992 (130): 41–68.

I. H. El-Sayed, X. Huang, and M. A. El-Sayed, "Selective laser photo-thermal therapy of epithelial carcinoma using anti-EGFR antibody conjugated gold nanoparticles," *Cancer Letters*, 2006, 239 (1):129–135.

S. J. Carter. Immunology. Tutorial Pharmacy. CBS Publications, 1986, 6th edn. :375-413.

I. Steinhauser, B. Sp¨ankuch, K. Strebhardt, and K. Langer, "Trastuzumab-modified nanoparticles: optimisation of preparation and uptake in cancer cells," *Biomaterials*, 2006, 27 (28):4975–4983.

J. Davies and L. Riechmann, "Antibody VH domains as small recognition units," *Bio/Technology*, 1995, 13 (5):475– 479.

J. J. Trill, A. R. Shatzman, and S. Ganguly, "Production of monoclonal antibodies in COS and CHO cells," *Current Opinion in Biotechnology*, 1995, 6 (5): 553–560.

J. M. Reichert, "Technology evaluation: lumiliximab, Biogen Idec," *Current Opinion in Molecular Therapeutics*, 2004, 6 (6): 675–683.

J. McCafferty, A. D. Griffiths, G. Winter, and D. J. Chiswell, "Phage antibodies: filamentous phage displaying antibody variable domains," *Nature*, 1990, 348 (6301): 552–554.

J. P.Mach, F. Buchegger,M. Forni, et al., "Use of radiolabeled monoclonal anti-CEA antibodies for the detection of human carcinomas by external photoscanning and tomoscintigraphy," *Immunol Today*, 1981 (2):239–249.

J. P.Mach, F. Buchegger,M. Forni, et al., "Use of radiolabeled monoclonal anti-CEA antibodies for the detection of human carcinomas by external photoscanning and tomoscintigraphy," *Immunol Today*, 1981 (2): 239–249.

K. J. Bentley, R. Gewert, and W. J. Harris, "Differential efficiency of expression of humanized antibodies in transient transfected mammalian cells," *Hybridoma*, 1998, 17 (6):559–567.

K. M. Tyner, S. R. Schiffman, and E. P. Giannelis, "Nanobiohybrids as delivery vehicles for camptothecin," *Journal of Controlled Release*, 2004, 95 (3):501–514.

Karl E. Hellstrom, I Hellsrrom, G. E. Goodman. Antibodies for Drug Delivery. Robinson Robinson's Controlled Drug Delivery, 2009, Published by Informa Healthcare USA, Inc. 2nd Edn. Vol. 20:623-653.

L. L. Green, "Antibody engineering via genetic engineering of the mouse: XenoMouse strains are a vehicle for the facile generation of therapeutic human monoclonal antibodies," *Journal of Immunological Methods*, 1999, 231 (1-2): 11– 23.

L. M. Weiner, "Monoclonal antibody therapy of cancer," *Seminars in Oncology*, 1999. 26 (5, supplement 14): 43– 51.

L. Zhu, M.-C. van de Lavoir, J. Albanese, et al., "Production of human monoclonal antibody in eggs of chimeric chickens," *Nature Biotechnology*, 2005, 23 (9):1159–1169.

M. Br¨uggemann and M. J. Taussig, "Production of human antibody repertoires in transgenic mice," *Current Opinion in Biotechnology*, 1997, 8 (4): 455–458.

M. Brennan, P. F. Davison, and H. Paulus, "Preparation of bispecific antibodies by chemical recombination of monoclonal immunoglobulin G1 fragments," *Science*,1985, 229 (4708): 81–83.

M. C. Garnett, "Targeted drug conjugates: principles and progress," *Advanced Drug Delivery Reviews*, 2001, 53 (2):171–216.

M. E. Hayes, D. C. Drummond, D. B. Kirpotin, et al., "Genospheres: self-assembling nucleic acid-lipid nanoparticles suitable for targeted gene delivery," *Gene Therapy*, 2006, 13 (7): 646–651.

M. Fujita, B.-S. Lee, N. M. Khazenzon, et al., "Brain tumor tandem targeting using a combination of monoclonal antibodies attached to biopoly(β-L-malic acid)," *Journal of Controlled Release*, 2007, 122 (3): 356–363.

M. J. Coloma, A. Hastings, L. A. Wims, and S. L. Morrison, "Novel vectors for the expression of antibody molecules using variable regions generated by polymerase chain reaction," *Journal of Immunological Methods*, 1992, 152 (1): 89–104.

M. J. Glennie and P. W. M. Johnson, "Clinical trials of antibody therapy," *Immunology Today*, 2000, 21 (8): 403– 410.

M. J. Glennie, H.M.McBride, A. T.Worth, and G. T. Stevenson, "Preparation and performance of bispecific F(ab'γ)2 antibody containing thioether-linked Fab'γ fragments," *The Journal of Immunology*, 1987, 139 (7): 2367–2375.

M. P. Reddy, C. A. S. Kinney, M. A. Chaikin, et al., "Elimination of Fc receptor-dependent effector functions of a modified IgG4 monoclonal antibody to human CD4," *The Journal of Immunology*, 2000, 164 (4):1925–1933.

M. R. Hurle and M. Gross, "Protein engineering techniques for antibody humanization," *Current Opinion in Biotechnology*, 1994, 5 (4): 428–433.

M. Valladares, I. Sanju´an, S. Magad´an, A. Molina, F. Sanju´an, and A´. Gonza´lez-Ferna´ndez, "H24: a human monoclonal antibody obtained from mice carrying human Ig genes, recognises human myeloid leukaemia and CD5- Non- Hodgkin's lymphoma," *Inmunologia*, 2004, 23 (1): 7–15.

Manuel Arruebo, M. Valladares, and A. G. Fernandez. Antibody-Conjugated Nanoparticles for Biomedical Applications. Journal of Nanomaterials, 2009, article ID 439389: 1-24.

N. Lonberg, L. D. Taylor, F. A. Harding, et al., "Antigenspecific human antibodies from mice comprising four distinct genetic modifications," *Nature*, 1994, 368 6474): 856–859.

Nicholas J Pokorny, J. I. Boutler-Bitzer, J. Chris H, Jack T. T., and Hung Lee. Recombinant antibodies for pathogen detection and immunotherapy. Handbook of Pharmaceutical Biotechnology, by Shayne Cox Gad. Published by Willey-Interscience, 2007:851-881.

P. Holliger and G. Winter, "Diabodies: small bispecific antibody fragments," *Cancer Immunology Immunotherapy*, 1997, 45 (3-4): 128–130.

P. J. Hudson, "Recombinant antibody fragments," *Current Opinion in Biotechnology*, 1998, 9 (4): 395–402.

P. S. Andersen, H. Orum, and J. Engberg, "One-step cloning of murine Fab gene fragments independent of IgH isotype for phage display libraries," *BioTechniques*, 1996, 20 (3): 340–342.

R. E. Bird, K. D. Hardman, J. W. Jacobson, et al., "Singlechain antigen-binding proteins," *Science*, 1998, 242 (4877): 423–426.

R. H. J. Begent, M. J. Verhaar, K. A. Chester, et al., "Clinical evidence of efficient tumor targeting based on single-chain Fv antibody selected from a combinatorial library," *Nature Medicine*, 1996, 2 (9):979–984.

R. K. Jain, "Transport of molecules across tumour vasculature," *Cancer Metastasis Reviews*, 1987 (6): 559–593.

R. Newman, J. Alberts, D. Anderson, et al., ""Primatization" of recombinant antibodies for immunotherapy of human diseases: a macaque/human chimeric antibody against human CD4," *Bio/Technology*, 1992, 10 (11):1455–1460.

S. D. Gorman and M. R. Clark, "Humanisation of monoclonal antibodies for therapy," *Seminars in Immunology*, 1990, 2 (6): 457–466.

S. Hu, L. Shively, A. Raubitschek, et al., "Minibody: a novel engineered anti-carcinoembryonic antigen antibody fragment (single-chain Fv-CH3) which exhibits rapid, highlevel targeting of xenografts," *Cancer*, 1996 (56):3055–3061.

S. L. Morrison, M. J. Johnson, L. A. Herzenberg, and V. T. Oi, "Chimeric human antibody molecules: mouse antigen binding domains with human constant region domains," *Proceedings of the National Academy of Sciences of the United States of America*, 1984, 81 (21): 6851–6855.

S. L.Morrison, "In vitro antibodies: strategies for production and application," *Annual Review of Immunology*, 1992 (10): 239–265.

S. Magad´an, M. Valladares, E. Suarez, et al., "Production of antigen-specific human monoclonal antibodies: comparison of mice carrying IgH/kappa or IgH/kappa/lambda transloci," *Biotechniques*, 2002, 33(3): 680–690.

T. Ashikaga, M. Wada, H. Kobayashi, et al., "Effect of the photocatalytic activity of TiO2 on plasmid DNA," *Mutation Research*, 2000, 466 (1):1–7.

T. J. Vaughan, A. J. Williams, K. Pritchard, et al., "Human antibodies with sub-nanomolar affinities isolated from a large non-immunized phage display library," *Nature Biotechnology*, 1996, 14 (3):309–314.

T. Suwa, S. Ozawa, M. Ueda, N. Ando, and M. Kitajima, "Magnetic resonance imaging of esophageal squamous cell carcinoma using magnetite particles coated with antiepidermal growth factor receptor antibody," *International Journal of Cancer*, 1994, 75 (4): 626–634.

The Project on Emerging Nanotechnologies, The Woodrow Wilson International Center for Scholars, 2008, http://www.Nanotech project.org/...

W. C. Earnshaw and S. R. Casjens, "DNA packing by the double-stranded DNA bacteriophages," *Cell*, 1980 (21): 319–331.

W. D. Huse, L. Sastry, S. A. Iverson, et al., "Generation of a large combinatorial library of the immunoglobulin repertoire in phage lambda," *Science*, 1989, 246 (4935): 1275–1281.

W. Kemmner, G. Moldenhauer, P. Schlag, and R. Brossmer, "Separation of tumor cells from a suspension of dissociated human colorectal carcinoma tissue by means of monoclonal antibody-coated magnetic beads," *Journal of Immunological Methods*, 1992, 147 (2): 197–200.

W. Zhou, C.-C. Chen, B. Buckland, and J. Aunins, "Fedbatch culture of recombinant NSO myeloma cells with high monoclonal antibody production," *Biotechnology and Bioengineering*, 1997, 55 (5): 783–792.

Wild M A, Xin H, Maruyama T, et al. Human antibodies from immunized donors are protective against anthrax atoxin *in-vivo. Nature Biotechnol*, 2003 (21):1305-1306

Beninati C, Arseni S., Mancuso G, et al. Protective immunizations against group B meningococci using anti idiotypic mimics of the capsular polysaccharide, *J. Immunol*, 2004 (172):2461-2468.

Mattews R C, Rigg G, Hodgetts S, et al. Preclinical assessment of the efficiency of mycograb, a human recombinant antibody against fungal HSP90. Antimicograb. Agents *Chemother,* 2003 (47): 2208-2216.

Li F, Dluzewewski A, Coley A M, et al. Phase-displayed peptides bind to the malarial protein apical membrane antigen-1 and inhibit the merozoite invasion of host erythrocytes. *J. Biol. Chem.*, 2002 (77):50303-50310.

Yoshida S, Matsuoka H, Luo E, et al. A single chain anti body fragment specific for the Plasmodium berghei ookinete protein Pbs21 confers transmission blockade in the mosquito midgut. *Mol. Biochem. Parasitol*, 1999 (104):195-204.

Villani M E, Roggero P, Gitti O, et al. Immunomodulation of *cucumber mosaic virus* infection by intrabodies selected in vitro from a stable frame work phase display library. *Plant Mol. Biol*, 2005 (58):305-316.

Kim Y J, Lebreton F, Kaiser C, et al. Isolation of *foot-and-mouth disease virus* specific bovine fragmentsa from phase display libraries. *J. Immunol. Meth.*, 2004 (286):155-166.

De Logu A. Williamson R A, Rozenshteyn R, et al. Characterization of a type-common human recombinant monoclonal antibody to *herpes simplex* virus with high therapeutic potential. *J. Clin. Microbiol.*, 1998 (36): 3198-3204.

Shaheen F, Duan L, Zhu M, et al. Targetting human immunodeficiency virus type-1 reverse transcriptase by intracellular expression of single chain variable fragments to inhibit early stages of the viral life cycle. *J. Virol*, 1996 (70):3392-3400.

Wu S C, Lin Y J, Chou J W, et al. Construction and characterzation of a Fab recombinant protein for *Japanese encephalitis virus* neutralization. *Vaccine*, 2004 (23): 163-171.

De carvalho N C, Williamson R A, Parren P W, et al. Neutralizing human Fab fragments against measles virus recovered by phase display. *J. Virol.*, 2002 (76): 251-258.

Ray K, Embleton M J, jaikhani B L, et al. Selection of a single chain variable fragments (scFv) against the glycoprotein antigen of the rabies virus from a human synthetic scfv phase display library and their fusion with Fc region of human IgG1. *Clin. Exp. Immunol.*, 2001 (125):94-101.

Higo-Moriguchi K, Ahahori Y, Iba Y, et al.. Isolation of human monoclonal antibodies that neutralize human rotavirus. *J. Virol*, 2004 (78): 3325-3332.

Kausmally L. Waalen K. Lobersli I, et al. Neutralizing human antibodies to varicella-zoster virus (VZV) derived from a VZV patient recombinant antibody library. *J. Gen. Virol.*, 2004 (85): 3493-3500.

Donofrio G, Heppner F L. Polymenidou M, et al. Paracrine inhibition of prion propagation by anti-PrP single chain Fv miniantibodies. J. Virol., 2005 (79): 8330-9338

Peretz D, Williamsom R A, Lanako k, et al. Antibodies inhibit prion propagation and clear cell cultures of prion infective. *Nature*, 2001 (412): 739-743.

Penttila T, Tammiruusu A, Liljestrom P, et al. DNA immunization followed by a viral vector booster in a *Chlamydia pneumoniae* mouse model. Vaccine, 2004 (22): 3386-3394

Sun J H, Yan Y X, Jiang J, et al. DNA imminization against very virulent infectious bursal disease virus with VP2-4-3 gene and chicken IL-6 gene. *J. Vet. Med.* B Infect. Dis. Vet. Public. Health. 2005 (52):1-7.

Lim K P, Li H, Nathan S. Expression and purification of a recombinant scFv towards the exotoxin of the pathogen, *burkholderia pseudomallei. J. Microbiol.*, 2004 (42):126-132.

www.becomehealthynow.com/article/bodyimmune/.....

www.becomehealthynow.com/body/immune/immune_system/glossary.shtmlo.....

www.en.wikipedia.org/wiki/file/: ....

Hong-Xuag H, Lei C, Cheng-Min W, et al. Expression of the recombinant fusion protein CP15-23 of Cryptosporidium parvum and its protective test. *J. Nanosci Nanotechnol.*, 2005 (5): 1292-1296

Melzer H, Baier k, Felici F, et al. Humoral immune response against proteiophospoglycan surface antigens of *Entamoeba histolytica* elicited by immunization with synthetic mimotope peptides. FEMS Immunol. *Med. Microbiol,* 2003 (37): 179-183.

Gonzalo R M, Rodriguez D, Garcia S A, et al. Enhanced CD8+ T cell response to HIV-1 env by combined immunization with influenza and vaccinia virus recombinants. *Vaccine,* 1999 (17): 887-892.

Simon B E, Cornell K A, Clark T R, et al. DNA vaccination protects mice against challenge with Listeria monocytogenes expressing the hepatitis C virus NS3 against protein. *Infect. Immun.*, 2003 (71): 6372-6380.

Sechi L A, Mara L. Cappai P, et al. Immunization with DNAvaccines encoding different mycobacterial antigens elicits a Th1 type immune response in lambs and protects against *Mycobacterium avium* subspecies *paratuberculosis* infection. *Vaccine*, 2006 (3): 229-235.

Xia Y J, Wen W H, Huang W Q, et al. Development of a phase displayed disulfide-stabilized Fv fragment vaccine against *Vibrio anguillarum*. *Vaccine*, 2005 (23): 3174-3180.

Ellmark P, Ghatnekar-Nilsson S, Meister A, Heinzelmann H, Montelius L, Wingren C and Borrebaeck CAK. Attovial-based antibody nanoarrays. *Proteomics,* 2009 (9): 5406-5413.

Andréasson U, Dictor M, Jerkeman M, Berglund M, Högerkorp C-M, Sundström C, Rosenquist R, Eden P, Borrebaeck CAK and Ek S. Identification of molecular targets associated with transformed diffuse large B cell lymphoma using highly purified tumour cells. *Am J Hematol,* 2009 (84): 803-808.

Broos S, Lundberg K, Akagi T, Kadowaki K, Akashi M, Greiff L, Borrebaeck CAK, Lindstedt M. Immunomodulatory nanoparticles as adjuvants and allergen-delivery system to human dendritic cells: Implications for specific immunotherapy. *Vaccine*, 2010 (28): 5075-5085.

Borrebaeck CAK and Wingren C. Design of high-density antibody microarrays for disease proteomics: Key technological issues. *J Proteomics,* 2009 (72): 928-935.

Persson H, Persson J, Danielsson L and Ohlin M. Charges drive selection of specific antibodies by phage display. *J Immunol Methods*, 2010 (353): 24-30.

7

Therapeutic Careers in Drug Delivery

Various strategies have been developed in the field of pharmaceutical research for targeted delivery of therapeutic agents to hepatic, renal, pulmonary, brain, blood and bone cells for different clinical or diagnostic purposes. Nevertheless, as mentioned above, relatively few products have reached the commercial drug market, and this rather interesting and yet complex and fascinating drug delivery method seems to be in its infancy stage. However, currently, extensive studies are being conducted by numerous research groups worldwide and we expect to see a broad range of these drugs becoming available to general public for treatment of different ailments in the coming years. In fact targeted drug delivery systems could be looked upon as the new generation of drug delivery systems, which could overtake the drug market in the near future. With the advent of the monoclonal antibody technology as well as the development of liposomes, resealed erythrocytes, neosomes and polymeric nanoparticle carriers, the drug targeting field enjoyed a welcoming expansion and the clinical applications. Monoclonal antibodies, liposomes, polymers, proteins and many other entities have all encountered a vast array of difficulties ranging from problems in the synthesis of the carriers and drug conjugates to unfavorable pharmacokinetics and toxicity. Furthermore, lack of knowledge on the anatomical and physiological barriers in the body has hampered their clinical application. However, many problems have been solved due to development in pharmaceutical formulation technology, rapid developments in molecular biology, cell biology and immunology led to a better understanding of the processes taking place *in-vivo* upon administration of carriers and conjugates.

Resealed Erythrocytes as Therapeutic Career

Introduction

The introduction of procedures for the transient opening of pores across the red blood cell membrane provides the extraordinary opportunity to manipulate red blood cell or erythrocytes for different biomedical applications. Erythrocytes have been extensively studied for their potential carrier capabilities for the delivery of drugs, chemicals or macromolecules and drug-loaded micro-spheres or nanoparticles. Such drug-loaded carrier erythrocytes are prepared simply by collecting blood samples from the organism of interest, separating erythrocytes from plasma, entrapping drug in the erythrocytes, and resealing the resultant cellular carriers, for which these careers are called as resealed erythrocytes.

Erythrocytes are biodegradable, can circulate for long period of time (months), have a large capacity and high percentage of drug encapsulations. Furthermore the morphological, immunological and biochemical properties of carrier erythrocytes are similar to those of native cells.

The most advantages of erythrocytes as career include: its biocompatibility, biodegradability with no generation of toxic products any undesired immune response against the loaded drug as a carrier and its degradation products, prevention of degradation of the loaded drug from inactivation by endogenous chemicals, the wide variety of chemicals that can be entrapped, and the modification of pharmacokinetic and pharmacodynamic parameters of drug as desired. Limitations include dose dumping, loading limitations of the particles beyond 600,000 Daltons in size.

7.1 Morphology and Physiology of Erythrocytes

Erythrocytes are the most abundant cells in the human body found to be (>5.4 million cells/mm^3) in a healthy male and (>4.8 million cells/mm^3) in a healthy female). These cells were described in human blood samples by Dutch Scientist Lee Van Hock in 1674. In the 19th century, Hope and Seyler; identified hemoglobin and its crucial role in oxygen delivery to various parts of the body. Erythrocytes are biconcave discs with an average diameter of 7.8μm and a thickness of 2.5 μm and a thickness. The flexible, biconcave shape enables erythrocytes to squeeze through narrow capillaries, which may be only 3 μm wide. Mature erythrocytes are quite simple in structure. They lack a nucleus and other organelles. Their plasma membrane encloses hemoglobin, a heme-containing protein that is responsible for carrying O_2 and CO_2 binding inside the erythrocytes. The main role of erythrocytes is to transport O_2 from the lungs to tissues and CO_2 produced in tissues back to lungs. Because a nucleus is absent, all the intracellular space is available for O_2 transport. Also, because mitochondria are absent and energy is generated anaerobically in erythrocytes, these cells do not consume any of the oxygen they are carrying.

Erythrocytes live only about 120 days because of wear and tear on their plasma membranes as they squeeze through the narrow blood capillaries. Worn-out erythrocytes are removed from circulation and destroyed in the spleen and liver (RES), and the breakdown products are recycled. The process of erythrocyte formation within the body is known as erythropoiesis. In a mature human being, erythrocytes are produced in red bone marrow under the regulation of a hemopoietic hormone called erythropoietin.

7.2 Source and Isolation of Erythrocytes

Various types of mammalian erythrocytes have been used for drug delivery, including erythrocytes of horse, mice, cattle, pigs, dogs, sheep, goats, monkeys, chicken, rats, and rabbits. To isolate erythrocytes, blood is to be collected in heparinized tubes by vein puncture. Fresh whole blood is typically used for loading purposes because of more entrapment efficiencies. Fresh whole blood after immediate collection centrifuged at 2500 rpm for 5 min., chilled to 4^0C in a refrigerated centrifuge. The erythrocytes (packed cells) are then harvested and washed with phosphate buffer saline of pH 7.4. The washed cells are suspended in buffer solutions at various hematocrit values as desired and are often stored in acid-citrate-dextrose buffer at 4^0C for as long as 48 h before use.

7.3 Methods of Drug Loading

Several methods can be used to load drugs or other bioactive compounds in erythrocytes, including physical (e.g., electrical-pulse method) osmosis-based systems, and chemical methods (e.g., chemical perturbation of the erythrocytes membrane). Irrespective of the method used, the optimal characteristics for the successful entrapment of the compound requires the drug to have a considerable degree of water solubility, resistance against degradation within erythrocytes, lack of physical or chemical interaction with erythrocyte membrane, size of the active substances must be within (5000 to 600,000 Dalton) and well defined pharmacokinetic and pharmacodynamic properties.

7.3.1 Hypotonic Hemolysis

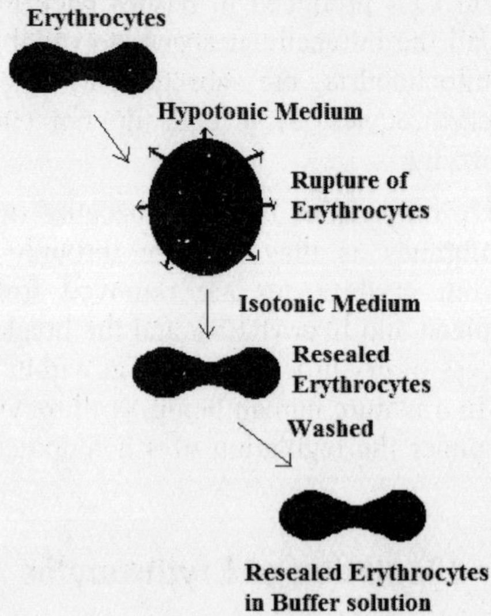

Fig. 7.1 Formulation of Resealed erythrocytes by Hypotonic Hemolysis Method.

This method is based on the ability of erythrocytes to undergo reversible swelling in a hypotonic solution. Erythrocytes have an exceptional capability for reversible shape changes with or without accompanying volume change and for reversible deformation under stress. An increase in volume leads to an initial change in the shape from biconcave to spherical (Fig 7.1)

The volume gain of 25-50% is permissible. The cell swells in the hypotonic solution (0.4% sodium chloride) at 0^0C temperature and after the increase in the volume above 50%, the membrane ruptures, releasing the cellular contents. At this point (just before cell lysis), some transient pores of 200-500 A^0 are generated on the membrane (the remnant is called an erythrocyte ghost) and at this point the drugs can be loaded in exchange of the inclusion compounds at equilibrium state. Increasing the ionic strength at 37^0C, membranes can be resealed by restoring the osmotic properties in isotonic conditions. Upon incubation, the cells resume their original biconcave shape and recover original impermeability.

7.3.2 Hypotonic Dilution

It is the simplest and fastest method, in which a volume of packed erythrocytes is diluted with 2-20 volumes of aqueous solution of a drug. The solution tonicity is then restored by adding a hypertonic buffer. Then the resultant mixture is centrifuged, the supernatant is discarded, and the pellet is washed with isotonic buffer solution. The major drawbacks of this method include low entrapment efficiency and a considerable loss of hemoglobin and other cell components. This reduces the circulation half life of the loaded cells. These cells are readily phagocytosed by RES macrophages and hence can be used for targeting RES organs. Hypotonic dilution is used for loading enzymes such as β-galactosidase and β -glucosidase, asparginase, arginase, and salbutamol, etc.

7.3.3 Preswell Dilution Technique

This method was developed by Rechsteiner in 1975 and was modified by Jenner et al. for drug loading. The technique is based upon initial controlled swelling in a hypotonic buffered solution. This mixture is centrifuged at low speed to recover the cells. The lysis point is detected by the disappearance of a distinct boundary between the cell fraction and the supernatant upon centrifugation. The tonicity of a cell mixture is restored at the lysis point by adding a calculated amount of hypertonic buffer. Then, the cell suspension is incubated at 37^0C to reseal the erythrocytes.

Drugs encapsulated in erythrocytes using this method include propranolol, asparginase, cyclophosphamide, cortisol-phosphate, methotrexate, metronidazol, levothyroxine, enalaprilat, and isoniazid, etc.

7.3.4 Hypotonic Dialysis

This method was first reported by Klibansky in 1959 and was used by Deloach and Ihler in 1977, for loading enzymes and lipids in which the drug mixture and erythrocyte solution is placed in a semi-permeable dialysis membrane, after tied with both the ends of the tube the tube placed in a swelling solution maintained to a temperature at 4^0C up to desired lysis time. The contents of the tube mixed with shaking after ward the tube placed in the isotonic phosphate buffer solution for resealing at the room temperature.

This method has been used for loading enzymes such as β-galactosidase, asparginase, inositol hexaphosphatase, gentamicin, adriamycin, pentamidine and furamycin, interlukin-2, and human recombinant erythropoietin, etc.

7.3.5 Isotonic osmotic Lysis

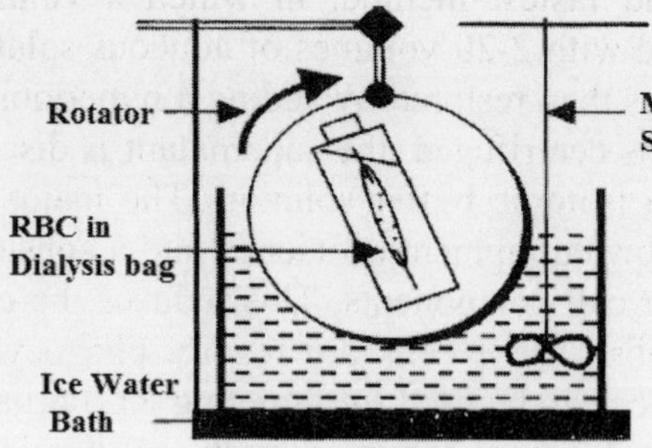

Fig. 7.2 Schematic of Isotonic Osmotic lysis method.

This method involves isotonic hemolysis of the erythrocytes in solutions of a substance with high membrane permeability; the solute will diffuse into the cells because of the concentration gradient. This process is followed by an influx of water to maintain osmotic equilibrium. Chemicals such as urea solution, polyethylene glycol, and ammonium chloride have been used for isotonic hemolysis. After the drug diffuses the suspending erythrocytes placed in an isotonic solution free from the above chemicals for resealing at 37^0C.

7.3.6 Chemical Perturbation of the Membrane

In 1973, Deuticke et al., showed that the permeability of erythrocyte membrane increases upon exposure to polyene antibiotic such as amphotericin B. In 1980, this method was used successfully by Kitao and Hattori to entrap the antineoplastic drug daunomycin in human and mouse erythrocytes. Lin et al., used halothane for the same purpose, however, these methods induce irreversible destructive changes in the cell membrane and hence are not very popular.

7.3.7 Electro-Insertion or Electroencapsulation

In 1973, Zimmermann tried an electrical pulse method to encapsulate bioactive molecules, which is known as electroporation. The method is based on the observation that electrical shock brings about irreversible changes in an erythrocyte membrane. In 1977, Tsong and Kinosita suggested the use of transient electrolysis to generate desirable membrane permeability for drug loading. The erythrocyte membrane is opened by a dielectric breakdown. Subsequently, the pores can be resealed by incubation at 37^{0}C in an isotonic medium.

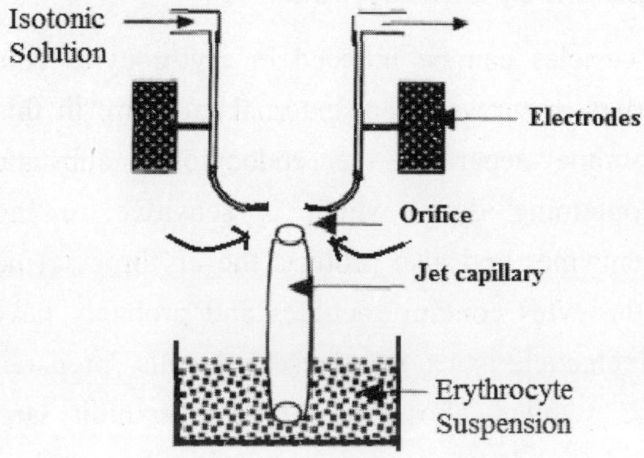

Fig. 7.3 Schematic of Formulation of Resealed erythrocytes by electroencapsulation method.

The procedure involves suspending erythrocytes in an isotonic buffer in an electrical discharge chamber, where a capacitor is used externally to discharge a definite voltage for a definite time interval through cell suspension to produce a square wave potential. The optimum intensity of an electric field is between 1-10 kW/cm and optimal discharge time is between 20-160 μs. The compound to be entrapped is added to the medium in which the cells are suspended from the commencement of the experiment. The characteristic pore diameter created in the membrane depends upon the intensity of electric field, the discharge time, and the ionic strength of suspending medium. The colloidal macromolecules contents of the cell may lead to cell lysis because of the increase in osmotic pressure.

One advantage of this method is a more uniform distribution of loaded cells in comparison with osmotic methods. The main drawbacks are the need for special instrumentation and the sophistication of the process. Entrapment efficiency of this method is >35% and the life span of the resealed cells in circulation is comparable with that of normal cells.

Various compounds such as sucrose, urease, methotrexate, isoniazid, human glycoprotein, DNA fragments, and latex particles of diameter 0.2 μm (4) can be entrapped within erythrocytes by this method.

7.3.8 Entrapment by Endocytosis

Intracellular vesicles can be induced in erythrocytes containing small molecules, drugs or viruses from external medium. In this method, the vesicle membrane separates the endocytosed substance from the cytoplasm containing drug, which is sensitive to inactivation by cytoplasmic enzymes and also protects the erythrocyte membrane. The resulting erythrocytes contain vacuoles and probably have different *in vivo* survival characteristics from resealed cells, prepared using other methods. The swollen ghosts so prepared exhibit larger endocytic vacuoles. The drug substances are entrapped in these endocytic vacuoles. Drug induced endocytosis is quite common and a variety of amphiphilic cations/drugs produce stomatocytosis in first step and then forms endocytic vacuole inside. This method is efficient for loading large particles such as viruses, enzymes and small molecules.

7.3.9 Loading by Lipid Fusion

Lipid vesicles containing a drug can be directly fused to human erythrocytes, which lead to an exchange with a lipid-entrapped drug. This technique was used for entrapping inositol monophosphate to improve the oxygen-carrying capacity of cells. However, the entrapment efficiency of this method is very low.

7.4 Characterization of Resealed Erythrocytes

The in vivo performance of resealed erythrocytes is affected to a great extent by their biological properties. Hence, *in-vitro* characterization forms an important part of their evaluation studies, which are summarized in Table 7.1.

Table 7.1 Characteristics and Evaluation methods of Resealed erythrocytes.

Parameter	Method/instrument used
I. Physical characterization	
Shape and surface morphology	Transmission electron microscopy, scanning electron microscopy, phase contrast microscopy, optical microscopy.
Vesicle size and size distribution	Transmission electron microscopy, optical microscopy.
Drug release	Diffusion cell and dialysis method
Drug content	Deproteinization of cell membrane followed by assay of resealed drug by spectrophotometric method at the wavelength of 540nm.
Surface electrical potential	Zeta potential measurement by Zeta meter
Surface pH	Using pH-sensitive probes, by pH meter
Deformability	Capillary rise method
II. Cellular characterization	
% Hb content	Deproteinization of cell membrane followed by hemoglobin assay, by hemoglobinometer.
Cell volume	Laser light scattering method
% Cell recovery	Neubaur's chamber, hematological analyzer
Osmotic fragility	Stepwise incubation with isotonic to hypotonic saline solutions followed by centrifugation and determination of drug content and hemoglobin assay
Osmotic shock	Dilution with distilled water and estimation of drug and hemoglobin by spectrophotometric method
Turbulent shock	Passage of cell suspension through 30- gauge hypodermic needle at 10 mL/min flow rate and estimation of residual drug and hemoglobin, vigorous shaking followed by hemoglobin estimation
Erythrocyte sedimentation rate	ESR methods
III. Biological characterization	
Sterility	Sterility test
Pyrogenicity	Rabbit method, LAL test
Animal toxicity	Toxicity tests

The morphology of erythrocytes decides their life span after administration. Shape change (deformability) is another factor that affects the life span of the cells. This parameter evaluates the ease of passage of erythrocytes through narrow capillaries and the RES. It determines the rheological behavior of the cells and depends on the viscoelasticity of the cell membrane, viscosity of the cell contents, and the cellular surface-to-volume ratio. The osmotic fragility of resealed erythrocytes is an indicator of the possible changes in cell membrane integrity and the resistance of these cells to osmotic pressure of the suspension medium. The test is carried out by suspending cells in media of varying sodium chloride concentration and determining the hemoglobin released. In most cases, osmotic fragility of resealed cells is higher than that of the normal cells because of increased intracellular osmotic pressure.

The turbulence fragility is yet another characteristic that depends upon changes in the integrity of cellular membrane and reflects resistance of loaded cells against hemolysis resulting from turbulent flow within circulation. Routine clinical hematological tests also can be carried out for drug-loaded cells, including mean corpuscular volume, and mean corpuscular hemoglobin content. Drug content of the cells determines the entrapment efficiency of the method used. The process involves deproteinization of packed i.e., loaded cells (0.5 mL) with 2.0 mL acetonitrile and centrifugation at 2500 rpm for 10 min. The clear supernatant is analyzed for the drug content.

The most important parameter for evaluation of resealed erythrocytes is the drug release pattern. Drug release involves the loss of cell membrane integrity indicating hemolysis. The rate of drug release is comparable to that of hemoglobin. This indicates that cell lysis is essential for drug release and drug can not be released by mere diffusion.

7.5 *In-Vitro* Storage

The success of resealed erythrocytes as a drug delivery system depends to a greater extent on their *in-vitro* storage. Preparing drug-loaded erythrocytes on a large scale and maintaining their survival and drug content can be achieved by using suitable storage methods, which include Hank's balanced salt solution and acid-citrate-dextrose at 4^0C in which cells remain viable in terms of their physiologic and carrier characteristics for at least 2 weeks. The addition of calcium-chelating agents or the purine nucleosides improve circulation survival time of cells upon injection.

Exposure of resealed erythrocytes to membrane stabilizing agents such as dimethyl sulfoxide, dimethyl,3,3-di-thio-bispropionamide, gluteraldehyde, toluene-2-4-diisocyanate followed by lyophilization or sintered glass filtration has been reported to enhance their stability upon storage. The resultant powder was stable for at least one month without any detectable changes. But the major disadvantage of this method is the presence of appreciable amount of membrane stabilizers in bound form that remarkably reduces circulation survival time. Other reported methods for improving storage stability include encapsulation of a prodrug that undergoes conversion to the parent drug only at body temperature high glycerol freezing technique, and reversible immobilization in alginate or gelatin gels.

7.6 *In-vivo* Life Span

The efficacy of resealed erythrocytes is determined mainly by their survival time in circulation upon injection. For the purpose of sustained action, a longer life span is required, although for delivery to target-specific RES organs, rapid phagocytosis and hence a shorter life span is desirable. The life span of resealed erythrocytes depends upon its size, shape, and surface electrical charge as well as the extent of hemoglobin and other cell constituents lost during the loading process.

The various methods used to determine *in-vivo* survival time include labeling of cells by ^{51}Cr or fluorescent markers such as fiuorescin isothiocyanate or entrapment of ^{14}C sucrose or gentamicin.

The circulation survival kinetics of resealed erythrocytes show typical bimodal behavior with a rapid loss of cells during the first 24 h after injection, followed by a slow decline phase with a half life on the order of days or weeks.

7.7 Applications of Resealed Erythrocytes

Resealed erythrocytes have several possible applications in various fields of human and veterinary medicine. Such cells could be used as circulating carriers to disseminate a drug within a prolonged period of time in circulation or in target-specific organs, including the liver, spleen, and lymph nodes. A majority of the drug delivery studies using drug-loaded erythrocytes are in the preclinical phase. In a few clinical studies, successful results were obtained. The mechanisms proposed for drug release include: passive diffusion; specialized membrane associated carrier transport; phagocytosis of resealed cells by macrophages of RES and subsequent accumulation of drug into the macrophage interior followed by slow release; accumulation of erythrocytes in lymph nodes upon subcutaneous administration followed by hemolysis to release the drug.

Routes of administration include intravenous, which is the most common, followed by subcutaneous, intraperitoneal, intranasal, and oral. Studies regarding the improved efficacy of various drugs given in this form in animal models have been published. Examples include: an enhancement in anti-inflammatory effect of corticosteroids in experimentally inflamed rats; increase in half life of isoniazid, levothyroxine, cytosine arabinoside and interlukin-2; prolongation of plasma half life of erythropoietin from 30 min to 35 h in mice; and can increase in mean survival time of mice with experimental hepatocarcinoma after injecting methotrexate loaded erythrocyte. A list of various bioactive agents encapsulated in erythrocytes along with their applications mentioned in the Table 7.2.

Table 7.2 Applications of Resealed Erythrocytes.

Bioactive Agent Encapsulated	Applications	References
Acetaldehyde dehydrogenase	Removal of blood acetaldehyde	Lagurre et al., 1991
Adenylosuccinate lyase	Enzyme replacement therapy	Salerno & Crifo, 1994)
Adriamycin	Cancer chemotherapy	Benatti et al., 1987
Albumin	Prolong release in circulation	Deloach et al., 1979
L-Asparagine	Cancer chemotherapy	Kravtzoff et al., 1991
Aspirin	Local prevention of thrombosis	Orekhova et al., 1990
β-galactosidase	Enzyme replacement therapy	Ihler et al., 1973
Brinase	Thrombolytic therapy	Flynn et al., 1994
Cortical-2-phosphate	Enzyme replacement therapy	Pitt et al., 1983
Deferroxime	Treatment of iron overload	Green et al., 1980
5-fluoro-2'-deoxyuridine	Cancer chemotherapy	Zocchi et al., 1988
Diclofenac sodium	Targeting to inflamed area	Jain & Vyas, 1994a
2'-3'-dideoxycystidine	Inhibition of HIV infection	Faternal et al., 1994,
Doxorubicin	Cancer Chemotherapy	Zocchi et al., 1991b
Glucocerebrosidase	Treatment of Goucher's disease	Dale et al., 1979, Zocchi et al., 1987
Heparin	Prevention of thromboembolism	Eichler et al., 1986a
Human recombinant interleukin-2	Immunotherapy	Mitchell et al., 1990
Inositol hexaphosphate	Inprovement of Oxygen Delivery	Ihler & Tosi, 1987, Ramo et al., 1994
Inulin	Prolonged release in circulation	Deloach et al., 1987a, Droleskey, 1982
Insulin	Treatment of Hyperglycemia	Fiddler et al., 1980; Bird et al., 1983
Luciferase	ATP monitoring	Vitvitsky, 1992
Meglumine antimonate	Targeting to RES	Jain & Jain, 1996a
Methotrexate	Cancer chemotherapy	Kruse et al., 1987a, b
Mycotoxin	Targeting to RES	Deloach et al., 1988a
Pentamidine	Treatment of leshmaniasis	Pei et al., 1994a
Propranolol	Prevention of hypertension	Alper & Irwin., 1987
Recombinant interleukin-2	Prolonged release in circulation	Mitchell et al., 1990
Urease	Enzyme deficiency	Ihler,1975

Advantages of this method include quantitative injection of materials into cells, simultaneous introduction of several materials into a large number of cells, minimal damage to the cell, avoidance of degradation effects of lysosomal enzymes, and simplicity of the technique. Disadvantages include a need for a larger size of fused cells, thus making them amenable to RES clearance; unpredictable effects on cell resulting from the co-introduction of various components. Hence, this method is limited to mainly cell biological applications rather than drug delivery.

Conclusion

The use of resealed erythrocytes looks promising for a safe and sure delivery of various drugs for passive and active targeting. However, the concept needs further optimization to become a routine drug delivery system. The same concept also can be extended to the delivery of biopharmaceuticals and much remains to be explored regarding the potential of resealed erythrocytes.

References

A.V. Gothoskar. Resealed Erythrocytes: A Review. *Pharma Technology,* 2004:140-158.

A.C. Guyton and J. E. Hall. Red Blood Cells,Anemia and Polycytemia," in *Textbook of Medical Physiology* (W.B. Saunders, Philadelphia, PA,1996, pp. 425–433. *Adv. Biosci.* (series), 1987 (67):11–15.

Al-Achi, A. and Boroujerdi, M. Pharmacokinetics and tissue uptake ofdoxorubicin associated with erythrocyte-membrane: erythrocytes-ghosts vs.erythrocytes-vesicles. *Drug Dev. Ind. Pharm.*, 1990 (16): 2199–2219.

Alpar, H.O. and Irwin, W.J. Some Unique Applications of Erythrocytes as Carrier Systems. *Adv. Biosci.* (Series), 1987 (67):1–9.

Alvarez, F.J., Jordán, J.A., Calleja, P., Lotero, L.A. , Olmos, G., Díez, J.C., Tejedor, M.C. Cross-Linking Treatment of Loaded Erythrocytes Increases delivery of Encapsu-lated Substance to Macrophages. *Biotechnol. Appl. Biochem.* 1998, 27 (2): 139–143.

Alvarez-Guerra, M. Nazaret, C. and Garay, R.P. The Erythrocyte Na, K, Cl, CoTransporter and Its Circulating Inhibitor in Dahl Salt- Sensitive rats. *J Hypertens*. 1998(10):1499–1504.

Bhaskaran, S. and Dhir, S.S. Resealed Erythrocytes as Carriers of Salbutamol Sulphate. *Indian J. Pharm. Sci*. 1995 (57): 240–242.

D.A. Lewis and H.O. Alpar, "Therapeutic Possibilities of Drugs Encapsulated in Erythrocytes," *Int. J. Pharm*. 1984 (22):137–146.

D.A. Lewis,"Red Blood Cells for Drug Delivery. *Pharm. J,* 1984 (233): 384–385.

De Loach, R. Encapsulation of Exogenous Agents in Erythrocytes and the Circulating Survival of Carrier Erythrocytes. *J. Appl. Biochem*., 1983, 5 (3): 149–157.

G.M. Iher, R.M. Glew, and F.W. Schnure, "Enzyme Loading of Erythrocytes," *Proc. Natl. Acad. Sci. USA,* 1973 (70): 2663–2666.

Gaudreault, B., Bellemare, Lacroix, J. Erythrocyte Membrane- Bound Daunorubicin as a Delivery System in Anticancer Treatment. *Anticancer Res.,* 1989, 9 (4):1201-5.

Gautam S, Barna B, Chiang T, Pettay J, Deodhar S. Use of resealed erythrocytes as delivery system for C-reactive protein (CRP) to generate macrophage-mediated tumoricidal activity. *J Biol Response Mod*. 1987, 6(3):346-54.

Gutierrez Millan, C., Zarzuelo Castaneda, A.,Sayalero Marinero, M.L., Lanao, J.M. Factors associated with the performance of carrier erythrocytes obtained by hypotonic dialysis. *Blood Cells Mol. Dis*., 2004 (33): 132–140.

H.O.Alpar and D.A. Lewis, "Therapeutic Efficacy of Asparaginase Encapsulated in Intact Erythrocytes," *Biochem. Pharmacol.* 1985 (34): 257–261.

Hamidi, M. and Tajerzadeh, H. Carrier Erythrocytes: An Overview. *Drug Delivery*, 2003 (10): 9–20.

Hamidi, M., Tajerzadeh, H. Carriererythrocytes: an overview. *Drug Deliv.,* 2003(10): 9–20.21.

Hamidi, M., Tajerzadeh, H., Dehpour, A. R., Ejtemaee-Mehr, S. ACE Inhibition in Rabbits Upon Administration of Enalaprilat-Loaded Intact Erythrocytes. *J. Pharm. Pharmacol.*, 2001 (53):1281–1289.

Hamidi, M., Tajerzadeh, H., Dehpour, A.R.,Rouini, M.R., Ejtemaee-Mehr, S. InVitro Characterization of Human Intact Erythrocytes Loaded by Enalaprilat. *Drug Delivery*, 2001 (8): 231–237.

Hamidi, M., Zarei, N., Zarrin, A. H., Mohammadi-Samani, S. Preparation and in vitro characterization of carrier erythrocytes for vaccine delivery. *Int. J. Pharm.* 2007 (338):70-78.

Ingrosso D, Cotticelli M G, D'Angelo S, Buro MD, Zappia V, Galletti P. Influence of osmotic stress on protein methylation in resealed erythrocytes. *Eur J Biochem.* 1997, 244(3):918-22.

Jain, S. and Jain, N.K. Engineered Erythrocytes as a Drug Delivery System. *Indian J. Pharm. Sci.*, 1997 (59): 275–281.

Jain, S., Jain, S. K., Dixit, V.K. Erythrocytes Based Delivery of Isoniazid: Preparation and In Vitro Characterization. *Indian Drugs*, 1995 (32): 471–476.

Jain, S., Jain, S.K., Dixit, V.K. Magnetically Guided Rat Erythrocytes Bearing Isoniazid: Preparation, Characterization, and Evaluation. *Drug Dev. Ind. Pharm.*, 1997 (23):999–1006.

Jain, S.K. and Vyas, S.P. Magnetically Responsive Diclofenac Sodium-Loaded Erythrocytes: Preparationand In Vitro Characterization, *J. Microencapsul*, 1994, 11 (2): 141–151.

Jaitely, V., Kanaujia, P., Venkatesan, N., Jain, S., Vyas, S.P. Resealed erythrocytes: carrier potentials and biomedical application. *Indian Drugs*, 1996 (33): 549–589.

Kravtzoff, R., Ropars, C., Laguerre, M.,Muh, J.P., Chassaigne, M. Erythrocytes as Carriers for L-Asparaginase:Methodological and Mouse In-Vivo Studies.J. Pharm. Pharmacol. 1990 (42): 473–476.

Lejeune, A., Poyet, P., Gaudreault, R.C.,Gicquaud, C. Nanoerythrosomes, A New Derivative of Erythrocyte Ghost: III. Is Phagocytosis Involved in the Mechanism of Action?. Anticancer Res., 1997 (17): 5A.

Lewis, D.A. Red Blood Cells for Drug Delivery. *Pharm. J.* 1985 (233): 384–385.

Li, L.H., Hensen, M.L., Zhao, Y.L., Hui, S.W. Electrofusion between Heterogeneous-Sized Mammalian Cells in aPellet: Potential Applications in Drug Delivery and Hybridoma Formation. *Biophys J.* 1996, 71 (1): 479–486.

Magnani, M. and Rossi, L. Erythrocyte Engineering for Drug Delivery and Targeting. *Biotechnol. Appl. Biochem.*, 1998 (28): 1–13.

Magnani, M., Bianchi, M., Rossi, L., Stocchi, V. Acetaldehyde Dehydrogenase-Loaded Erythrocytes as Bioreactors for Removal of Blood Acetaldehyde, Alcoholism. *Clin. Exp. Res.* 1989 (13): 849–859.

Mangal, P.C., and Kaur, A. Electroporation of Red Blood Cell Membrane and its Use as a Drug Carrier System, *Ind. J. Biochem. Biophys*. 1991, 28 (3):219-221.

Mitchell, D. H., James, G.T., and Kruse , C.A. Bioactivity of Electric Field-Pulsed Human Recombinant Interleukin-2 and Its Encapsulation into Erythrocyte Carriers. *Biotechnol. Appl. Biochem.*, 1990, 12 (3): 264–275.

Moorjani, M., Lejeune, A., Gicquaud, C.,Lacroix, J., Poyet, P., Gaudreault, R.C. Nanoerythrosomes, A New Derivative of Erythrocyte Ghost II: Identification of the Mechanism of Action, *Anticancer Res.* 1996, 16 (5A): 2831–2836.

N. Talwar and N.K. Jain,"Erythrocytes as Carriers of Primaquin Preparation: Characterization and Evaluation," *J. Controlled Release,* 1992 (20):133–142.

Pei , L., Omburo, G., Mc Guinn, W.D., Petrikovics, I. , Dave, K., Raushel, F.M.,Wild, J.R., DeLoach, J.R., Way, J.L. Encapsulation of Phosphotriesterase withinMurine Erythrocytes. *Toxicol. Appl. Pharmacol.* 1994, 124 (2): 296–301.

Pitt, E., Johnson, C.M., Lewis, D.A., Jenner, D. A., Offord, R. E. Encapsulation of Drugs in Intact Erythrocytes: An Intravenous Delivery System. *Biochem. Pharmacol.* 1983 (22): 3359–3368.

Price, R. J., Skyba, D.M., Kaul, S., Skalak, T.C. Delivery of Colloidal Particles and Red Blood Cells to Tissue through Micro-vessel Ruptures Created by Targeted Micro-bubble Destruction with Ultrasound, *Circulation*, 1998, 98 (13): 1264–1267.

Rossi, L. and Magnani, M. (1996).Encapsulation,Metabolism, and Release of2-Fluoro-Ara-AMP from HumanErythrocytes. *Biochim. Biophys. Acta.* 1291(2): 149–154.

Rossi, L., Serafini, S., Pierige, F., Antonelli,A., Cerasi, A., Fraternale, A., Chiarantini, L.and Magnani, M. Erythrocyte-based delivery. *Exp. Opin. Deliv.,*2005 (2): 311–322.

S. Jain and N. K. Jain. Resealed Erythrocytes as Drug Carriers. Controlled and Novel Drug Delivery. 1st Edn, CPS Publishers, 2008: 256-291.

S. Jain and N.K. Jain, "Engineered Erythrocytes as a Drug Delivery System," *Indian J. Pharm. Sci.*1997 (59): 275–281.

S. J. Updike and R. T. Wakamiya, "Infusion of Red Blood Cell-Loaded Asparaginase in Monkey," *J. Lab. Clin.Med.* 1983 (101): 679–691.

S.P. Vyas and R.K. Khar, *Resealed Erythrocytes in Targeted and Controlled Drug Delivery: Novel Carrier Systems* (CBS Publishers and Distributors, India, 2002: 87–416.

Schrier, S. L., Junga, I. , Johnson, M. Energized Endocytosis in Human Erythrocyte Ghosts. *J. Clin. Invest.* 1975, 56 (1): 8–22.

Tajerzadeh, H. and Hamidi, M. Evaluation of the Hypotonic Preswelling Method for Encapsulation of Enalaprilat in Human Intact Erythrocytes. *Drug Dev Ind. Pharm.*, 2000 (26): 1247–1257.

Talwar, N. and Jain, N.K. Erythrocytes as Carriers of Primaquin Preparation: Characterization and Evaluation. *J. Controlled Release*, 1992 (20): 133–142.

Talwar, N., and Jain, N.K. Erythrocytes as Carriers of Metronidazole: In-Vitro Characterization. *Drug Dev. Ind. Pharm.*, 1992 (18):1799–1812.

U. Zimmermann, *Cellular Drug-Carrier Systems and Their Possible Targeting In Targeted Drugs*, EP Goldberg, Ed., John Wiley & Sons, New York, 1983:153–200

V. Jaitely et al., "Resealed Erythrocytes: Drug Carrier Potentials and Biomedical Applications," *Indian Drugs*, 1996 (33): 589–594.

V. S. Gopal, A. Ranjith Kumar, A. N. Usha, A. Karthik, N. Udupa. Effective drug targeting by Erythrocytes as Carrier Systems. *Current Trends in Biotechnology and Pharmacy,* 2007, 1 (1): 18-33.

Vyas, S. P. and Dixit, V. K. Pharmaceutical Biotechnology-1 (CBS Publishers & Distributors, New Delhi),1999:655.

Vyas, S. P. and Khar, R. K. Resealed Erythrocytes in Targeted and Controlled Drug Delivery: Novel Carrier Systems, CBS Publishers and Distributors, India, 2002: 387-416.

Jagadale V.L., Aloorkar N.H., Dohe G.H. and Gehlot M.V. Resealed Erythrocytes: An Effective Tool in Drug Targeting. *Indian J. Pharm. Educ. Res.*, 2009, 43(4): 375-383.

7A
Liposome as Therapeutic Drug Career

Introduction

Liposomes have been widely investigated since 1970 as drug carriers for improving the delivery of therapeutic agents to specific sites in the body. As a result, numerous improvements have been made, thus making this technology potentially useful for the treatment of certain diseases in the clinics. The range of medical applications of liposomes extends from chemotherapy of cancer; fungal infections to vaccines and most recently to gene therapy. The success of liposomes as drug carriers has been reflected in a number of liposome-based formulations, which are commercially available or are currently undergoing clinical trials. The current pharmaceutical preparations of liposome-based therapeutic systems mainly result from our understanding of lipid-drug interactions and liposome disposition mechanisms. The insight gained from clinical use of liposome drug delivery systems can now be integrated to design liposomes that can be targeted on tissues, cells or intracellular compartments with or without expression of target recognition molecules on liposome membranes.

7A.1 Liposomes

Liposomes are lyotropic liquid crystals composed of relatively biocompatible and biodegradable materials and consist of an aqueous core entrapped in one or more bilayered natural and/or synthetic lipids. Drugs with widely varying lipophilicities can be encapsulated in liposomes either in the phospholipid bilayer and/or entrapped in the aqueous core. The most common phospholipid used is phosphatidylcholine (PC) molecule. PC is an amphipathic molecule in which a glycerol bridge links a pair of hydrophobic acyl-hydrocarbon chains, with a hydrophilic polar head groups, phospho choline (Fig. 7A.1).

Formulation of drugs in liposomes has provided an opportunity to enhance the therapeutic indices of various agents mainly through the alteration of bio-distribution. They are versatile drug carriers, which can be used to control retention of entrapped drugs in the presence of biological fluids, controlled vesicle residence in the systemic circulation or other compartments in the body and enhanced vesicle uptake by target cells. Liposomes composed of natural lipids are biodegradable, biologically inert, weakly immunogenic, produce no antigenic or pyrogenic reactions and possess limited intrinsic toxicity. Therefore, drugs encapsulated in liposomes are expected to be transported without rapid degradation and minimum side effects to the recipients.

The molecules of PC are not soluble in water and in aqueous media, they align themselves closely in planner bilayer sheets in order to minimize the unfavorable action between the bulk aqueous phase and the long hydrocarbon fatty chain. Such interactions are completely eliminated, when the sheets fold on themselves to form closed sealed vesicle (Fig. 7A.2). It is proposed that the double fatty acid chain gives the molecule an overall tubular shape, which leads to a spherical micellar structure (Fig. 7A.1).

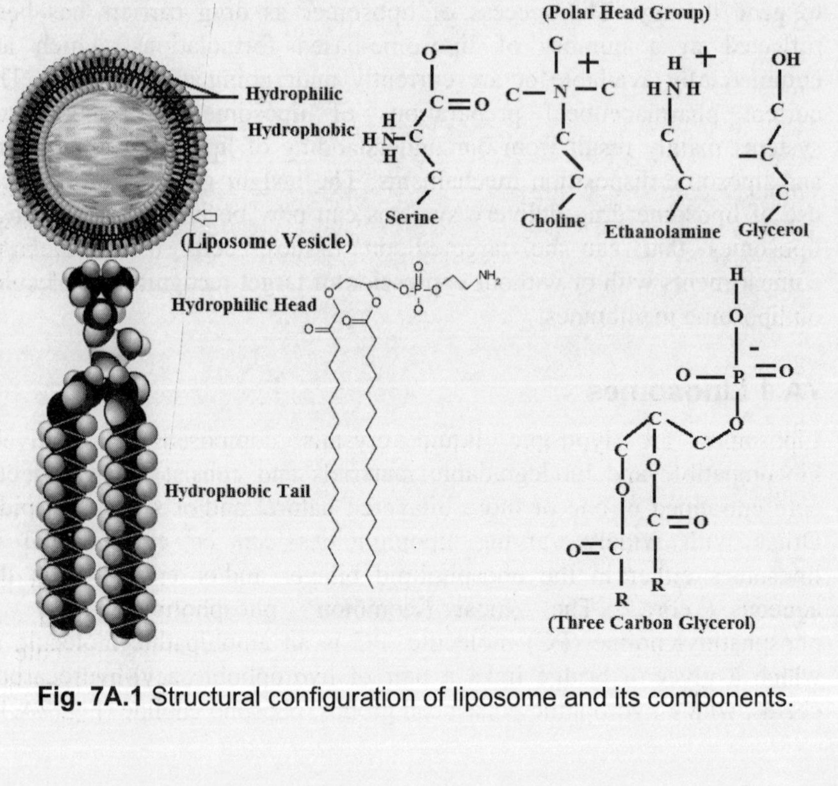

Fig. 7A.1 Structural configuration of liposome and its components.

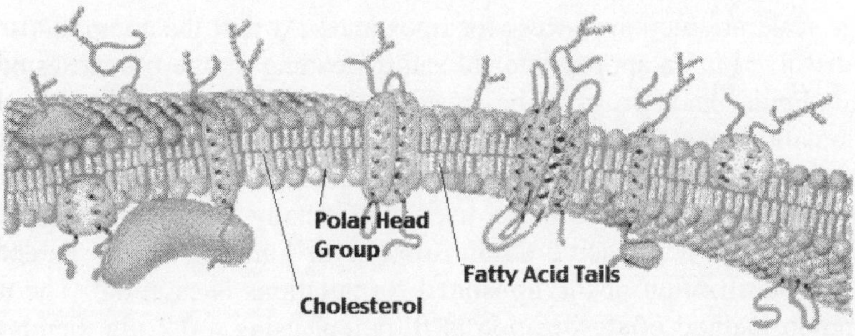

Fig. 7A.2 Schematic representation of Cell membrane of Liposome)

7A.2 Structural Properties of Liposomes

The diameter of liposomes ranges typically from 50 to 5000 nm. Liposomes are self-assembling structures that resemble cell membranes. Small unilamellar vesicles (SUV) have a single bilayer and a diameter of 50 to 200 nm, while large unilamellar vesicles (LUV) have a diameter range from 200 to 800 nm. Multilamellar vesicles (MLV) are composed of several concentric lipid bilayers and their diameter ranges from 500 to 5 000 nm. The type of liposomes depends on the preparation method and the lipid composition. Physical properties of the liposomes can be readily modified. Size, charge, rigidity, and surface properties can be changed by using different liposome preparation methods and lipid compositions. Liposomes with a net surface charge can be prepared by using charged lipids, such as phosphatidyl-glycerol (PG). Many liposomes exhibit structural phase transition from the gel phase to the liquid crystalline phase, when temperature is increased. Such phase transition increases the permeability of the liposomal membrane to the drug, and the transition temperature (T) can be adjusted by varying the lipid composition. In general, longer alkyl chains increase T and saturated lipids exhibit higher T than unsaturated ones. Phosphatidyl-ethanolamines (PE) lipids have higher T compared to PC with same acyl chains. Cholesterol is often used to increase the liposomal stability. Also other materials, such as polymers can be incorporated into the liposomes.

7A.3 Preparation of Liposomes

7A.3.1 Large Scale Production

The industrial manufacturing procedure of liposomes is complex and consists of several steps. Figure 7A.3, provides an overview on a standard large scale production process for liposomes. At first the compounds will be dissolved in an appropriate solvent depending on the properties of the used lipids and drugs. The residual solvent concentration in the formulation must be reduced by an elimination step before continuing the processing of the liposomal formulation. A variety of hydration and homogenization steps can be used and finally, the sterilization and stabilization. For invasive administration in humans by the parenteral route a sterilization of the liposomal formulations is essential. The most common method of sterilization is filtration, using a 0.2 µm membrane. Also gamma-irradiation, heat sterilization of the end product by autoclaving at 121°C or steam sterilization has been used for the sterilization of liposomes.

Organic solvents are often employed in liposomal production processes to solubilize, either the lipids or the drug. The elimination of such organic solvents and the removal of non-encapsulated drugs are often required for the purification of the final product. In several cases this represents the limiting factor for scalability of the process. Additionally, to avoid stability problems like sedimentation or leakage of liposomes over the storage time, manufacturing for uniform liposomal size distributions by homogenization is particularly important. For certain applications it is necessary to achieve certain liposome sizes e.g. multi-lamellar (MLV), small uni-lamellar (SUVs), large uni-lamellar (LUVs) or multi-vesicular vesicles (MVVs). The effect of liposome size can be important in respect of drug loading or the circulation time accumulation behavior after application to the patient.

Drug loading can be achieved either passively (i.e. the drug is to be entrapped during liposome formation) or actively (i.e. after liposome formation). Hydrophobic drugs, such as: amphotericin-B. taxol or kanamycin, can be directly incorporated into liposomes during vesicle formation, and the extent of uptake and retention is governed by drug-lipid interactions. Trapping efficiencies of 100% are often achievable, but this is dependent on the solubility of the drug in the liposome membrane. Passive encapsulation of water-soluble drugs relies on the ability of liposomes to trap aqueous buffer containing a dissolved drug during vesicle formation. Trapping efficiencies (generally <30%) are limited by the trapped volume contained in the liposomes and drug solubility. Alternatively, water-soluble drugs that have protonizable amine functions can be actively entrapped by employing pH gradients, which can result in increasing trapping efficiencies approaching 100%.

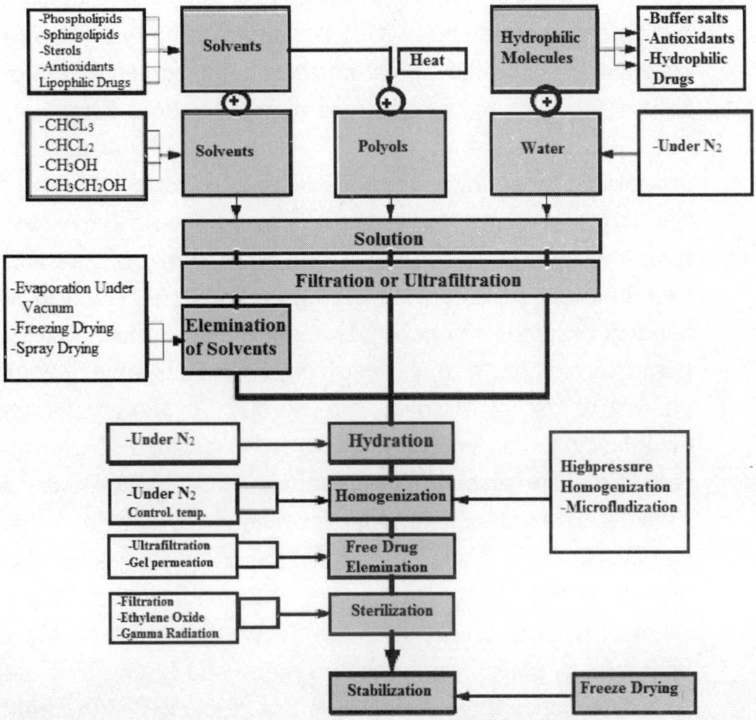

Fig. 7A.3 Schematic for a Large-Scale production of Liposome.

Ref: Redzinik et at., 1995

7A.3.2 Preparation of Liposome by Lipid Hydration Method

Properties of lipid formulations can vary depending on the composition (cationic, anionic or neutral lipid species), however, the same preparation method can be used for all lipid vesicles regardless of composition. The general elements of the procedure involve preparation of the lipid for hydration, hydration with agitation, and sizing to a homogeneous distribution of vesicles.

(a) ***Preparation of Lipid for Hydration*:** To obtain a clear lipid solution preparing liposomes with mixed lipid composition, the lipids must first be dissolved and mixed in an organic solvent to assure a homogeneous mixture of lipids. Usually this process is carried out using chloroform or chloroform: methanol for complete mixing of lipids. Typically lipid solutions are prepared at 10-20mg lipid/ml organic solvent, although higher concentrations may be used if the lipid solubility and mixing are acceptable. Once the lipids are thoroughly mixed in the organic solvent, the solvent is removed to yield a lipid film. For small volumes of organic solvent (<1mL), the solvent may be evaporated using a dry nitrogen or argon stream in a fume hood. For larger volumes, the organic solvent should be removed by rotary evaporation yielding a thin lipid film on the sides of a round bottom flask. The lipid film is thoroughly dried to remove residual organic solvent by placing the vial or flask on a vacuum pump overnight. If the use of chloroform is objectionable, an alternative is to dissolve the lipid(s) in tertiary butanol or cyclohexane. The lipid solution is to be transferred to containers and frozen by placing the containers on a block of dry ice or swirling the container in a dry ice-acetone or alcohol (ethanol or methanol) bath. Care should be taken when using the bath procedure that the container can withstand sudden temperature changes without cracking. After freezing completely, the frozen lipid cake is placed on a vacuum pump and lyophilized until dry (1-3 days depending on volume). The thickness of the lipid cake should be no more than the diameter of the container being used for lyophilization. Dry lipid films or cakes can be removed from the vacuum pump, the container close tightly and taped, and stored frozen until ready to hydrate.

(b) **Hydration of Lipid Film/Cake:** Hydration of the dry lipid film/cake is accomplished simply by adding an aqueous medium to the container of dry lipid and agitating. The temperature of the hydrating medium should be above the gel-liquid crystal transition temperature (Tc or Tm) of the lipid with the highest Tc before adding to the dry lipid. After addition of the hydrating medium, the lipid suspension should be maintained above the Tc during the hydration period. For high transition lipids, this is easily accomplished by transferring the lipid suspension to a round bottom flask and placing the flask on a rotary evaporation system without a vacuum. Spinning the round bottom flask in the warm water bath maintained at a temperature above the Tc of the lipid suspension allows the lipid to hydrate in its fluid phase with adequate agitation. Hydration time may differ slightly among lipid species and structure, however, a hydration time of 1 hour with vigorous shaking, mixing, or stirring is highly recommended. It is also believed that allowing the vesicle suspension to stand overnight (aging) prior to downsizing makes the sizing process easier and improves the homogeneity of the size distribution. Aging is not recommended for high transition lipids as lipid hydrolysis increases with elevated temperatures. The hydration medium is generally determined by the application of the lipid vesicles. Suitable hydration media include distilled water, buffer solutions, saline, and nonelectrolytes such as sugar solutions. Physiological osmolality (290mOsm/kg) is recommended for *in-vivo* applications. Generally accepted solutions, which meet these conditions, are 0.9% saline, 5% dextrose, and 10% sucrose. During hydration some lipids form complexes unique to their structure. Highly charged lipids have been observed to form a viscous gel when hydrated with low ionic strength solutions. The problem can be alleviated by addition of salt or by downsizing the lipid suspension. Poorly hydrating lipids such as phosphatidylethanolamine have a tendency to self aggregate upon hydration. Lipid vesicles containing more than 60 mol% phosphatidylethanolamine form particles having a small hydration layer surrounding the vesicle. As particles approach one another there is no hydration or repulsion to repel the approaching particle each other and the two membranes fall into an energy, where they adhere and form aggregates. The aggregates settle out of solution as large flocculates which will disperse on agitation but reform upon sitting, which results in production of large, multilamellar vesicle (LMV) with each lipid bilayer separated by a water layer. The spacing between lipid layers is dictated by composition with poly-hydrating layers being closer together than highly charged layers which separate based on electrostatic repulsion. Once a stable, hydrated LMV suspension has been produced, the particles can be downsized by a variety of techniques, including sonication or extrusion.

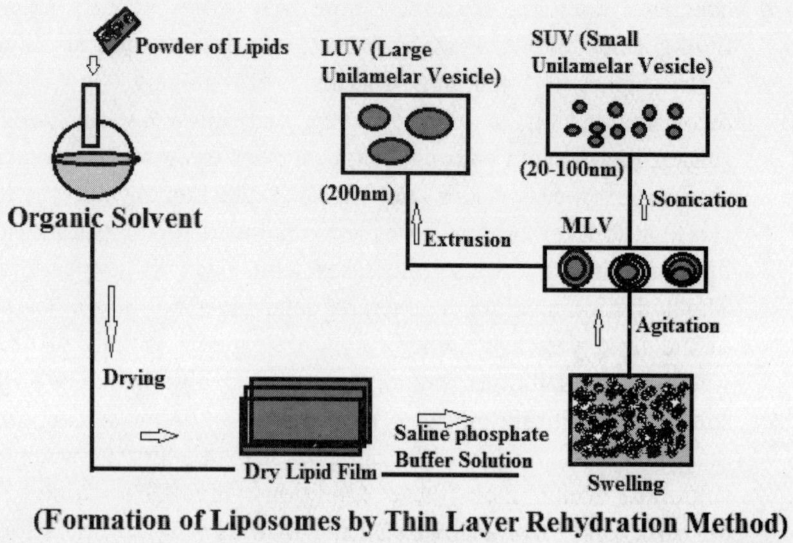

Fig. 7A.4 Formation of Liposome by Thin Layer Rehydration Method.

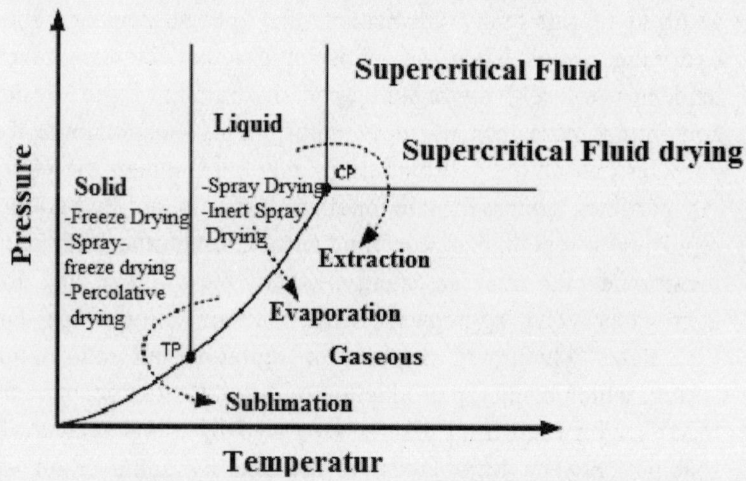

Fig. 7A.5 Schematic pressure-temperature diagram with tripple point (TP) and the different phases. The investigating drying pathways from the liquid phase over solid (freeze-drying, spray freeze-drying, percolative vacuum drying) and supercritical phase (supercritical fluid drying) and the direct evaporation (spray-drying and intent spray drying)

(c) Sizing of Lipid Suspension

(i) ***Sonication*:** Disruption of LMV suspensions using sonic energy (sonication) typically produces small, unilamellar vesicles (SUV) with diameters in the range of 15-50nm. The most common instrumentation for preparation of sonicated particles is bath and probe tip sonicators. Cup-horn sonicators can successfully produced SUV. Probe tip sonicators deliver high energy input to the lipid suspension but suffer from overheating of the lipid suspension causing degradation. Sonication tips also tend to release titanium particles into the lipid suspension which must be removed by centrifugation prior to use. For these reasons, bath sonicators are the most widely used instrumentation for preparation of SUV. Sonication of an LMV dispersion is accomplished by placing a test tube containing the suspension in a bath sonicator (or placing the tip of the sonicator in the test tube) and sonicating for 5-10 minutes above the Tc of the lipid. The lipid suspension should begin to clarify to yield a slightly hazy transparent solution. The haze is due to light scattering induced by residual large particles remaining in the suspension. These particles can be removed by centrifugation to yield a clear suspension of SUV. Mean size and distribution is influenced by composition, concentration, temperature, sonication time, volume, and sonicator tuning. Since it is nearly impossible to reproduce the conditions of sonication, size variation between batches produced at different times is not uncommon. Also, due to the high degree of curvature of these membranes, SUV are inherently unstable and will spontaneously fuse to form larger vesicles when stored below their phase transition temperature.

(ii) ***Extrusion*:** Lipid extrusion is a technique in which a lipid suspension is forced through a polycarbonate filter with a defined pore size to yield particles having a diameter near the pore size of the filter used. Prior to extrusion through the final pore size, LMV suspensions are disrupted either by several freeze-thaw cycles or by prefiltering the suspension through a larger pore size (typically 0.2µm-

1.0μm). This method helps prevent the membranes from fouling and improves the homogeneity of the size distribution of the final suspension. As with all procedures for downsizing LMV dispersions, the extrusion should be done at a temperature above the Tc of the lipid. Attempts to extrude below the Tc will be unsuccessful as the membrane has a tendency to foul with rigid membranes which cannot pass through the pores. Extrusion through filters with 100nm pores typically yields large unilamellar vesicles (LUV) with a mean diameter of 120-140nm. Mean particle size also depends on lipid composition and is quite reproducible from batch to batch.

7A.3.3 Preparation of Multi-lamellar Liposomes (MLV)

Following is a brief outline of the protocol for lyophilizing (freeze-drying) lipid mixtures for preparation of liposomes: Dissolve the lipids in chloroform, combine the lipids in the appropriate ratio, carefully evaporate the organic solvent using a dry nitrogen stream, resuspend the lipid mixture in cyclohexane. If the mixture is not completely soluble in the cyclohexane, add a small amount of ethanol (1-2% of the cyclohexane volume). Do not use too much ethanol, as the solution will not freeze with excessive ethanol presence. Freeze the cyclohexane solution using dry ice. Quickly place the frozen mixture on a high vacuum system (lyophilization system). The sample should remain frozen until it is completely dry (if the sample begins to thaw, indicate either the vacuum is not strong enough or there is too much ethanol present). A thawed sample will not produce a white powder and it may bump or foam out of the vial. Leave the sample on the vacuum system for 3-5 hours (depending on how much cyclohexane was used) or until the sample is completely dry (the vial should not feel cold to the touch or smell of cyclohexane, when removed from the vacuum).

(*Note*: This produces a dry, white powder which readily suspends in water. An alternative method is to resuspend the lipid film produced by evaporating the chloroform using the appropriate aqueous buffer).

Suspend the lipid mixture in the aqueous buffer (buffer temperature should be above the phase transition of the lipid) and allow the mixture to hydrate above the transition temperature of the lipid for 30-60 minutes (vortexing occasionally). This procedure yields large, multilamellar vesicles (LMV or MLV). (For water soluble compounds to be entrapped, the same protocol is followed except the compound is dissolved in the aqueous buffer used to reconstitute the dry lipid and the external compound, which are not encapsulated is remove by gel filtration).

Extrusion of multilamellar liposomal suspensions using membranes with a pore size >0.2μm does not produce unilamellar liposomes. Liposomes produced with larger pore membranes will yield a polydisperse suspension of multilamellar liposomes. Unilamellar liposomal suspensions with a low polydispersity can only be prepared with membranes having a pore size of ≤0.2μm.

7A.4 Stability of Liposomal Formulation

More recent investigations revealed an advantage of homogenous unilamellar vesicles in the size range of 50 to 200 nm. Liposome stability decreases with increasing size and shows an optimum at a size between 80 and 200 nm. The selection of an appropriate size is always a compromise between loading efficiency, which increases with liposome size, and the resulting decline in stability. Physical and chemical stability of liposomes can be strengthened by choosing a homogenous size, the optimum encapsulation efficiency and by the addition of diverse excipients like antioxidants (e.g. α- or γ- tocopherhol) or chelating agents [(e.g. like ethylenediaminetetraacetic acid (EDTA) or diethylene triamine pentaacetic acid (DTPA)]. Furthermore, by the optimization of the size and the formulation conditions, liquid formulations can be stable for several years like DauXome®, Doxil® or Myocet®. With the addition of appropriate cryoprotectants and lyoprotectants, liquid formulations can also be frozen or lyophilized to enhance the stability. As already mentioned, freeze-drying is exclusively employed as dehydration process for industrial manufacturing and stabilization of liposomal products. Besides the commercial Paclitaxel containing liposomal formulations AmBisome® and Visudyne® are freeze-dried.

7A.5 Characterization of Liposomes

The liposomes are to be characterized for its physical properties such as size, shape, %ge of drug entrapment, percentage drug release and chemical composition such as contents of lipid materials, quantity and purity of starting material, etc. to know the integrity and stability of the liposomes, which are summarized in the Table 7A.1.

Table 7A.1 Characterization of Liposomes

Parameters	Methods/Instruments used
I. Physical Parameters	
Size and its distribution	Scanning Electron Microscope (SEM); Fluorescent Microscope; Transmission Electro Microscope (TEM)
Translational diffusion coefficient	Proton Corellation Spectroscopy (PCS) (measured to determine the hydrodynamic radious)
Size distribution	Gel Electrophoresis
Surface Charges	Electrophoresis; Zeta meter
Percent capture Entrapment	Mini column centrifugation for separation followed by assay; Protamine aggregation followed by centrifugation and assay
Entrapment volume	Water replacement technique and NMR method
Lamellarity	The number of bilayer can be measured by Freeze Electron microscopy and p-NMR technique,
Drug Release	*In-vitro* diffusion cell followed by assay of drug released with respect to time.
II. Chemical properties	
Quantitative determination of Phospholipids	Barlett assay or Stewart Assay (is a complexometric method with ammonium ferro- cyanate)
Phospholipid hydrolysis	Separation by HPLC and assayed by UV-Spectrophotometry
Phospholipid Oxidation	GLC; UV absorbance method; iodometric method
III. Pharmacodynamic properties	By *in-vivo* assay

7A.6 Drug Release from Liposomes

After the delivery to the target tissue, drug must be released from liposomes in the extra-cellular space or within the cells in order to induce the desired therapeutic activity.

Accumulation of liposomes into the target tissue does not necessarily improve therapeutic effect, if the drug release from carrier is hampered. The release depends mainly on the lipid composition of the liposomes and physical properties of the drug molecule. Lipid composition defines the physical phase of the lipid bilayer structure. For example, phospholipid-based liposomes in the liquid crystalline state release the drug more efficiently than liposomes in solid (gel) state. Cholesterol can be used for enhanced drug retention by improving the packing density of phospholipids. Drug release from liposomes takes place by diffusion through the liposomal bilayer according to a concentration gradient. Diffusion is dependent on the molecular mass and the polarity of the drug. As the drug release from liposomes is crucial for the therapeutic effect, fast and effective release at the target site is need, but on the other hand, the liposomes must remain intact in the bloodstream prior to reaching the target tissue.

Functional liposomes that respond to a specific factor at the target site can be used for site specific drug release. For example, the elevated temperature in solid tumors has been utilized for the triggered drug delivery. Temperature-sensitive liposomes remain intact at the normal body temperature, but become more permeable and release the drug in the higher temperature of the tumor tissue. Likewise, the acidic pH of the endosomes can be used for pH-sensitive drug release from the liposomes. The pH-sensitive liposomes are endocytosed by the cells, and thereafter, they fuse with the endovacuolar membrane and release their contents to the cytoplasm.

7A.7 Pharmacokinetics of Liposomes

One of the major limitations in the systemic parenteral use of liposomes is their fast elimination from the bloodstream by the reticuloendothelial system (RES). Liposomes are opsonized by plasma proteins followed by recognition and uptake by macrophages mainly in the liver and spleen. Certain hydrophilic polymer coatings are used to extend the circulation time of liposomes. Flexible polymer chains grafted to the liposome surface form a protective barrier and reduce recognition and uptake by the macrophages. As a result of the stearic stabilization, the liposomes circulate longer in the bloodstream, providing enhanced delivery to the target tissues and improved efficiency of the drug. The gold standard for polymer coating is poly (ethylene glycol) (PEG). PEG coating prolongs the blood circulation time of the liposomes significantly. In mice, half-life of the liposomes increased 10-fold, and the accumulation in the liver and spleen (elimination) decreased to 50 %. Also other hydrophilic polymers, such as 2-(N hydroxypropyl) methacrylamide, polyvinyl alcohol and poly (oxazoline) have been shown to prolong the plasma half-life of liposomes. Liposomes must be able to extravasate from the systemic blood circulation in order to reach the target tissues. In general, small liposomes (> 100 nm in diameter) can escape from the vascular bed easier than larger ones. Liposomes can easily escape the blood stream in the leaky sinusoidal vessels of the liver and spleen, but they cannot permeate across the tight endothelia in the brain and retinal capillaries. Drug distribution to the target tissue can be increased with liposomes. When the drug is administered in the form of long-circulating liposomes, the extended half-life of the drug enhances its accumulation into the tumour by enhanced permeability and retention (EPR) effect. The physiology of solid tumors differs from normal tissues: tumors promote extensive angiogenesis with hyper-vasculature, higher vascular permeability, and impaired lymphatic system. Thus, liposomes can be passively targeted into the tumors with long-circulating liposomes.

7A.8 Cellular Internalization of Liposomes

The interaction between liposomes and target cells is important for efficient drug delivery. Early studies on liposome-cell interactions suggested that liposomes fuse with the cell membrane. According to present understanding, liposomes are internalized via endocytosis. It has been shown that more than 90% of the cell-associated liposomes were adsorbed on the cell surface and endocytosed in several different cell lines. The rate of liposome uptake through endocytosis varies between cell lines. Usually liposomes reach the endosomes after 30 min to 3 h of incubation. Naturally, different cell types have different liposome-binding sites, which are likely to have several types of receptors, utilized to bind liposomes on the cell surface. Endocytosis may direct the liposomes into the lysosomes that have high catalytic enzyme activity. This can be a problem, if the drug is degraded by the enzymes (e.g. oligonucleotides, proteins). In that case liposomes should preferably be release the drug to the cytoplasm to avoid their targeting to lysosomes.

7A.9 Routes of Administration

Liposomes have been administered using different routes, such as oral, topical, and pulmonary route. More recently, delivery of peptides, proteins, siRNA and DNA with liposomes has been investigated, as well as the liposomal vaccine formulations and use of liposomes in diagnostic imaging being established. Examples of routes of administration and pharmaceutical applications of liposomes are shown in Table 7A.2.

Table 7A.2 Routes of Administration of Liposomes.

Routes of Administration	Applications
Intravenous drug delivery	Improved delivery of doxorubicin for ovarian cancer; Improved delivery of daunorubicin to solid tumors; Increased plasma concentration of vincristine
Oral drug delivery	Enhanced bioavailability of fenofibrate; Enhanced bioavailability and therapeutic efficacy of 5-fluorouracil; Stability of liposomes in gastrointestinal tract
Topical drug delivery	Various applications of liposomes in topical drug delivery; Skin permeation enhancement; Improved drug penetration; Enhanced dermal and transdermal drug penetration
Pulmonary drug delivery	Improved efficiency of camptothecin against human cancer; xenografts; Liposomal aerosol of paclitaxel for lung cancer; Aerosolized liposomal amphotercin B for the treatment of invasive pulmonary aspergillosis.
Ocular drug delivery	Enhanced delivery of immunosuppressive agent, FK506, to ocular tissues; Increased corneal absorption of acyclovir; Various applications of liposomes in ophtalmology
Peptide and protein delivery (Oral route)	Oral delivery of peptide drug; Improved therapeutic effect of interleukin-2 against hepatic metastases; Intracellular delivery of cytosolic exopeptidase prolidase
Liposomal vaccination	Liposomes as immunological adjuvants; Enhanced immune response by DNA vaccination
Nucleic acid delivery	Efficient plasmid transfection into mammalian cells; Targeted gene delivery for cancer treatment; Cationic liposomes as non-viral vectors for gene therapy; In vivo delivery of siRNA; Improved stability and intracellular delivery of oligonucleotides
Diagnostic imaging	Delivery of imaging agents for gamma- and magnetic resonance imaging; Enhanced magnetic resonance imaging with immunoliposomes; Immunoliposomes for molecular imaging
Liposomes for active drug targeting	Tumour-targeted delivery of antibody-incorporated Liposomes; Folate-mediated cellular targeting of macromolecules; Transferrin-mediated tumor targeting of cytostatics
Liposomes for triggered drug release	Temperature-sensitive liposomes for tumor treatments; Improved drug delivery with pH-sensitive liposomes; Light-triggered drug release from liposomes

7A.10 Clearance of Liposome from the Body

Immediately after i.v. injection, liposomes become coated by proteins circulating in the blood. Some of these proteins compromise the integrity of the lipid bilayer causing rapid leakage of liposome contents. Others promote recognition and subsequent elimination of liposomes from the blood. For example, liposomes composed of unsaturated lipids such as Ethanolamine Phosphatidyl choline (EPC) rapidly lose their membrane integrity through lipid transfer to lipoproteins and disintegrate. This process involves insertion of ApoA1, an apolipoprotein found predominantly in the high-density lipoprotein fraction, into the lipid bilayer. Other proteins called opsonins, mark liposomes for removal through phagocytic cells. Examples of opsonins include components of the complement system (C3b, iC3b), IgG_2, glycoprotein and fibrinolectin. The removal of foreign matter including liposomes is carried out by the mononuclear phagocyte system (MPS), in particular the resident macrophages of the liver (Kupffer cells), spleen, lung and bone marrow. The bulk of the injected liposomes accumulate in the liver and spleen

7A.11 Therapeutic Applications

Active targeting of drug-loaded liposomes to a specific location in the body has evident potential for enhanced drug delivery. Active targeting is achieved by modifying the liposome surface with targeting molecules such as proteins, peptides, aptamers or small molecules that act as ligands for cell surface receptors. The rationale of this approach is that the target receptor or antigen is expressed on the target cells at much higher levels than elsewhere. Therefore, the liposomes will be preferably distributed to the target cells.

Antibody-mediated targeting is the most extensively studied method for homing the liposomes to the target tissue. The liposome surface is modified with monoclonal antibodies that recognize the antigen expressed on the target cell surface (Park et al. 2001, Kim et al. 2009, Simard, Leroux 2009). Antibodies have high specificity and affinity to their targets. Antibody-functionalized liposomes can be also used to target diagnostic and imaging agents to tissues (Kozlowska et al. 2009).

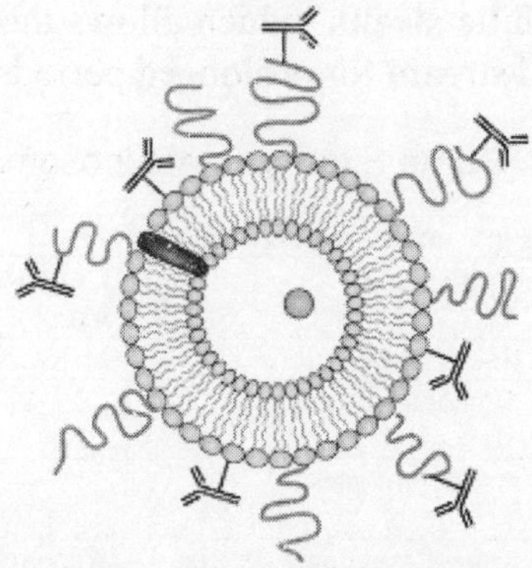

Fig. 7A.6 Schematic illustration of immunoliposome. The surface of liposome is coated with monoclonal antibodies and protective polymer (Modified from Torchilin 2005).

Various receptors that are over expressed on the surface of the cancer cells can be used for active liposome targeting. Epidermal growth factor receptor (EGFR) is over expressed in many tumour cells and has been used for liposome targeting (Kim et al. 2009). Likewise, vasoactive intestinal peptide (VIP) (Dagar et al. 2003) and folate-mediated liposome targeting have been extensively studied. Folate receptors are over-expressed by many tumour cells, and folate-derivatised liposomes are internalized through folate receptor mediated endocytosis (Leamon, Low 1991). This system has also been utilized in targeted delivery of DNA (Hofland et al. 2002) and oligonucleotides (Leamon, Cooper & Hardee 2003) to the tumor cells. Transferrin-coupled liposomes have also been used, since transferring receptors are over-expressed on many tumor cells (Maruyama et al. 2004, Hatakeyama et al. 2004). Aptamers, i.e. oligonucleotide or peptide-based molecules that bind to specific target molecule can be also used for targeting. For example, vascular endothelial growth factor (VEGF) –specific aptamers have been incorporated into liposomes for enhanced inhibition of angiogenesis (Willis et al. 1998). Some examples mentioned in Table 7A.3.

Doxorubicin (DXR) is a potent antineoplastic agent, which is active against a wide range of human cancer including lymphomas, leukemia and solid tumors. However, administration of this drug produces acute toxicity in the form of bone marrow depression, alopecia and oral ulceration. DXR entrapped in liposomes shows reduced non-specific toxicity and maintains or enhances anticancer effect. DXR hydrochloride constitutes the first liposomal product (DoxilTM) to be licensed in the United States. Surface grafted methoxypolyethylene glycol (MPEG) provides the hydrophilic stealth, which allows the DoxilTM liposomes to circulate in the blood stream for prolonged periods.

Table 7A.3 Some Examples of Liposomal Applications

Examples of Drugs in Liposomal Formulations			
Drug Application	**Application**	**Commercial name**	**Composition of liposomes**
Amikacin	Bacterial infections	MiKasome	HSPC/CH/DSPG
Adriamycin (doxorubicin)	Stomach cancer	---	DPPC/CH
Ampicilin Listeria	monocytogenes	---	CH/PC/PS 5:4:1 CH:DSPC:DPPG
Annamycin	Kaposi's sarcoma, Breast cancer, Leukemia	Annamycin	Liposomes
Amphotericin B	Systemic fungal infections	AmBisome	HSPC/CH/DSPG
All–*trans*-retinoic acid	Acute promyelocytic leukemia, Lymphoma, Prostate cancer	ATRAGEN	Liposome
1-β-D-Arabinofuranozidecytozine	Leukemia	----	SM/PC/CH 1:1:1
Ciprofloxacine	*Pseudomonas aeruginosa*	-----	DPPC
Clodronate	Macrophage suppression	-----	PC/CH
Cyclosporin	Immunosuppressor	----	PC/CH
Chloroquine	Malaria	-----	PC/PG/CH 10:1:5
Doxorubicin	Cancers	Doxil	HSPC/CH/PEG–DSPE
Daunorubicin	Cancers	DaunoXome	DSPC/CH
Ganciclovir	Cytomegalovirus retinitis	----	Liposomes
Interleukin 2	Immunostimulant	----	DMPC
Methotrexate	Cancers	----	DPPC/PI 18:2w/w
Nystatin	Systemic fungal infections	NYOTRAN	Liposomes
Platinum drugs e.g., cisplatin	Mezotelioma	PLATAR	Liposomes
Lurtotecan	Cancers	NX 211	Liposomes
Oligonucleotides against *c–myc*	Cancers	INXC–6295	Liposomes TCS
Ribavirin	Herpes simplex	----	Liposomes
Vincristin	Cancer, Lymphoma	Onco TCS	Liposomes TCS
Abbreviations: HSPC, hydrogenated soya phosphatidylcholine (hydrogenated soya lecithin); CH, cholesterol; DSPG, distearoylphosphatidylglycerol; DPPG, dipalmitoyl phosphatidyl glycerol; DMPC, dimyristoyl phosphatidyl choline; DPPC, dipalmitoyl phosphatidyl choline; DSPC, distearoyl phosphatidyl choline; PC, phosphatidyl choline; PS, phosphatidyl serine; SM, sphingo myelin; PI, phosphatidyl inositol; DCP, dicetylphosphate; PEG-DSPE, polyethylene glycol-phosphatidyl ethanolamine derivative; liposomes TCS, commercial composition of liposome-forming lipids; PEG-liposomes, liposomes modified with components containing a PEG (polyethylene glycol)			

For targeting on the solid tumor tissue, the Fab' fragment of 21B2 antibody and transferrin pendant type immunoliposomes (Fab'-PFG-ILP and TF-PEG-ILP) were used, which showed a low RES uptake and a long circulation time, and resulted in enhanced accumulation of the liposomes in the solid tumor. TF-PEG-ILP was internalized into tumor cells with receptor- mediated endocytosis after extravasation into tumor tissue. The pendant type immunoliposomes can escape from the gaps between adjacent endothelial cells and openings at the vessel termini during tumor angiogenesis by passive convective transport much higher than ligand directed targeting. Active targeting on tumor tissue with the pendant type immunoliposomes is particularly important for many highly toxic anticancer drugs in cancer chemotherapy.

References

Abraham, S. A., Waterhouse, D. N., Mayer, L. D., Cullis, P. R., Madden, T. D. and Bally, M. B., "The liposomal formulation of doxorubicin", *Liposomes, Pt E,* 2005 (391): 71-97.

Kim, I.Y., Kang, Y.S., Lee, D.S., Park, H.J., Choi, E.K., Oh, Y.K., Son, H.J. & Kim, J.S., "Antitumor activity of EGFR targeted pH-sensitive immunoliposomes encapsulating gemcitabine in A549 xenograft nude mice", *Journal of Controlled Release,* 2009, 140, (1): 55-60.

Papahadjopoulos, D. & Benz, C.C., "Tumor targeting using anti-her2 immunoliposomes", *Journal of Controlled Release,* 2001, 74 (1-3): 95-113.

Hofland, H.E.J., Masson, C., Iginla, S., Osetinsky, I., Reddy, J.A., Leamon, C.P., Scherman, D., Bessodes, M. & Wils, P., "Folate-targeted gene transfer in vivo", *Molecular Therapy,* 2002, 5 (6): 739-744.

Leamon, C.P. & Low, P.S., "Delivery of Macromolecules into Living Cells - a Method that Exploits Folate Receptor Endocytosis", *Proceedings of the National Academy of Sciences of the United States of America,* 1991, 88 (13): 5572-5576.

Leamon, C.P., Cooper, S.R. & Hardee, G.E., "Folate-liposome-mediated antisense oligodeoxynucleotide targeting to cancer cells: Evaluation in vitro and in vivo", *Bioconjugate chemistry,* 2003, 14 (4): 738-747.

Dagar, S., Krishnadas, A., Rubinstein, I., Blend, M.J. & Onyuksel, H., "VIP grafted sterically stabilized liposomes for targeted imaging of breast cancer: in vivo studies", *Journal of Controlled Release,* 2003, 91 (1-2): 123-133.

Simard, P. & Leroux, J.C., "pH-sensitive immunoliposomes specific to the CD33 cell surface antigen of leukemic cells", *International journal of pharmaceutics,* 2009, 381 (2): 86-96.

Willis, M.C., Collins, B., Zhang, T., Green, L.S., Sebesta, D.P., Bell, C., Kellogg, E., Gill, S.C., Magallanez, A., Knauer, S., Bendele, R.A., Gill, P.S. & Janjic, N., "Liposome anchored vascular endothelial growth factor aptamers", *Bioconjugate chemistry,* 1998, 9 (5): 573-582.

Hatakeyama, H., Akita, H., Maruyama, K., Suhara, T. & Harashima, H., "Factors governing the in vivo tissue uptake of transferrin-coupled polyethylene glycol liposomes in vivo", *International journal of pharmaceutics,* 2004, 281 (1-2): 25-33.

Kozlowska, D., Foran, P., MacMahon, P., Shelly, M.J., Eustace, S. & O'Kennedy, R., "Molecular and magnetic resonance imaging: The value of immunoliposomes", *Advanced Drug Delivery Reviews,* 2009, 61 (15): 1402-1411.

Kozubek, A. & Tyman, J.H.P., Resorcinolic lipids, the natural non-isoprenoid phenolic amphiphiles and their biological activity. *Chem. Rev.* 1999 (99): 1- 26.

Arkadiusz Kozubek, Jerzy Gubernator, Ewa Przeworska and Maria Stasiuk. Liposomal drug delivery, a novel approach: PLARosomes. Acta Biochiminica Polonica. 2000, 47(3): 639–649.

Assadullahi, T.P., Hider, R.C. & McAuley, A.J., Liposome formation from synthetic polyhydroxyl lipids. *Biochim. Biophys. Acta,* 1991(1083): 271-276.

Bangham AD, Standish M M, Watkins J C: Diffusion of univalent ions across the lamellae of swollen phospholipids. *J. Mol. Biol.,* 1965(13): 238-252.

Torchilin, V.P., "Recent advances with liposomes as pharmaceutical carriers", *Nature Reviews Drug Discovery,* 2005, 4(2): 145-160.

Bangham, A.D., Standish, M.M. and Watkins, J.C., Diffusion of univalent ions across lamellae of swollen phospholipids. *J. Mol. Biol.* 1965(13): 238-252.

Budker, V., Gurevich, V., Hagstrom, J.E., Bortzov, F. & Wolff, J.A., pH-Sensitive cationic liposomes: A new synthetic virus-like vector. *Nature Biotechnol.,* 1996(14) 760-764.

Chen, H.M., Torchilin, V. & Langer, R., Polymerized liposomes as potential oral vaccine carriers: Stability and bioavailability. *J. Control. Rel.,* 1996 (42): 263-272.

Choquet, C.G., Patel, G.B., Beveridge, T.J. & Sprott, G.D., Stability of pressure extruded liposomes made from archaebacterial ether lipids. *Appl. Microbiol. Biotechnol.* 1994 (42): 375-384.

Cistola, D.P., Atkinson, D., Hamilton, J.A. & Small, D.M., Phase behavior and bilayer properties of fatty acids:hydrated 1:1 acid- soaps. *Biochemistry,* 1986 (25): 2804-2812.

Clary, L., Verderone, G., Santaella, C. & Vierling, P., Membrane permeability and stability of liposomes made from highly fluorinated double-chain phosphocholines derived from diaminopropanol, serine or ethanolamine. *Biochim. Biophys. Acta,* 1997(1328): 55-64.

Coukell, A.J. & Spencer, C.M., Polyethylene glycol-liposomal doxorubicin: A review of its pharmacodynamic and pharmacokinetic properties, and therapeutic efficacy in the management of AIDS-related Kaposi's sarcoma. *Drugs,* 1997 (53): 520-538.

Crowe, J.H., Crowe, L.M., Preservation of liposomes by freeze-drying, in: Liposome Techn. (2nd Ed.), Ed. Gregoriadis, G., CRC Press, 1993 (1): 229-252.

Danilo D. Lasic. Liposomes in Gene Delivery. Published in 1997 by CRC Press LLC.

Engberts, J.B.F.N. and Hoekstra, D., Vesicle-forming synthetic amphiphiles. *Biochim. Biophys. Acta,* 1995 (1241): 323-340.

Frank Szoka, Jr. and Demetrios Papahadjopoulos, "Comparative Properties and Methods of Preparation of Lipid Vesicles (Liposomes)", *Ann. Rev. Biophys. Bioeng.,* 1980 (9): 467-508.

Frederiksen, L., Anton, K., van Hoogvest, P., Keller, H.R., Leuenberger, H., Preparation of liposomes encapsulating water-soluble compounds using supercritical carbon dioxide, J. Pharm. Sci., 1997, 86(8): 921-928.

Gabizon, A., Goren, D., Cohen, R. & Barenholz, Y., Development of liposomal anthracyclines: From basics to clinical applications. *J. Control. Rel.,* 1998(53): 275-279.

Gadras, C., Santaella, C. & Vierling, P., Improved stability of highly fluorinated endocytosis of conventional and sterically stabilized liposomes. *Biochemistry,* 1999 (37): 12875- 12883.

Goldbach, P., Borchart, H., Stamm, A., Spray-Drying of Liposomes for a Pulmonary Administration. II. Retention of Encapsulated Materials, *Drug. Dev. Ind. Pharm.*, 1993, 19(19): 2623-2636.

Kikuchi, H., Yamauchi, H., Hirota, S., A spray-drying method for mass-production of liposomes, *Chemical and Pharmaceutical Bulletin*, 1991, 39(6): 15-22-1527.

Kulkarni, S.B., Betageri, G.V., Singh, M., Factors affecting microencapsulation of drugs in liposomes, *J. Microencaps.*, 1995, 12(3): 229-246.

Lasic, D.D., Novel application of liposomes, Trends Biotechnol., 1998, 16(7): 307-321.

Lauri Paasonen. External Signal-Activated Liposomal Drug Delivery Systems. Academic dissertation. Division of Biopharmaceutics and Pharmacokinetics, Centre for Drug Research Faculty of Pharmacy University of Helsinki, Finland, 2010

Litzinger, D.C., Buiting, A.M.J., van Rooijen, N., Huang, L., Effect of liposome size on the circulation time and intraorgan distribution of amphipathic poly (ethylene glycol)-containing liposomes, Biochim. Biophy. *Acta, Biomem.* 1994, 1190(1): 99-107.

Lo, Y-L., Tsai, J-C., Kuo, J-H., Liposomes and disaccharides as carries in spray-dried powder formulations of superoxide dismutase, *J. Cont. Rel.*, 2004, 94(2-3): 259-272.

Lu, D., Hickey, A.J., Liposomal dry powders as aerosols for pulmonary delivery of proteins, *AAPS PharmSciTech*, 2005, 6(4): E641-648.

Michael Wiggenhorn.Scale-Up of Liposome Manufacturing: Combining High Pressure Liposome Extrusion with Drying Technologies. Dissertation zur Erlangung des Doktorgrades der Fakultät für Chemie und Pharmazie der Ludwig-Maximilians Universität München, Juli 2007.

Miller, C.R., Bondurant, B., McLean, S.D., McGovern, K.A. & OBrien, D.F., Liposome-cell interactions in vitro: Effect of liposome surface charge on the binding and phospholipid-based vesicles in the presence of bile salts. *J. Control. Rel.,* 1998 (57): 29-34

Nagayasu, A., Uchiyama, K. & Kiwada, H., The size of liposomes: A factor which affects their targeting efficiency to tumors and therapeutic activity of liposomal antitumor drugs. *Adv. Drug Deliv. Rev.* 1999 (40): 75-87.

Needham, D., T. J. McIntosh, and E. Evans, Thermomechanical and transition properties of dimyristoylphosphatidylcholine/cholesterol bilayers. *Biochemistry,* 1988(27): 4668-4673.

New, R.R.C., Preparation of Liposomes; Liposomes a practical approach, Ed. New, R. R. C., IRL Press, 1990: 33-104.

Norbert Maurer, David B Fenske & Pieter R Cullis. Developments in liposomal drug delivery systems, A Review. *Expert Opin. Biol. Ther.,* 2001, 1(6):1-25

Ostro, M.J., Industrial application of liposomes: what does that mean?, in: Liposomes as Drug Carriers, Ed. Gregoriadis, G., Wiley&Sons,1988:844-862.

P. Goyal, Parveen goyal, kumud goyal, Sengodan gurusamy vijaya kumar, Ajit singh, Om Prakash Katare, Dina nath mishra. Liposomal drug delivery systems – Clinical applications, *Acta Pharm.,* 2005 (55): 1–25.

Patel, G.B. & Sprott, G.D., Archaeobacterial ether lipid liposomes (Archaeosomes) as novel vaccine and drug delivery systems. *Crit. Rev. Biotechnol.*, 1999(19): 317-357.

Pector, V., Caspers, J., Banerjee, S., Deriemaeker, L., Fuks, R., ElOuahabi, A., Vandenbranden, M., Finsy, R. & Ruysschaert, J.M., Physico-chemical characterization of a double long-chain cationic amphiphile (Vectamidine) by microelectrophoresis. *Biochim. Biophys. Acta,* 1998 (1372): 339-346.

Redziniak, G., Perrier, P., Marechal, C., Liposomes at the industrial scale; in: Liposomes as tools in basic research and industry, Eds. Philippot, J.R., Schuber, F., CRC Press, 1995: 59-67.

Sanjay K. Jain and N. K. Jain. Liposomes as Drug Carriers. Controlled and Novel Drug Delivery, 1st Edn., CBS Publications, 2008: 304-352.

Scherphof, G.L. and Kamps, J.A.A.M., Receptor versus non-receptor mediated clearance of liposomes. *Adv. Drug Deliv. Rev.*, 1998 (32): 81-97.

Sharma, A. & Sharma, U.S., Liposomes in drug delivery: Progress and limitations. *Int. J. Pharm.*, 1997 (154): 123-140.

Sharma, A., Mayhew, E., Bolcsak, L., Cavanaugh, C., Harmon, P., Janoff, A., Bernacki, R.J., Activity of Paclitaxel liposome formulations against human ovarian tumor xenografts, *Int. J. Cancer*, 1997, 71(1): 103-107.

Sprott, G.D., Tolson, D.L. & Patel, G.B., Archaeosomes as novel antigen delivery systems. *FEMS Microbiol. Lett.* 1997 (154): 17-22.

Straubinger, R,M., Sharma, A., Murray, M., Mayhew, E., Novel Taxol formulations: Taxol – containing liposomes, *J. Nat. Cancer Inst. Monographs*, 1993 (15): 69-78.

Tardi, C., Drechsler, M., Bauer, K.H., Brandl, M., Steam sterilization of vesicular phospholipid gels, *Int. J. Pharm*., 2001, 217(1-2): 161-172.

Thompson, D.H., Gerasimov, O.V., Wheeler, J.J., Rui, Y.J. and Anderson, V.C., Triggerable plasmalogen liposomes: Improvement of system efficiency. *Biochim. Biophys. Acta,* 1996 (1279): 25-34.

Vemuri, S., Rhodes, C.T., Preparation and characterization of liposomes as therapeutic delivery systems: a review, *Pharm. Acta Helv*., 1995, 70(2): 95-111.

Walde, P., Namani, T., Morigaki, K., Hauser, H., Formation and Properties of Fatty Acid Vesicles (Liposomes), in: Liposome Technol., 3rd Ed., Ed, Gregoriadis, G., informa, 2007(1): 1-20.

Wiseman, L.R., Spencer, C.M., Paclitaxel. An update of its use in the treatment of metastatic breast cancer and ovarian and other gynaecological cancers, Drugs and aging, 1998, 12(4): 305-334.

Woodle, M.C. and Papahadjopoulos, D., Liposome preparation and size characterization. *Methods Enzymol.* 1998 (171): 193-217.

www.avantilipids.com/image/preparationofliposome.gif

Yamauchi, K., Doi, K. & Kinoshita, M., Archaebacterial lipid models: Stable liposomes from 1-alkyl-2-phytanyl-*sn*-glycero- 3-phosphocholines. *Biochim. Biophys. Acta,* 1996 (1283): 163-169

Zeisig, R., Arndt, D., Stahn, R. & Fichtner, I., Physical properties and pharmacological activity *in vitro* and *in vivo* of optimised liposomes prepared from a new cancerostatic alkylphospholipid. *Biochim. Biophys. Acta,* 1998(1414): 238-248

Zuidam, N.J., Lee, S.S., Crommelin, D.J.A., Sterilization of liposomes by heat treatment, *Pharm. Res.,* 1993, 10(11): 1591-1596.

Zuidam, N.J., Lee, S.S.L., Crommelin, D.J.A., Gamma-irradiation of non-frozen, frozen, and freezedried liposomes, *Pharm. Res*., 1995, 12(11): 1761-1768.

7B
Niosomes as Therapeutic Drug Career

Introduction

Paul Ehrlich, in 1909, initiated the era of development for targeted drug delivery when he envisaged a drug delivery mechanism that would target directly to diseased cell. Since then, numbers of carrier were utilized to carry drug at the target organ/tissue, which include immunoglobulins, serum proteins, synthetic polymers, liposomes, microspheres, erythrocytes, niosomes, etc. Among different carriers liposomes and niosomes are well established drug delivery systems used for target specific drug delivery.

7B.1 Niosomes

Niosomes are non-ionic surfactant vesicles of microscopic lamellar structure formed on admixture of non-ionic surfactant of the alkyl or dialkyl polyglycerol ether class and cholesterol with subsequent hydration in aqueous media. In niosomes, the vesicles forming amphiphile with a non-ionic surfactant such as Span-40, Span-60 and/or Tween-80, which is usually stabilized by addition of cholesterol and a small amount of anionic surfactant such as dicetyl phosphate.

Structurally niosomes are similar to liposomes, in that they are also made up of a bilayer (Fig 7B.1)

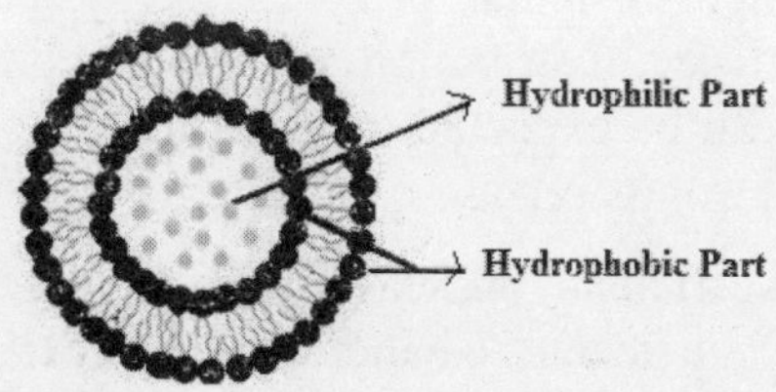

Fig. 7B.1 Schematic representation of a noisome.

However, the bilayer in the case of niosomes is made up of non ionic surface active agents rather than phospholipids as seen in the case of liposomes. Most surface active agents when immersed in water yield micellar structures; however some surfactants can yield bilayer vesicles which are niosomes. Niosomes may be unilamellar or multilamellar depending on the method used to prepare them. The niosome is made of a surfactant bilayer with its hydrophilic ends exposed on the outside and inside of the vesicle, while the hydrophobic chains face each other within the bilayer.

7B.2 Comparison of Niosome Vs Liposome

Niosomes are different from liposomes in that they offer certain advantages over liposomes. Liposomes face problems such as: they are expensive; their ingredients like phospholipids are chemically unstable because of their predisposition to oxidative degradation, they require special storage and handling and purity of natural phospholipids. Because of above mentioned drawbacks of liposomes, alternative nonionic surfactants such as monoalkyl or dialkyl polyoxyethylene ether vesicles have been investigated for entrapping hydrophilic and hydrophobic solutes. Niosomes are unilamellar or multilamellar vesicles capable of entrapping hydrophilic and hydrophobic drugs. Cholesterol, 5-cholesten-3ß-ol is also used in combination with nonionic surfactant for the formation of stable niosomes.

Niosomes are prepared from uncharged single-chain surfactant and cholesterol, whereas liposomes are prepared from double-chain phospholipids (neutral or charged). Handjani-Vila *et al.,* were first to report the formation of vesicular system on hydration of mixture of cholesterol and a single-alkyl chain non-ionic surfactant.

Niosomes behave *in-vivo* like liposomes, prolonging the circulation of entrapped drug and altering its organ distribution and metabolic stability. Encapsulation of various anti neoplastic agents in these carrier vesicles has been shown to decrease drug induced toxicity, increasing the anti-tumor efficacy. Such vesicular drug carrier systems alter the plasma clearance kinetics, tissue distribution, metabolism and cellular interaction of the drug. They can be expected to target the drug to its desired site of action and/or to control its release.

The entrapment efficiency increases with increase in the concentration and lipophilicity of surfactant. Chandraprakash *et al.,* made methotrexate loaded non-ionic surfactant vesicles using lipophilic surfactants like Span 40, Span 60 and Span 80 and found that Span 60 (HLB = 4.7) gave highest percent entrapment while Span 85 (HLB = 9.8) gave least entrapment. They also observed that as HLB value of surfactant decreased, the mean size was reduced.

7B.3 Method of Preparation of Niosomes

7B.3.1 Ether Injection Method

This method provides a means of making niosomes by slowly introducing a solution of surfactant dissolved in diethyl ether into warm water maintained to a temperature of 60°C. The surfactant mixture in ether is injected through 14 gauge needle into an aqueous solution of material. Vaporization of ether leads to formation of single layered vesicles. Depending upon the conditions used the diameter of the vesicle range from 50 to 1000 nm.

7B.3.2 Hand Shaking Method (Thin Film Hydration Technique)

The mixture of vesicles forming ingredients like surfactant and cholesterol are dissolved in a volatile organic solvent (diethyl ether, chloroform or methanol) in a round bottom flask. The organic solvent is removed at room temperature (20°C) using rotary evaporator leaving a thin layer of solid mixture deposited on the wall of the flask. The dried surfactant film can be rehydrated with aqueous phase at 0-60°C with gentle agitation. This process forms typical multilamellar niosomes. Thermo sensitive niosomes were prepared by evaporating the organic solvent at 60°C and leaving a thin film of lipid on the wall of rotary flash evaporator. The aqueous phase containing drug was added slowly with intermittent shaking of flask at room temperature followed by sonication.

Sonication

A typical method used for the production of the vesicles is by sonication of solution. In this method an aliquot of drug solution in buffer is added to the surfactant/cholesterol mixture in a 10ml glass vial. The mixture is probe sonicated at 60°C for 3 min. using a sonicator with a titanium probe to yield niosomes.

Micro Fluidization

Micro fluidization is a recent technique used to prepare unilamellar vesicles of defined size distribution. This method is based on submerged jet principle in which two fluidized streams interact at ultra high velocities, in precisely defined micro channels within the interaction chamber. The impingement of thin liquid sheet along a common front is arranged such that the energy supplied to the system remains within the area of niosomes formation, which results in a uniform, small size with better reproducible niosomes.

Extrusion: Mixture of surfactant, cholesterol and dicetyl phosphate in chloroform is made into thin film by evaporation. The Extrusion is a technique in which the above mixture is forced through a polycarbonate filter with a defined pore size to yield particles having a diameter near the pore size of the filter used. Prior to extrusion through the final pore size, LMV suspensions are disrupted either by several freeze-thaw cycles or by pre-filtering the suspension through a larger pore size (typically 0.2μm-1.0μm). This method helps prevent the membranes from fouling and improves the homogeneity of the size distribution of the final suspension.

7B.3.3 Reverse Phase Evaporation Technique (REV)

Cholesterol and surfactant (1:1) are dissolved in a mixture of ether and chloroform. An aqueous phase containing drug is added to this and the resulting two phases are sonicated at 4-5°C. The clear gel formed is further sonicated after the addition of a small amount of phosphate buffered saline (PBS). The organic phase is removed at 40°C under low pressure. The resulting viscous niosome suspension is diluted with PBS and heated on a water bath at 60°C for 10 min. to yield niosomes.

7B.3.4 Trans Membrane pH Gradient Drug Uptake Process

Surfactant and cholesterol are dissolved in chloroform. The solvent is then evaporated under reduced pressure to get a thin film on the wall of the round bottom flask. The film is hydrated with 300ml citric acid buffer of pH 4.0, by vortex mixing. The multilamellar vesicles are frozen and thawed 3 times and later sonicated. To this niosomal suspension, aqueous solution containing 10mg/ml of drug is added and vortexed. The pH of the sample is then raised to 7.0-7.2 with 1M disodium phosphate. This mixture is later heated at 60°C for 10min. to give niosomes.

7B.3.5 The "Bubble" Method

In this method niosomes are formed without use of organic solvents. The bubbling unit consists of round bottomed flask with three necks positioned in water bath to control the temperature. Water cooled reflux and thermometer is positioned in the first and second neck and nitrogen supply through the third neck. Cholesterol and surfactant are dispersed together in this buffer (pH 7.4) at 70°C, the dispersion mixed for 15 seconds with high shear homogenizer and immediately afterwards "bubbled" at 70°C using nitrogen gas resulting niosomes.

7B.3.6 Formation of Niosomes from Proniosomes

Another method of producing niosomes is to coat with water soluble carrier such as sorbitol with surfactant. The coating process consists of covering of water soluble particle within a thin film of dry surfactant. This preparation is termed "Proniosomes". The niosomes are formulated by the addition of aqueous phase at T > Tm and brief agitation.

T = Temperature.

T_m = mean phase transition temperature.

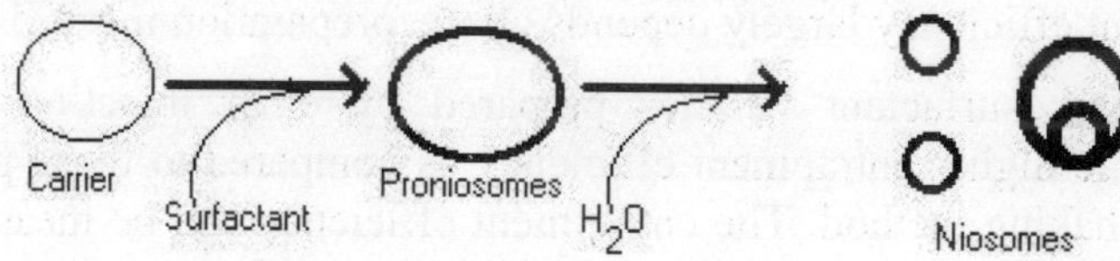

Fig. 7B.2 Formation of niosome from proniosomes.

7B.3.7 Formation of Niosomes by Supercritical Carbon Dioxide ($scCO_2$) Fluid

The $scCO_2$ apparatus for niosome preparation consisted of two parts in which one is the volume view cell that has two glass windows on both sides and other is the high pressure pump for feeding CO_2 gas. The temperature inside the volume view cell was measured by a Platinum (Pt) resistance thermometer. The pressure was measured with a strain gauge. The mixtures in volume view cell were continuously mixed by a magnetic stirrer. The total amount of surfactant mixed with cholesterol together with drug was added into the view cell. The temperature in the cell was raised to 60°C and the CO_2 gas was introduced into the view cell. The pressure and temperature in the view cell were maintained at 200 bar and 60±1°C, respectively. After continuation of stirring for 30 mins, the pressure will be released to get the niosomal dispersions.

7B.4 Separation of Unentrapped Drug

The removal of unentrapped solute from the vesicles can be accomplished by various techniques, which include: Dialysis using dialysis tubing against phosphate buffer or gel-filtration through a sephadex-G-50 column and eluted with phosphate buffered saline or by

Centrifugation in which suspension is centrifused and the supernatant is separated and resuspended to obtain a niosomal suspension free from unentrapped drug.

7B.5 Characterization of Niosomes

7B.5.1 Determination of Shape and Size

The shape and size of the niosome can be determined by electron microscopic measurement using Scanning Electron Microscope (SEM) or Transmission Electron Microscope (TEM).

7B.5.2 Entrapment Efficiency

Entrapment efficiency largely depends on the preparation method.

Non-ionic surfactant vesicles prepared by ether injection method demonstrate higher entrapment efficiency as compared to those prepared by hand shaking method. The entrapment efficiency can be measured by dialysis or ultracentrifugation methods. The niosome entrapped drug could be separated from the free drug by dialysis or centrifugation method. The dialysate or the supernatant liquid can be measured against phosphate buffer saline using Ultra violet spectrophotometer and the absorbance (A) of the corresponding blank niosome can be measured under the same condition. The concentration of free drug could be obtained from absorbance difference based on standard curve. The entrapment efficiency of the drug is defined as the ratio of the mass of niosome associated drug to the total mass of drug.

The amount of entrapped drug can be determined by lysis of the vesicles with absolute ethanol. The vesicle dispersion can be mixed with an equal volume of ethanol to give a clear solution. The amount of drug can be measured by a High Performance Liquid Chromatography method. The percentage of drug encapsulated was calculated from the ratio of the drug in the vesicles to the total amount of drug in the aqueous suspension.

Using gel-filtration chromatography the entrapped charged drug can be separated from free drug using a column packed with 10 ml Sephadex G-50 fine or medium.

The column can be eluted with PBS (pH 6.6). The eluted fractions is to be diluted with a ratio of 1:1 (v/v) with l-propanol, to disrupt the NSVs. The solutions must be analyzed for charged drug with absorption spectrometry.

7B.5.3 Drug Release Studies

The release of drug from niosomes can be determined using the membrane diffusion technique. Suspend an accurately measured amount of drug niosomal formulation in 1ml phosphate buffer saline and transferred to a glass tube covered at its lower end by a soaked cellulose membrane. Suspend the glass tube in the dissolution flask of a dissolution apparatus containing 750ml phosphate buffer saline and rotate it at 50 rpm. Keep the temperature at 37ºC. Draw the aliquots of the dialysate at predetermined time interval and replenish immediately with the same volume of fresh simulated fluid. Analyze the withdrawn samples using spectrophotometer.

7B.5.4 Physical Stability Study

Physical stability study is required to investigate the leaching of drug from niosomes during storage. To know the stability, seal the prepared niosomes in 20ml glass vials and store at a temperature of 2-8ºC for a period of 60 – 90 days, withdraw samples from each batch at definite time intervals and determine the residual amount of the drug in the vesicles after separation from un-entrapped drug by ultracentrifugation or dialysis method.

7B.5.5 Surface Charges Determination

Vesicle surface charge can be estimated by measurement of particle electrophoretic mobility and is expressed as the zeta potential by using Zeta meter following Henry equation as

$$z = mE4ph/ S$$

Where, z = zeta potential, mE = electrophoretic mobility, h= viscosity of the medium, S= dielectric constant.

7B.5.6 Measurement of Elasticity Value

Elasticity of the empty and drug loaded (both conventional and elastic) niosomes can be carried out by the extrusion measurement through polycarbonate membrane filter with pore size of 50nm (Millipore, USA) at constant pressure (2.5 bar). The elasticity of the vesicles can be expressed in terms of deformability index according to the equation: Deformability Index = j (rv/rp) 2

Where, j is the weight of dispersion, which was extruded in 10min through a polycarbonate filter of 50nm pore size, rv is the size of the vesicle after extrusion and *r*p is the pore size of the filter membrane.

7B.5.7 Measurement of Electron Density Profile of the Bilayer

Small angle X-ray diffraction measurement (**SAXS**) is a very valuable method for obtaining information on the distribution of the number of bilayer in vesicles and for obtaining information on the electron density profile of the bilayer.

7B.5.8 *In-Vivo* Evaluations of Niosomes

After i.v. administration the niosomes are filtered basing on their size and passed through the lungs to the body, which can be cleared by the liver and spleen. At the cellular level it can interact with the cell by fusion of the outer bilyer with the plasma membrane or by stable adsorption to the cell surface either by non-specific or by ligand bound or by endocytosis or transfer through lipid molecules of the membrane. The *in-vivo* evaluation can be done by bioassay in animal or human model using appropriate analytical methods for drug determination after administration with respect to time to know the bio-availability and distribution of the drugs from niosomal formulations.

7B.6 Biomedical Applications

Niosomes as Drug Carriers

A number of workers have reported the preparation, characterization and use of niosomes as drug carriers. It has been found that the niosomal encapsulation of **Methotrexate** and **Doxorubicin** increases drug delivery to the tumor and tumoricidal activity of the drug. Raja Naresh *et al*, reported the anti-inflammatory activity of niosome encapsulated **Diclofenac sodium** in arthritic rats. It was found that the niosomal formulation prepared by employing a 1:1 combination of Tween-85 elicited a better consistent anti-inflammatory activity for more that 72 hrs after administration of single dose. Baillie *et al,* investigated the encapsulation and retention of entrapped solute 5,6-carboxy fluorescence (CF) in niosomes. They observed that stable vesicles could not be formed in the absence of cholesterol but were more permeable to entrapped solute. The physical characteristics of the vesicles were found to be dependent on the method of production. Parthasarthi *et al*, prepared niosomes of **vincristine sulfate** which had lesser toxicity and improved anticancer activity. Jagtap and Inamdar prepared niosomes of **Pentoxifylline** and studied the *in-vivo* bronchodilatory activity in guinea pigs. The entrapment efficiency was found to be 9.26 ± 1.93% giving a sustained release of drug over a period of 24 hrs. Carter *et al,* reported that multiple dosing with **sodium stilbogluconate** loaded niosomes was found to be effective against parasites in the liver, spleen and bone marrow as compared to simple solution of sodium stibogluconate. Azmin *et al.*, reported the preparation and oral as well as intravenous administration of **Methotrexate** loaded niosomes in mice. They observed significant prolongation of plasma levels and high uptake of Methotrexate in liver from niosomes as compared to free drug solution. Chandraprakash *et al*, reported the formation and pharmacokinetic evaluation of **Methotrexate** niosomes in tumor bearing mice. Cable *et al*, modified the surface of niosomes by incorporating polyethylene alkyl ether in the bilayered structure. They compared the release pattern and plasma level of Doxorubucin in niosomes and Doxorubucin mixed with empty niosomes and observed a sustained and higher plasma level of doxorubicin from niosomes in mice. D' Souza *et al*, studied absorption of **Ciprofloxacin** and **Norfloxacin** when administered as niosome encapsulated inclusion complexes. Namdeo *et al*, reported the formulation and evaluation of **Indomethacin** loaded niosomes and showed that therapeutic effectiveness increased and simultaneously toxic side effect reduced as compared with free Indomethacin in paw oedema bearing rats.

Targeting of Bioactive Agents

The cells of **RES** preferentially take up the vesicles. The uptake of niosomes by the cells can also be achieved by circulating serum factors known as opsonins, for prolonged period of time. Such localized drug accumulation been exploited in treatment of animal tumors known to metastasize to the liver and spleen and in parasitic infestation of liver.

It has been suggested that the antibodies such as immunoglobulins seem to bind quite readily to the lipid surface, thus offering a convenient means for targeting of drug as carrier.

In Neoplasia: Doxorubicin, the anthracyclic antibiotic with broad spectrum anti tumor activity, shows a dose dependant irreversible cardio toxic effect. Niosomal delivery of this drug to mice bearing S-180 tumor increased their life span and decreased the rate of proliferation of sarcoma. Niosomal entrapment increased the half-life of the drug, prolonged its circulation and altered its metabolism. Intravenous administration of methotrexate entrapped in niosomes to S-180 tumor bearing mice resulted in total regression of tumor and also higher plasma level and slower elimination.

In Leishmaniasis: Baillie *et al*, reported increased sodium stibogluconate efficacy of niosomal formulation and that the effect of two doses given on successive days was found to be additive to treat leishmaniasis.

Delivery of peptide drugs: Yoshida *et al*, investigated that the oral delivery of 9- Desglycinamide, 8-Arginine vasopressin entrapped in niosomes in an in-vitro intestinal loop model and reported that stability of peptide increased significantly.

Immunological application of niosomes: Niosomes have been used for studying the nature of the immune response provoked by antigens. Brewer and Alexander [41] have reported niosomes as potent adjuvant in terms of immunological selectivity, low toxicity and stability.

Niosomes as carriers for Hemoglobin: Niosomal suspension shows a visible spectrum superimposable career for free hemoglobin. Vesicles are permeable to oxygen and hemoglobin dissociation curve can be modified similarly to non-encapsulated hemoglobin.

Transdermal Delivery of Drugs by Niosomes

Slow penetration of drug through skin is the major drawback of transdermal route of delivery. An increase in the penetration rate has been achieved by transdermal delivery of drug incorporated in niosomes. Jayraman *et al*, has studied the topical delivery of **erythromycin** from various formulations including niosomes or hairless mouse. From the studies of confocal microscopy, it was seen that non-ionic vesicles could be formulated to target pilosebaceous glands.

Other Applications

Sustained Release

Azmin *et al*, suggested the role of liver as a depot for **methotrexate** after niosomes are taken up by the liver cells. Sustained release action of niosomes can be applied to drugs with low therapeutic index and low water solubility since those could be maintained in the circulation via niosomal encapsulation.

Localized Drug Action

Drug delivery through niosomes is one of the approaches to achieve localized drug action, because of their size and low penetrability through epithelium and connective tissue keeps the drug localized at the site of administration. Localized drug action results in enhancement of efficacy of potency of the drug and at the same time reduces its systemic toxic effects, for example: **antimonials** encapsulated within niosomes are found to be taken up by mononuclear cells resulting in localization of drug, increase in potency and hence decrease strength of dose and toxicity.

References

Aggarwal D , Kaur IP .Improved pharmacodynamics of timolol maleate from amucoadhesive niosomal ophthalmic drug delivery system. *Int. J. Pharm.* 2005 (290): 155–159.

Alsarra IA, Bosela AA, Ahmed SM, Mahrous GM .Proniosomes as a drug carrier for transdermal delivery of ketorolac. *Eur. J. Pharm. Biopharm.* 2005 (59):485-490.

Arunothayanun P, Bernard MS, Craig DM , Uchegbu IF , Florence AT .The effect of processing variables on the physical characteristics of non-ionic surfactant vesicles (niosomes) formed from a hexadecyl diglycerol ether. *Int. J. Pharm.* 2000 (201): 7–14.

Attia IA, El-Gizawy SA, Fouda MA, Donia AM. Influence of a Niosomal Formulation on the Oral Bioavailability of Acyclovir in Rabbits. *AAPS Pharm Sci Tech.*, 2007, 8(4): Article 106. DOI: 10.1208/ pt0804106

Azmin M.N., Florence A.T., Handjani-Vila R.M., Stuart J.F.B., Vanlerberghe G., and Whittaker J.S. The effect of non-ionic surfactant vesicle (niosome) entrapment on the absorption and distribution of methotrexate in mice. *J. Pharm. Pharmacol.* 1985 (37): 237–242.

Baillie A.J., Coombs G.H. and Dolan T.F. Non-ionic surfactant vesicles, niosomes, as delivery system for the anti-leishmanial drug, sodium stribogluconate, *J. Pharm. Pharmacol.* 1986 (38): 502-505.

Baillie A.J., Florence A.T., Hume L.R, Rogerson A., and Muirhead G.T. The preparation and properties of niosomes-non-ionic surfactant vesicles. *J. Pharm Pharmacol.* 1985, 37(12):863–868

Biju SS, Talegaonkar S, Mishra PR, Khar RK. Vesicular systems: An overview. *Ind. J. Pharm. Sci.,* 2006 (68):141-53.

Bouwstra JA, Gooris GS, Bras W, Talsma H.Small angle X-ray scattering: possibilities and limitations in characterization of vesicles. *Chemistry and Physics of Lipids*. 1993 (64): 83-98

Bouwstra JA, Vanhal DA, Hans EJ, Hofland, Hans EJ. Preparation and characterization of nonionic surfactant vesicles. Colloids and Surfaces: *A Physicochemical and Engineering Aspects*. 1997:123-124.

Brewer J.M. and Alexander J.A. The adjuvant activity of non-ionic surfactant vesicles (niosomes) on the BALB/c humoral response to bovine serum albumin. *Immunology*. 1992, 75 (4): 570-575.

Carafa M , Santucci E, Alhaique F, Coviello T , Murtas E, Riccieri FM,Lucania G, Torrisi MR . Preparation and properties of new unilamellar non-ionic: ionic surfactant vesicles. *Int. J. Pharm.* 1998, (160):51–59.

Chandraprakash K.S., Udupa N., Umadevi P. and Pillai G.K. Formulation and evaluation of Methotrexate niosomes. *Ind. J. Pharm. Sci.* 1992, 54 (5): 197.

Chandraprakash K.S., Udupa N., Umadevi P. and Pillai G.K. Pharmacokinetic evaluation of surfactant vesicles containing methotrexate in tumor bearing mice. *Int. J. Pharma*. 1990, R1-R3: 61.

Chauhan S. and Luorence M.J. The preparation of polyoxyethylene containing non-ionic surfactant. vesicles. *J. Pharm. Pharmacol*. 1989, (41): 6.

D' Souza R., Ray J., Pandey S. and Udupa N. Niosome encapsulated ciprofloxacin and norfloxacin BCD complexes. *J. Pharm. Pharmcol*. 1997, 49(2): 145-149.

Devaraj GN, Parakh SR , Devraj R, Apte SS, Rao BR , Rambhau D. Release Studies on Niosomes Containing Fatty Alcohols as Bilayer Stabilizers Instead of Cholesterol. *J. Coll. Interf. Sci*. 2002 (251): 360–365.

Don A., Van H., Joke A.B. and Hans E. Non ionic surfactant vesicles containing estradiol for topical application. Ph.D. thesis. Centre for drug research. 1997: 330-339

Fang JY, Hong CT, Chiu WT, Wang YY. Effect of liposomes and niosomes on skin permeation of Enoxacin. *Int. J. Pharm*. 2001 (21): 61–72.

Gayatri Devi S., Venkatesh P. and Udupa N. Niosomal sumatriptan succinate for nasal administration. *Int. J. Pharm. Sci*. 2000, 62(6): 479-481.

Ghada A, Nashwa EG.Niosome- Encapsulated gentamicin for ophthalmic controlled delivery. *AAPS Pharm.Sci. Tech*. 2008, 9(3): 740-747

Guinedi AS, Nahed DM, Samar M, Rania MH.Preparation and evaluation of reversephase evaporation and multilamellar niosomes as ophthalmic carriers of acetazolamide. *Int. J. Pharm*. 2005 (306): 71–82.

Gupta P N , Mishra V, Rawat A , Dubey P , Mahor S, Jain S , Chatterji DP, Vyas SP Non-invasive vaccine delivery in transfersomes , niosomes and liposomes: a comparative study. *Int. J. Pharm*. 2005 (293): 73–82.

Handjani-Vila R.M., Ribier A., Rondot B. and Vanlerberghe G. Dispersions of lamellar phases of non-ionic lipids in cosmetic products. *Int. J. Cosmetic Sci*. 1979 (1):303–314.

Hao Y, Zhao F, Li N, Yang Y, Li K .Studies on a high encapsulation of colchicine by a noisome System. *Int. J. Pharm*. 2002 (244):73–80.

Honga M, Zhu S, Jiang Y, Tang G, Pei Y. Efficient tumor targeting of hydroxycamptothecin loaded PEGylated niosomes modified with transferrin. *J. Control. Release*., 2009 (133): 92–102.

Hu C. and Rhodes D.G. Proniosomes: a novel drug carrier preparation. *Int. J. Pharm*. 1999 (185): 23-35.

Hua W, Liu T .Preparation and properties of highly stable innocuous niosome in Span 80/PEG 400/H2O system. Colloids and Surfaces A: Physicochem. *Eng. Aspects*. 2007 (302):377–382.

Hunter C.A., Dolan T.F., Coombs G.H. and Baillie A.J. Vesicular systems (niosomes and liposomes) for delivery of sodium stibogluconate in experimental murine visceral leishmaniasis. *J.Pharm. Pharmacol*. 1988, 40(3): 161-165.

Jagtap A. and Inamdar D. Study of antiparkinson's activity of plain and niosomal pentoxifylline. *Ind. J. Pharm. Sci*. 2001, 63(1): 49–54.

Kato K, Walde P, Koine N, Ichikawa S, Ishikawa T , Nagahama R, Ishihara T, Sujii T,? Shudou M, Omokawa Y, Kuroiwa T. Temperature-Sensitive Nonionic Vesicles Prepared from Span 80(Sorbitan Monooleate). *Langmuir*. 2008 (24): 10762-10770

Liu T, Guo R, Hua W, Qiu J. Structure behaviors of hemoglobin in PEG 6000/Tween 80/Span 80/H2O niosome system. Colloids and Surfaces A: Physicochem. *Eng. Aspects*. 2007 (293):255–261.

Mahmoud M, Sammour OA, Mohammed A, Hammad NA. Effect of some formulation parameters on flurbiprofen encapsulation and release rates of niosomes prepared from proniosomes. *Int. J. Pharm.* 2008 (361):104–111

Malhotra M. and Jain N.K. Niosomes as Drug Carriers. *Indian Drugs* (1994), 31(3):81 86.

Manconi M, Valenti D, Sinico C, Lai F, Loy G, Anna MF.Niosomes as carriers for tretinoin II. Influence of vesicular incorporation on tretinoin photostability. *Int. J. Pharm.* 2003 (260): 261–272.

Manosroi A, Chutoprapat R, Masahiko A, Manosroi J. Characteristics of niosomes prepared by supercritical carbon dioxide (scCO2) fluid . *Int. J. Pharm.* 2008 (352): 248-255.

Manosroi A, Jantrawut P, Manosroi J.Anti-inflammatory activity of gel containing novel elastic niosomes entrapped with diclofenac diethylammonium. *Int. J. Pharm.*, 2008 (360): 156–163.

Murdan S, Gregoriadis G, Florence AT, Sorbitan monostearate/ polysorbate 20 organogels containing niosomes: delivery vehicle for antigens? *Eur. J. Pharm. Sci.*, 1999 (8):177–185

Muzzalupo R, Tavano L, Trombino S ,Cassano R , Picci N , Mesa CL .Niosomes from alpha,omega-trioxyethylene-bis(sodium 2-dodecyloxy-propylenesulfonate): Preparation and characterization . Colloids and Surfaces B: *Biointerfaces.*, 2008 (64):200–207.

Namdeo A., Mishra P.R., Khopade A.J. and Jain N.K. Formulation and evaluation of niosome encapsulated indomethacin. *Indian Drugs*.1999, 36(6): 378-380.

N. Udupa. Niosome as Drug Careers. Controlled and Novel drug Delivery, by N. K. Jain, 1st Edn, 2008: 292-303.

Palozza P, Muzzalupo R, Trombino S, Valdannini A, Picci N . Solubilization and stabilization of beta -carotene in niosomes: delivery to cultured cells. *Chemistr and Physics of Lipids.* 2006 (139): 32–42.

Parthasarathi G., Udupa N., Umadevi P. and Pillai G.K. Niosome encapsulated of vincristine sulfate: improved anticancer activity with reduced toxicity in mice. *J. Drug Target.* 1994, 2(2): 173-182.

Rakesh P. Patel. Niosome: A Unique Drug Delivery System. Latest Reviews, Vol. 5(1), 2007; www.pharmainfo.net/latest-reviews

Raja Naresh R.A., Chandrashekhar G., Pillai G.K. and Udupa N. Antiinflammatory activity of Niosome encapsulated diclofenac sodium with Tween -85 in Arthitic rats. *Ind. J. Pharmacol.* 1994 (26):46-48.

Rogerson A., Cummings J., Willmott N. and Florence A.T. The distribution of doxorubicin in mice following administration in niosomes. *J Pharm Pharmacol.* 1988, 40(5): 337–342.

Schreief H, Bouwstra J.Liposomes and niosomes as topical drug carriers: dermal and transdermal drug delivery. *J. Control. Release.* 1994 (30): 1-15.

Sheena I.P., Singh U.V., Kamath R., Uma Devi P., and Udupa N. Niosomal withaferin A, with better tumor efficiency. *Indian J. Pharm. Sci.* 1998, 60(1):45-48.

Stafford S., Baillie A.J. and Florence A.T. Drug effects on the size of chemically defined non-ionic surfactant vesicles. *J. Pharm.Pharmacol.* 1988, 40(suppl.): 1-26

Suzuki K. and Sokan K. The Application of Liposomes to Cosmetics. *Cosmetic and Toiletries.* 1990; 105: 65-78.

Tabbakhian M , Tavakoli N, Jaafari MR , Daneshamouz S. Enhancement of follicular delivery of finasteride by liposomes and niosomes1. *In vitro* permeation and *in vivo* deposition studies using hamster flank and ear models. *Int. J. Pharm.* 2006 (323): 1–10.

Uchegbu IF, Florens AT. Non-ionic surfactant vesicles (Niosomes): Physical and pharmaceutical chemistry. *Advances in Colloid and Interface Science.* 1995 (58):1-55.

Vanhal DA, Jeremiasse E, Vringer TD, Junginger HE , Bouwstra JA .Encapsulation of lidocaine base and hydrochloride into non-ionic surfactant vesicles (NSVs) and diffusion through human stratum comeum *in vitro* . *Eur. J. Pharm. Sci.* 1996 (4):147-157.

Vijay D. Waghand Onkar J. Desmukh. Niosomes as ophthalmic drug delivery systems: A review. *J Pharm Res.*, 2010, 3(7): 1558-1563.

Vora B, Khopade AJ, Jain NK. Proniosomes based transdermal delivery of levonorgestrel for effective contraception. *J. Control. Release.* 1998 (54): 149–165.

Vyas SP, Khar SP. Controlled drug Delivery Concepts and advances. First edition, Vallabh Prakashan, New Delhi, 2002: 315-382.

Weissman G., Bloomgarden D., Kaplan R., Cohen C., Hoffstein S., Collins T., Gotlieb A. and Nagle D. A general method for the introduction of enzymes, by means of immunoglobulin-coated liposomes, into lysosomes of deficient cells. *Proc. Natl. Acad. Sci.*1975, 72(1): 88-92.

Yoshida H, Lehr CM , Kok W , Junginger HE, Verhoef JC, Bouwstra JA .Niosomes for oral delivery of peptide drugs . *J. Control. Release*. 1992 (21):145-154

Yoshioka T, Sternberg B, Florence AT. Preparation and properties of vesicles (niosomes) of sorbitan monoesters (Span 20, 40, 60 and 80) and a sorbitan trimester (Span 85). *Int. J. Pharm*. 1994 (105): 1-6.

Yoshioka T., Sternberg B., Moody M. and Florence A.T. Niosomes from Span surfactants: Relations between structure and form. *J. Pharm. Pharmcol*. Supp. 1992 (44): 1044.

8

Oral Controlled Release Systems

Introduction

Controlled drug delivery dominates over the conventional drug delivery systems in the sense to alter the pharmacokinetic and pharmacodynamic parameters of the active therapeutic moieties by adopting novel drug delivery technology or by modifying the molecular structure and/or physiological parameters of the drug through a selected route of administration. An appropriately designed controlled-release drug delivery system can have a major advantage towards solving the problems by targeting the drug to a specific organ or tissue and controlling the rate of drug delivery to the target tissue at a predetermined rate at the predetermined rate of time.

This correctly suggests that there are sustain release system that can not be considered controlled release system, in fact controlled release indicate controlled to release at a particular site and/or at a particular rate for a particular time period. In general, the goal of a sustained release dosage from is to maintain therapeutic blood or tissue levels of drug for an extended period this is usually accomplished by attempting to obtain zero-order release from the dosage form. How ever new and more sophisticated controlled release and sustained release delivery systems are constantly being developed and tested for their therapeutic potentials.

Sustained release, sustained action, prolonged action, controlled release, extended action, timed release, depot and repository dosage forms are terms used to identify drug delivery system that are designed to achieve prolonged therapeutic effect by continuously releasing medication over an extended period of time after administration of a single dose.

The primary objectives of controlled drug delivery are to ensure safety and to improve efficiency of drugs as well as patient compliance. This achieved by better control of plasma drug levels and frequent dosing. For conventional dosage forms, only the dose and dosing interval can vary and for each drug there exists a therapeutic window of plasma concentration, below which therapeutic effect is insufficient, and above which toxic side effects are elicited. This is often defined as the ratio of median lethal dose (LD 50) to median effective dose (ED50)

Systems that are designed as prolonged release can also be considered as an attempt to achieve sustained-release delivery. Repeat action tablets are an alternative method of sustained release in which multiple doses of drug are loaded within a dosage form, and each dosage is related to a periodic interval. Delayed release systems, in contrast may not be sustaining, science often function of these dosage forms is to maintain the drug within the dosage form for some time before release. Successful fabrication of sustained release products is usually difficult and involves consideration of physicochemical properties of drug, pharmacokinetic behavior of drug, route of administration, disease state to be treated, and most important is that the placement of the drug in dosage form, which will provide the desired temporal and spatial delivery pattern for the drug.

During the last two decades there has been remarkable increase in interest in controlled release drug delivery system. This has been due to various factor viz. the prohibitive cost of developing new drug entities, expiration of existing international patents, discovery of new polymeric materials suitable for prolonging the drug release, and the improvement in therapeutic efficiency and safety achieved by these delivery systems.

8.1 Modified Release Drug Product

The term modified release drug product is used to describe products that alter the timing and/or the rate of release of the drug substance.

Types of Modified Release Drug Products

***Extended Release Dosage Forms*:** A dosage form that allows at least a two fold reduction in dosage frequency as compared to that of drug presented as an immediate release form.

Sustained release: It includes any drug delivery system that achieves slow release of drugs over an extended period of time not particularly at a pre-determined rate.

Controlled release: It includes any drug delivery system from which the drug is delivered at a predetermined rate over a long period.

***Delayed Release Dosage Forms*:** A dosage form releases a discrete portion of drug at a time or times other than promptly after administration, Ex: Enteric coated dosage forms.

Targeted Release Dosage Forms: A dosage forms that releases drug at /near the intended physiological site of action. Targeted release dosage forms may have extended release characteristics.

Repeat Action Dosage Forms: It is a type of modified release dosage forms, which is designed to release one dose or drug initially followed by a second dose of drug at a latter time.

***Prolonged Action Dosage Forms*:** It is designed to release the drug slowly and to provide a continuous supply of drug over an extended period.

Advantages of Sustained/Controlled Release Dosage Forms include reduction in dosing frequency; reduced fluctuations in plasma drug levels; increased patient compliance; more uniform effect; and patient compliance.

Disadvantages of Sustained/Controlled Release Dosage Forms include unpredictable or poor *in-vitro* and *in-vivo* correlation; dose dumping; reduced potential for dosage adjustment; poor systemic availability in general.

Criteria to be met by drug proposed to be formulated in controlled release dosage forms include: a) Desirable half-life, b) High therapeutic index, c) Small dose, d) Desirable absorption and solubility characteristics, e) Desirable absorption window and f) protection from first past effect.

8.2 Types Controlled Release Systems:

Diffusion controlled

- Reservoir
- Matrix
- Reservoir and monolithic

Dissolution controlled

- Encapsulation
- Matrix

Water penetration controlled

- Osmotically controlled
- Swelling controlled

Chemically controlled

- Erodible systems
- Drug covalently linked with polymer

Hydro gels

- Chemically controlled
- Swelling controlled
- Diffusion controlled
- Environment responsive

Ion-exchange resins

- Cationic exchange
- Anionic exchange

8.3 The Basic Mechanisms of Controlled / Sustained Release System

8.3.1 Diffusion controlled systems

Diffusion systems are characterized by release rate of drug, which is dependant on its diffusion through inert water insoluble membrane

There are basically two types of diffusion devices.

(i) Reservoir devices

(ii) Matrix devices

8.3.2 Reservoir Devices

Reservoir Devices are those in which a core of drug is surrounded by polymeric membrane. The nature of membrane determines the rate of release of drug from system.

The process of diffusion is generally described by a series of equations governed by **Fick's first law** of diffusion.

$$J = -D\frac{DC}{DX} \quad \text{......(1)}$$

Where, J: is the flux of drug across the membrane given in units of amount / area time, D: is diffusion coefficient of drug in membrane in units of area / time. This reflects on the drug molecule's ability to diffuse through the solvent and is dependent on the factors such as molecular size and charge i.e., dc/dt: represents rate of change in concentration C relative to a distance X in the membrane.

The law states that the amount of drug passing across a unit area are proportional to the concentration difference across that plane.

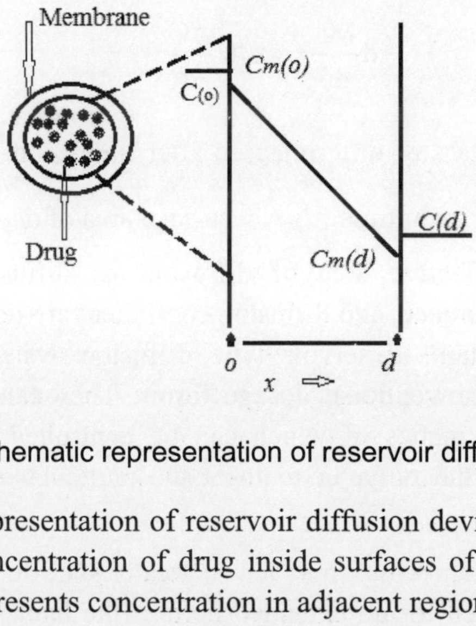

Fig. 8.1 Schematic representation of reservoir diffusion device

Schematic representation of reservoir diffusion device C_m *(o)*, and C_m *(d)* represent concentration of drug inside surfaces of membrane and *C (o)* and *C (d)* represents concentration in adjacent regions.

It is assumed that the drug on the both side of membrane is in equilibrium with its respective membrane surface (Fig. 8.1). Therefore the concentration just inside the membrane surface can be related to the concentration in the adjacent region, which can be expressed by following expression.

$$K = \frac{Cm(o)}{C(d)} \text{ at } x = 0 \qquad .....(2)$$

$$K = \frac{Cm(d)}{C(d)} \text{ at } x = d \qquad .....(3)$$

where, K = partition coefficient.

If we consider K & D are constants then equation (1) becomes,

$$J = DK\frac{\Delta C}{d} \quad .....(4)$$

where, Δ_c is the concentration difference across the membrane and d is path length of diffusion.

The simplest system to consider is that of slab, where drug release is from only one surface as shown in Figure 1, in this case equation (4) becomes:

$$d\frac{Mt}{dt} = ADK\frac{\Delta C}{d} \quad ....(5)$$

where, Mt = Mass of drug released after time t; $d\frac{Mt}{dt}$, is the steady state drug release rate at time 't'; A : surface area of device.

A constant effective area of diffusion is: diffusion path length, concentration difference, and diffusion coefficient are required to obtain a release rate constant. Reservoir type diffusion systems have several advantages over conventional dosage forms. They can offer zero order release of drug, kinetics of which can be controlled by changing the characteristics of the polymer to meet the particular drug and therapy conditions.

Plot (Fig. 8.2) showing approach to steady state for reservoir device that has been stored for an extended period (the burst effect curve) and for device that has been freshly made (the time lag curve)

Common methods used to develop reservoir type of devices include micro encapsulation of drug particles and press coating of tablets containing drug cores. The drug release generally involves combination of dissolution and diffusion with dissolution being process that controls the release rate. Examples of some materials used for designing reservoir type device include: hardened gelatin, ethyl or methyl cellulose, polyhydroxy- methacrylate, hydroxypropyl-methylcellulose, polyhydroxy -methacrylate, polyvinyl acetate, and various waxes, which may be used alone or in combination.

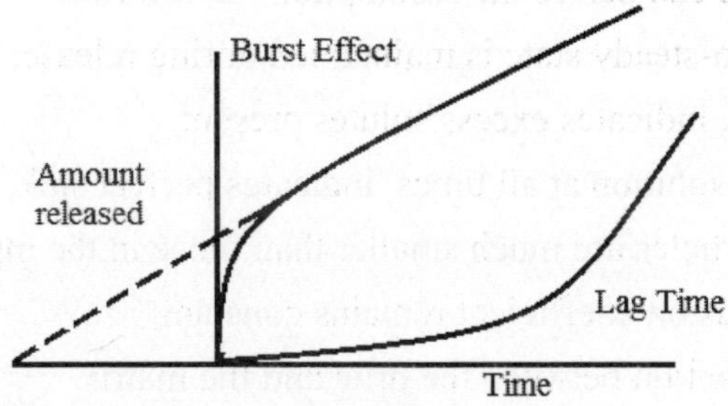

Fig. 8.2 Plot showing approach to steady state for reservoir device; the burst effect curve; and the time log cuve.

8.3.3 Matrix devices

A matrix device, as the name implies, consists of drug dispersed homogenously in an insoluble matrix. The rate of drug release depends on the drug diffusion but not on the solid dissolution. The appropriate equation describing drug release from the system gas been derived by **T. Higuchi**.

$$Q = \left[D\varepsilon / \tau \left(2A - \varepsilon Cs \right) C_s t \right]^{1/2} \qquad .....(6)$$

where, Q = weight in grams of drug released per unit surface area; D = diffusion coefficient of drug in the release medium at time = t; ε = porosity of the matrix; Γ = tortuosity of the matrix; Cs = solubility of the drug in the release medium; and A = concentration of drug in the tablet, expressed as g/ml.

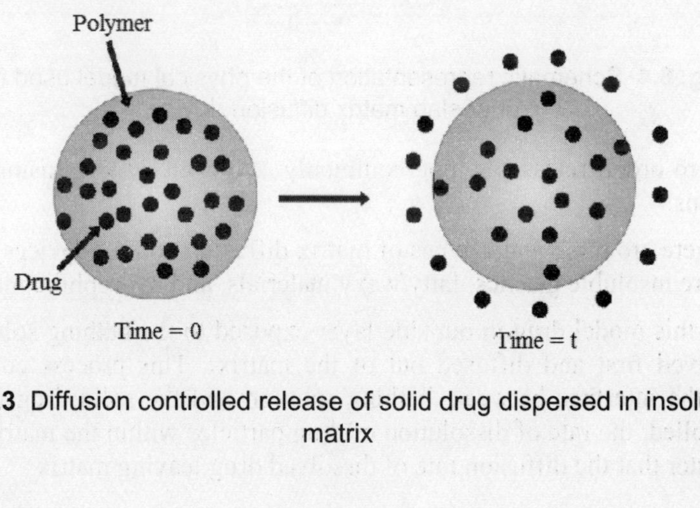

Fig. 8.3 Diffusion controlled release of solid drug dispersed in insoluble matrix

From eq. (6), one can derive the assumptions as follows:

- A pseudo-steady state is maintained during release.
- A >> Cs, indicates excess solutes present.
- C = 0 in solution at all times, indicates perfect sink.
- Drug particles are much smaller than those in the matrix.
- The diffusion coefficient remains constant.
- No interaction between the drug and the matrix

For the purpose the eq. (6) can be reduced to,

$$Q = kt^{1/2} \quad .....(7)$$

Therefore, a plot of amount of drug released versus the square root of time should be linear, if the drug release from the matrix is of diffusion controlled, which depends on: initial concentration of the drug in the matrix; drug solubility; porosity, tortuosity, leaching solvent composition; and the polymer system making up matrix.

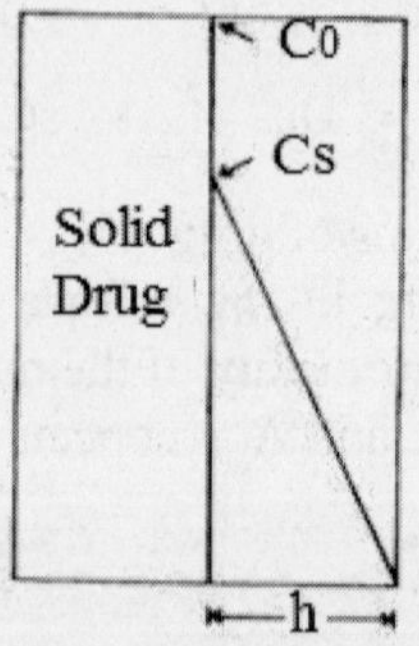

Fig. 8.4 Schematic representation of the physical model used for a planer slab matrix diffusion device.

Zero-order release is not commonly achieved *in-vivo* using these systems.

There are three major types of matrix diffusion control devices, which are: are insoluble plastics, fatty/waxy materials, and hydrophilic matrices.

In this model drug in out side layer exposed to the bathing solution is dissolved first and diffused out of the matrix. This process continues with the interface between bathing solution and the solid drug moving controlled, the rate of dissolution of drug particles within the matrix must be faster that the diffusion rate of dissolved drug leaving matrix.

8.3.4 Diffusion and dissolution Controlled Systems

The characteristics of swelling fluid uptake and the drug dissolution phenomena greatly affect the drug release since they both may represent the rate-limiting step for the macroscopic release. It can be assumed that the incoming swelling fluid affects the drug molecules motion by ruling the drug diffusion coefficient in the polymeric network, by altering the drug flux through the density gradient arising in the matrix during swelling, and by making possible drug dissolution. Although the bulk flow contribution to the global drug flux may be not negligible, the analysis can be confined to those systems characterized by either a low swelling degree or low swelling kinetics.

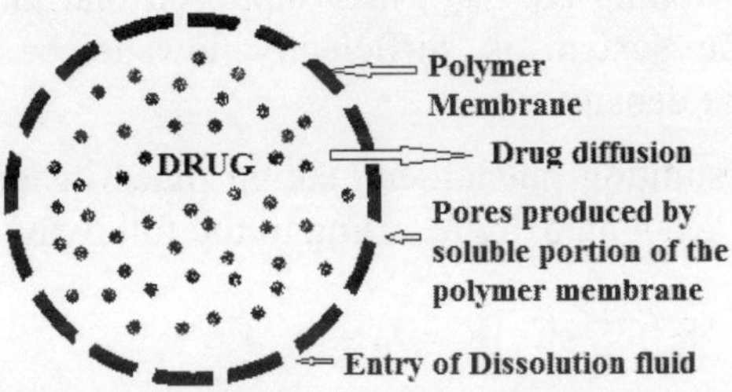

Fig. 8.5 Dissolution and Diffusion Controlled Release system

Even if the drug diffusion coefficient D_d depends on the swelling fluid concentration C and on drug concentration C_d, it can be assumed that only the first dependency plays an important role in determining D_d. Among the plethora of equations expressing the Dd dependence on C, a suitable choice is represented by the **Mackie and Meares equation**, which requires only the knowledge of the polymer volume fraction φ, which, in turn, depends on the solvent concentration C. This equation reads:

$$\frac{D_d}{D_{d0}} = \frac{(1-\varphi_i)^2}{(1+\varphi_i)^2} \qquad .....(8)$$

in which

$$\varphi_i = 1 - \frac{C_i}{\rho_s} \qquad .....(9)$$

where, D_{d0} is the drug diffusion coefficient in the pure swelling fluid, ρ_s is the swelling fluid density and i, represents the ith volume element into which the matrix can be ideally subdivided in. The mathematical expression for the density profile developing inside the swelling matrix may be formulated resorting to a swelling fluid mass balance made on the ith matrix volume element as:

$$\rho_{mi} = \frac{4_{\rho_p} + \frac{\rho_s\left(3C_i + C_{i-1}\right)}{2\rho_s - \left(C_i + C_{i-1}\right)} + \frac{\rho_s\left(3C_i + C_{i+1}\right)}{2\rho_s - \left(C_i + C_{i+1}\right)}}{4 + \frac{3C_i + C_{i-1}}{2\rho_s - \left(C_i + C_{i-1}\right)} + \frac{2C_i + C_{i+1}}{2\rho_s - \left(C_i + C_{i+1}\right)}} \qquad .....(10)$$

where, ρ_{mi} is the matrix density in the ith matrix volume element, ρ_p is polymer density, and C_{i-1}, C_i and C_{i+1} represent the swelling fluid concentration in the (i-1)th, ith and (i+1)th matrix volume elements, respectively. In writing eq. (10), it is supposed that the amount of drug contained in the system is sufficiently low to be neglected while computing matrix density.

The drug dissolution phenomena taking place inside the matrix may be conveniently accounted, by resorting to the following equation:

$$\frac{f\,C_{dd}}{f\,t} = -K\left(C_s - C_d\right)K = K^1S \qquad .....(11)$$

where, t is time, K is the dissolution constant, K^1 is the dissolution constant per unit area, S is the dissolution area (i.e., the surface of the solid drug particle contained in the ith matrix volume element), C_{dd} is the solid drug concentration, Cs is the drug solubility in the swelling fluid. Of course, eq.(11) takes place only in the presence of solid drug and if the local swelling fluid concentration is sufficiently high to allow significant dissolution of the drug. Assuming that, during dissolution, S is constant with time, K remains also constant. This implies that the solid drug particles filling the matrix can be considered as a very thin slab. If it is not the case, S and K are time dependent and, for spherical particles, K depends on time t according to:

$$K = 4\pi R\left(t\right)^2 K^I \qquad .....(12)$$

Being R(*t*), the time dependent particle radius.

In order to keep the model closer to reality, a phase transition upon dissolution has to be considered. This transition, although not affecting the drug diffusion coefficient, alters the drug solubility and a suitable model for this modification may be that proposed by **Nogami** as:

$$\frac{dC_s}{dt} = K_r\left(C_s^f - C_s\right) \quad .....(13)$$

where, Kr is the crystallization constant and C_s^f is the drug solubility at the end of the transition.

8.3.5 Erosion type Matrix system

Polymer erosion can take place because of chemical and physical reactions. Under particular physiological conditions, the hydrolysis of eventually present water-labile bonds incorporated in the polymer can cause chain breaking. Moreover, erosion can also be due to enzyme attack and chemical reactions on particular polymeric chains sites.

On the contrary, in physically cross-linked matrices, erosion is due to chain disentanglement, induced by the matrix swelling fluid and the hydrodynamic conditions imposed in the release environment. Obviously, polymer characteristics play a very important role in erosion kinetics. Erosion can develop according to two different mechanisms:

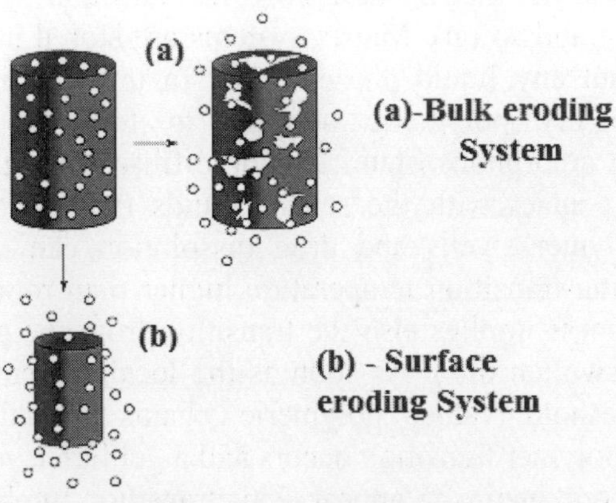

Fig. 8.6 Erodible Matrices

(a) surface or heterogeneous erosion, and (b) bulk or homogeneous erosion.

A mathematical model describing erosion induced by chemical reaction is reported by Thombre, Joshi and Himmelstein. These authors focussed their attention on polymer degradation from poly (ortho ester)-based delivery systems containing an acid-producing species to accelerate matrix hydrolysis. Assuming a slab geometry and perfect sink conditions, they proposed that firstly, water (A) moves in the matrix hydrolyzing the acid generation of compound (B) (for example an acid anhydride). Then, the generated acid (C) catalyzes polymer degradation according to two steps: (i) formation of an unstable intermediate ester (D*) and ii) reaction with water to final polymer degradation products

$$A+B \rightarrow C;\ C+D \rightarrow D^*;\ D^*+A \rightarrow E$$

The diffusion of the compounds due to the degradation process is ruled by the following equation as:

$$\frac{\partial C_i}{\partial t} = \frac{\partial}{\partial X}\left(D_i\left(X,t\right)\frac{\partial C_i}{\partial X}\right) + V_i \qquad \text{.....(14)}$$

where, C_i and D_i are the concentration and diffusion coefficient of the diffusing species respectively; X is the space coordinate; V_i is the net sum of synthesis and degradation rate of species I, at time t.

8.3.5.1 Release Mechanism from Matrices

Drug release kinetics may be affected by many factors such as polymer swelling, polymer erosion, drug dissolution/diffusion characteristics, drug distribution inside the matrix, drug/polymer ratio and system geometry (cylinder, sphere and so on). Matrix systems are stored in dry, shrunken state, i.e. without any liquid phase inside. In this condition, the drug, present in the dry polymeric network in form of microcrystals, nanocrystals or amorphous state, cannot diffuse through the network meshes. Upon contact with the release fluids (water or physiological media), the polymer swells and drug dissolution can take place (for polymers showing transition temperature higher than room temperature, the swelling process implies also the transition from the glassy dry state to the rubbery swollen one). As soon as the local solvent concentration exceeds a threshold value, polymeric chains unfold so that the glassy/rubbery polymer transition occurs and a gel like layer, surrounding the matrix dry core, begins to appear. This transition implies a molecular rearrangement of polymeric chains that tend to reach a new equilibrium condition as the old one was altered by the presence of the incoming solvent. The time required for this rearrangement depends on the relaxation time tr of the given polymer/solvent system which, in turn, is a function of both solvent concentration and temperature. If tr is much lower than the characteristic time of diffusion td of the solvent (defined as the ratio of the solvent diffusion coefficient at the equilibrium and the square of a characteristic length), then solvent adsorption may be described by means of Fick's law with a concentration dependent diffusion coefficient. On the contrary, if tr is much greater than td, then a Fickian solvent adsorption with constant diffusivity takes place.

However, in both cases, the diffusion of the drug molecule in the swelling network may be described by Fick'law with a non-constant diffusion coefficient, and the macroscopic drug release is defined as Fickian. When tr ≈ td, solvent adsorption does not follow Fick's law of diffusion. In such case, the macroscopic drug release becomes anomalous or non-Fickian. Thus, solvent absorption and drug release depends also on the polymer/solvent couple visco-elastic properties.

The glassy - rubbery transition enormously increases polymer chains mobility, so that network mesh enlarges and the drug can dissolve and diffuse through the gel layer.

While microcrystals dissolution does not show particularly interesting aspects, dissolution of nanocrystals and amorphous drug exhibits a peculiar behavior. Indeed, as solubility depends on crystal size, both of them, due to their small dimension (amorphous state can be considered as a crystal of vanishing dimensions), are characterized by a higher solubility in aqueous medium with respect to the infinitely large crystal, here identifiable with microcrystal (significant solubility improvements due to reduced dimensions, usually takes place for crystals radius smaller than 10 nm). Unfortunately, nanocrystals and amorphous state do not represent a thermodynamically stable condition and in the presence of the incoming solvent, they tend to merge to form the original macrocrystal. Accordingly, upon dissolution, nanocrystals and amorphous state tend to come back to the macrocrystal condition and yet dissolved drug undergoes a re-crystallisation with unavoidable solubility reduction. Dissolved drug re-crystallisation can take place inside the matrix as well as in the release environment and this phenomenon can be characterised by different recrystallisation constants. Other than this the average solubility and bioavailability may increase explaining the term 'activated drug' for matrix systems containing nanocrystals or amorphous state, provided that a proper polymer choice is made. In the dry, shrunken state, nanocrystals and amorphous state are fully stabilised by the polymeric network, which may reduce the recombination of the drug molecules into macrocrystals. This action is mainly due to both the polymer–drug interactions and the physical presence of the polymeric chains. Indeed, drug macrocrystals can form on the condition in which the dry network meshes be sufficiently wide.

Of course, drug diffusion through the swelling network depends on polymer/drug physical and chemical characteristics; adsorption / desorption phenomenon, which can take place on polymer chains during diffusion; diffusivity; and mesh size.

Additionally, drug diffusion can be heavily affected by a particular aspect of matrix topology. If wide network meshes are defined as accessible sites for the diffusing drug and small network meshes as forbidden sites, the entire network can be seen as a percolative network (fractal network). If forbidden sites approach a threshold value, it can be demonstrated that the diffusion on percolative (fractal) networks differs a lot from diffusion in non-fractal networks and release kinetics is different. Drug distribution in the matrix (i.e., drug concentration profile) can heavily affect drug release kinetics.

A very clear way for the representation of polymer swelling/erosion and drug delivery from matrix systems is observed in three main fronts appears during the release process (Fig. 8.7). The eroding front separates the release environment from the matrix (it moves outwards when swelling kinetics is predominant on the erosion process, while it moves inwards in the opposite case) and its position depends on the combination of release environment, hydrodynamic conditions and on matrix cross-linking strength.

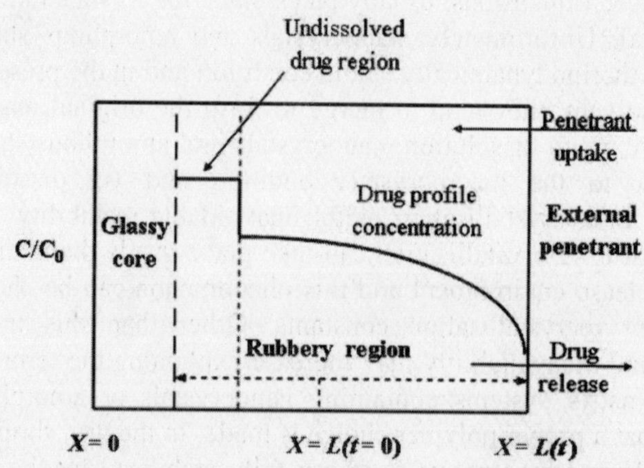

Fig. 8.7 The external fluid uptake gives origin to the formation of three fronts: the swelling front, the diffusion front and the erosion front.

Finally, it is also worth mentioning that matrix geometry (planar, spherical, cylindrical, and so on) highly influences drug release kinetics.

8.3.6 Osmotic Controlled Release Systems

In this type of system, osmotic pressure is the driving force that generates constant drug release (Fig. 8.8). The system can be fabricated by applying a semepermeable membrane around a core of an osmotically active drug or a core of an osmotically inactive drug in combination with an osmotically active salt. A delivery orifice is drilled in each system by laser or by a high-speed mechanical drill. The optimum size of the orifice can be calculated by the equation:

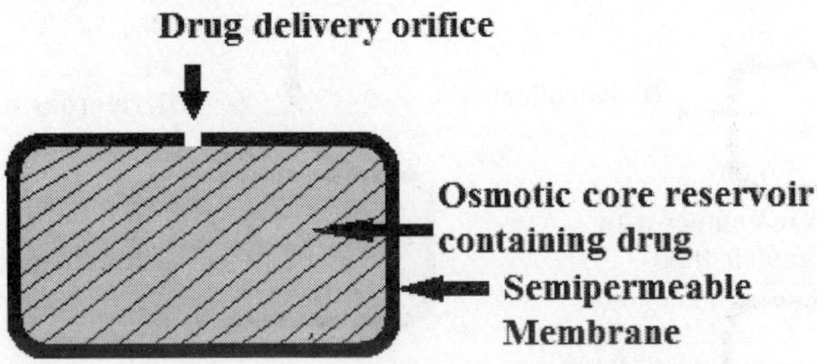

Fig. 8.8 The elementary osmotic pump

$$A_s = (LV/t)\,(8\pi)\,(\square/P)^{1/2} \qquad .....(15)$$

where, A_s is the cross sectional area of the orifice. V/t is the volume released per unit time, L is the diameter of the orifice, π is 3.4, η is the viscosity of the solution moving from the inside to the outside of the device, and P is the hydrostatic pressure difference.

When, an osmotic system exposed to water or body fluid, water flow inside because of pressure difference across the semipermeable membrane. Under this osmotic pressure gradient, the volume flow of water into the core reservoir, dV/dt, is expressed as:

$$dV/dt = (Ak/h)\,(\Delta\pi \sim \Delta P) \qquad .....(16)$$

where, A, k, and h are the area, membrane permeability, and thickness, respectively, $\Delta\pi$ is the osmotic pressure difference. The drug pumps out of the orifice at a controlled rate, dM/dt, which is equal to the volume flow rate of water into the core by the drug concentration, Cs. Thus the rate of delivery will be governed by the equation as:

$$dM/dt = (dV/dt)\,Cs \qquad .....(17)$$

In principle the drug delivery from the osmotic controlled systems follows zero-order rate until the concentration of the osmotically active agent is below saturable solubility.

8.3.7 Altered Density Systems

This approach involve in altering the density of the formulations using high or low density polymer or pellet to increase the GI residence of the drug for prolonged drug delivery.

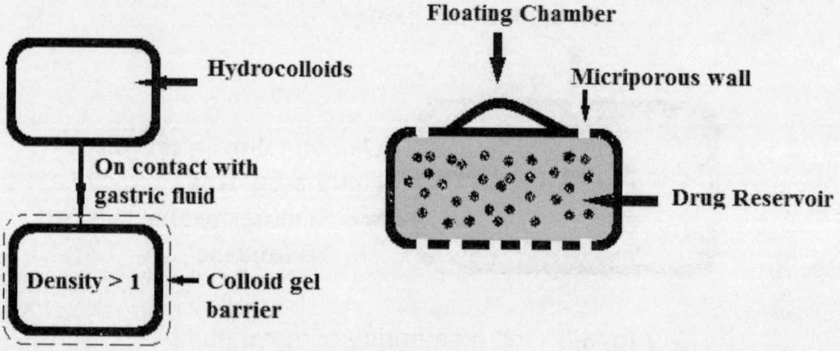

Fig. 8.9 Floating or Buoyant system (Tablet or Capsule)

(a) High-density approach:

In this approach, the density of the pallets must exceed that of normal stomach content and be therefore at least 1.4, while preparing such systems, drug can be coated on a heavy core, mixed with barium sulphate, titanium dioxide, iron powder, and zinc oxide to formulate high density systems, which after administration sink to the bottom of the stomach and are entrapped in the folds of the antrum and withstand the peristaltic waves of the stomach. Sedimentation has been employed as a retention mechanism for high density systems.

(b) Low-density approach:

This approach is used to retain the drugs in the stomach by increasing gastric residence of the drug and is called as **gastro retentive drug delivery** systems (**GRDDS**), which helps in sustaining the drug delivery for longer period of time. One such approach can be achieved by making **floating drug delivery** systems (**FDDS**) or also called as buoyant systems, in which the drugs mixed with swellable polymers such as hydroxyl-ethyl-cellulose, hydroxyl-propyl-cellulose, hydroxyl-propyl-methyl-cellulose to prepare the matrices and compressed to get the tablet. On contact with gastric fluid, the tablet forms a water impermeable colloid gel barrier around its surface and maintains a bulk density less than one, there by allowing it to remain floating in the gastric fluid.

Another application of buoyant system follows the incorporation of a gel filled flotation chamber into a microporous compartment, which housed a reservoir, as shown in the Fig. 9. The presence of apertures along the top and bottom walls of the system, allows the gastric fluids to enter inside to dissolve the drug. Its peripheral walls are sealed to prevent the passage of undissolved drugs.

8.3.8 Ion-Exchange Resins

It is based on the formation of drug resin complex in which an ionic solution is kept in contact with ion-exchange resins. The drug from this complex gets exchanged in gastrointestinal tract and released with excess of Na^+ and Cl^- present in gastrointestinal tract as:

$$\text{Resin } [N(CH_3)]^+ X^- + Z^- \longrightarrow \text{Resin } [[N(CH_3)]^+ Z + X^- \quad .....(18)$$

or,

$$\text{Resin } (SO_3^-) A^+ + B^+ \longrightarrow \text{Resin } (SO_3^-) B^+ + A^+ \quad .....(19)$$

where, X^- and A^+ are drug ions.

These systems generally utilize resin compounds of water insoluble cross-linked polymer. They contain salt – forming functional group in repeating positions on the polymer chain. The rate of drug diffusion out of the resin is controlled by the area of diffusion, diffusion path length and rigidity of the resin which is function of the amount of cross linking agent used to prepare resins. The release rate can be further controlled by coating the drug resin complex by micro-encapsulation process. The resins used include amberlite indion, polysterol resins and others.

8.3.9 pH– Independent Formulations

The gastrointestinal tract present some unusual features for the oral route of drug administration with relatively brief transit time through out the gastrointestinal tract, which constraint the length of prolongation. Further the chemical environment (variable in pH ranges from 7 in the mouth, 1 to 3 in the stomach and 6 to 8 in the small intestine) throughout the length of gastrointestinal tract is a constraint on dosage form design. Since most drugs are either weak acids or weak bases, the release from sustained release formulations is pH dependent. However, buffers such as salts of amino acids, citric acid, phthaleic acid, phosphoric acid or tartaric acid can be added to the formulation, to help in maintaining a constant pH, thereby rendering pH independent drug release. A buffered controlled release formulation is prepared by mixing a basic or acidic drug with one or more buffering agent, granulating with appropriate pharmaceutical excipients and coating with gastrointestinal fluid permeable film forming polymer. When gastrointestinal fluid permeates through the membrane, the buffering agents adjust the fluid inside to suitable constant pH thereby rendering a constant rate of drug release e.g. propoxyphene in a buffered controlled release formulation, which significantly increase reproducibility.

8.4 Factors Governing the Design of SR /CR Formulations

8.4.1 Physico-Chemical Properties

8.4.1.1 Molecular Size and Diffusivity

A drug must diffuse through a variety of biological membranes during its time course in the body. In addition to diffusion through these biological membranes, drugs in many extended-release systems must diffuse through a rate-controlling polymeric membrane or matrix. The ability of a drug to diffuse in polymers, so-called its diffusivity (diffusion coefficient D), is a function of its molecular size (or molecular weight). For most polymers, it is possible to relate log D empirically to some function of molecular size as:

$$\log D = -S\nu \log U + K\nu = -sM \log M + Km \qquad .....(20)$$

Where, ν is the molecular volume, M is molecular weight, $S\nu$, sM, $K\nu$ and Km are constants. Thus the value of D, is related to the size and shape of the cavities as well as size and shape of drugs.

8.4.1.2 Aqueous Solubility

The fraction of drug absorbed into the portal blood is a function of the amount of drug in the solution in the G.I tract, which depends on the intrinsic solubility of the drug

For a drug to be absorbed, it must dissolve in the aqueous phase surrounding the site of administration and then partition into the absorbing membrane. The aqueous solubility of a drug influences its dissolution rate, which in turn establishes its concentration in solution and hence, the driving force for diffusion across membranes.

Dissolution rate is related to aqueous solubility, as shown by the **Noyes-Whitney equation**, under sink conditions as:

$$dC/dt = kD\ A.Cs$$

where, dC/dt is the dissolution rate, kD is the dissolution rate constant, A is the total surface area of the drug particles, and Cs is the aqueous saturation solubility of the drug. The dissolution rate is constant only if "A" remains constant, but the important point to be noted that at the initial state the rate of dissolution is directly proportional to Cs. Therefore, the aqueous solubility of a drug can be used as a first approximation of its dissolution rate. Drugs with low aqueous solubility have low dissolution rates and usually suffer from oral bioavailability problems. The aqueous solubility of weak acids or bases is governed by the pKa of the compound and the pH of the medium. For a weak acid

$$St = S_0(1+Ka/[H+]) = S_0(1+10pH\text{-}pKa) \qquad .....(21)$$

where, St is the total solubility (both the ionized and unionized forms) of the weak acid, S_0 is the solubility of the unionized form. Ka is the acid dissociation constant, and [H+] is the hydrogen ion concentration in the medium. Similarly, for a weak base

$$St = S_0(1+[H+]/Ka) = S_0(1+10pKa\text{-}pH) \qquad .....(22)$$

where, St is the total solubility (both the conjugate acid and free base forms) of the weak base, S_0 is the solubility of the free-base, and Ka is the dissociation constant of the conjugate acid, Equations 20 and 21 predict the total solubility of a weak acid or base with a given pKa, which can be affected by the pH of the medium. Taking to the consideration of equations 21 and 22, a pH-partition hypothesis can be established relative to drug absorption, which simply states about the unionized form of the drug in the stomach (pH = 1 to 2), their absorption will be excellent in such an acidic environment. On the other hand, weakly basic drugs exist primarily in the ionized form (conjugate acid) at the same site, and their absorption will be poor. In the upper portion of the small intestine, the pH is more basic (pH = 5 to 7), and the reverse will be expected for weak acids and bases.

The Bio-pharmaceutical Classification System (BCS) allows estimation of likely contribution of three major factors such as solubility, dissolution and intestinal permeability, which reflects on the oral drug absorption.

Classification of drugs according to BCS

Class I High solubility-High permeability

Class II Low solubility-High permeability

Class III High solubility-Low permeability

Class IV Low solubility-Low permeability

High solubility: Largest dose dissolves in water over a pH range 1-8

High permeability: Extent of absorption is > 90% for Class III and poor for Class IV drug candidates.

In general, extreme aqueous solubility of a drug is undesirable for formulation of an extended-release product. A drug with very low solubility and a slow dissolution rate will exhibit dissolution-limited absorption and yield an inherently sustained blood level. In most instances, formulation of such a drug into an extended release system may not provide considerable benefits over conventional dosage forms. Even, if a poorly soluble drug was considered to be a candidate for formulation of an extended-release system. Any system relying on diffusivity of the drug through a polymer as the rate-limiting step in the drug release would be suitable for sustained drug delivery. However, for a drug with very high solubility and a rapid dissolution rate, may decrease its dissolution rate and slow down its absorption.

The pH dependent solubility, particularly in the physiological pH range would be another problem for S.R/C.R formulation, because of the variation in the pH throughout the gastro intestinal tract and hence variation in dissolution rate. Ex: Phenytoin.

Examples of drugs which are poor candidates for S.R/C.R release systems:

Drugs, limited in the absorption by their dissolution rates are: Digoxin, Warfarin, Griseofulvin, and Salicylamide; Drugs poorly soluble in the intestine are: Diazepam, Diltiazem, Cinnarizine, Chlordiazepoxide, and Chlorpheniramine; Drugs having lower solubility in stomach: Furosemide.

8.4.1.3 pKa / Ionization Constant

The pKa is a measure of the strength of an acid or a base. The pKa used to determine the charge on a drug molecule at any given pH. Drug molecules are active only in the undissociated state and also unionized molecules cross these lipoidal membranes much more rapidly than the ionized species. It has been found that the amount of drug exists in unionized form, is a function of dissociation constant of that drug and the pH of fluid at absorption site. For a drug to be absorbed, it must be in unionized form at the site of absorption. Drugs which exist in ionized form at the absorption site are poor candidates for sustained/controlled dosage forms.

8.4.1.4 Partition Coefficient

Partition coefficient influences not only the permeation of drug across the biological membranes but also diffusion across the rate controlling membrane or matrix. After administration and before elimination a drug must have to diffuse through a variety of biological membranes that act primarily as lipid-like barriers. A major criterion in evaluation of the ability of a drug to penetrate these lipid membranes (i.e., its membrane permeability) is its measurement of apparent oil/water partition coefficient, defined as

$$K = C(O)/C(W)$$

where, C(O) is the equilibrium concentration of all forms of the drug in an organic phase at equilibrium, and C(W) is the equilibrium concentration of all forms in an aqueous phase In general, drugs with extremely large values of K are very oil-soluble and will partition into membranes quite readily.

It has been observed that the value of K at which optimum activity is observed is approximately 1000/1 in n-octanol/water. Drugs with a partition coefficient that is higher or lower than the optimum are, in general, poorer candidates for formulation into extended release dosage forms.

8.4.1.5 Release Rate and Dose

The rate of release can best be illustrated by the following simple kinetic scheme as:

The absorption pool represents a solution of the drug at the site of absorption, and the terms *Kr*, *Ka* and *Ke* are first order rate constants for drug release, absorption, and overall elimination, respectively. Immediate release from a conventional dosage form implies that *Kr* >> Ka or, alternatively, the absorption of drug across a biological membrane, such as the intestinal epithelium, is the rate limiting step in delivery of the drug to its target area. For non immediate-release dosage forms, *Kr* << *Kw* i.e., release of drug from the dosage form is the rate-limiting step.

The ideal goal in designing an extended-release system is to deliver drug to the desired site at a rate according to the needs of the body (i.e., a self-regulated system based on feedback control), which, implies that the rate of delivery must be independent of the amount of drug remaining in the dosage form and constant over time. The release pattern should follow zero-order kinetics, following equation:

$$K^0_r = \text{Rate In} = \text{Rate Out} = K_r\, C_d\, V_e$$

where, K^0_r is the zero-order rate constant for drug release (amount/time), *ke* is the first-order rate constant for overall drug elimination (time^{-1}), C_d is the desired drug level in the body (amount/volume), and *Vd* is the volume of the space in which the drug is distributed. The values of *ke, Cd, and Vd* need to calculate $K^0_{r,}$ which can be obtained from appropriately designed single-dose pharmacokinetic studies. It already has been mentioned that a perfectly invariant drug blood or tissue level versus time profile is the ideal starting goal of an extended release system, which can be achieved by a maintenance dose, which releases its drug content following zero-order kinetics.

However, satisfactory approximations of a constant drug level can be obtained by suitable combinations of the initial dose and a maintenance dose that releases its drug following a first-order process. The total dose for such a system can be calculated by:

$$W = D_i + (K_e C_d / K_r V_d)$$

where, K*r* is the first-order rate constant for drug release (time^{-1}), and K_e, C_d, and V_d are as defined previously. To maintain drug blood levels within the therapeutic range over the entire time course of therapy, most extended-release drug delivery systems are, like conventional dosage forms with multiple rather than single doses. For an ideal extended-release system that releases drug by zero-order kinetics, the multiple dosing regimens are analogous to that of a constant intravenous infusion (Alfonso R., 2002).

8.4.2 Biological Factors

8.4.2.1 Absorption

The rate, extent, and uniformity of absorption of a drug are important factors when considering its formulation into an extended release system. The critical factors incase of oral administration is Kr << Ka, assuming that the transit time of drug through the absorptive area of gastrointestinal tract is between 9-12 hours and the maximum absorption half-life should be within 3-4 hours. This corresponds to a minimum absorption rate constant Ka value of 0.17-0.23/hr necessary for about 80-95% absorption over a 9-12hr transit time

Gilberts et al., 2001, stated that for a drug with a very slow rate of absorption (Ka<<0.17/hr), the first order release rate constant Kr is less than 0.17/hr, results in unacceptable poor bioavailability in many patients. Therefore, slowly absorbed drug will be difficult to be formulated into extended release systems, where the criterion Kr<<Ka must be met (Rudnic and Schawartz., 2000). If the drug irregularly absorbed, because of variable absorptive surface of gastrointestinal tract, design of the sustained/controlled release product would be more difficult. Ex: The oral anticoagulant such as Dicoumarol, Iron, etc. Also, drugs absorbed by active transport system are unsuitable for sustained / controlled drug delivery system such as: Methotrexate, Enalapril, Riboflavin, Pyridoxine, 5-Fluorouracil,5-Bromo uracil, Nicotinamide, Fexofenadine, and Methyl-dopa, etc.

8.4.2.2 Distribution

The distribution of a drug into vascular and extra vascular spaces in the body is an important factor in the overall elimination kinetics, which depends on the volume of distribution and barriers in transportation. Apparent volume of distribution and ratio of drug in tissue to plasma (T/P) concentration are used to describe the distribution characteristics of a drug, which plays major roles in determining the dosage regimen. For drugs which have apparent volume of distribution higher than real volume of distribution, such drugs are extensively bound to extra vascular tissues, for example: **chloroquine**. For such drugs the elimination half life is decreased, for which those drug leaves the body gradually provided drug elimination rate is controlled by the release of drug from tissue binding sites and the concentration of the drug released from the tissues exceed the threshold level or within the therapeutic range resulting sustained the drug release.

8.4.2.3 Metabolism

The metabolic product may be active or inactive materials to be eliminated. If the metabolite is an active one, that will again reflect on the fixation of dosage regimen basing on its rate of elimination and recycling.

There are two areas concerning metabolism, which significantly restrict SR product design such as: if a drug upon chronic administration is capable of either inducing or inhibiting enzyme synthesis, it will be a poor candidate for a S.R/C.R product because of the difficulty in maintaining uniform blood levels of that drug; and if there is a variable blood level of a drug through either intestinal (or tissue) metabolism or through first pass effect, which may make formulation of SR dosage form difficult, because of saturation, the fraction of the drug loss will be dose dependent, resulting significant reduction in bioavailability. Examples of drugs unsuitable to formulate SR formulations because of:

- Fluctuating drug blood levels due to intestinal metabolism upon oral dosing: Salicylamide, Isoproterenol, Chlorpromazine, Clonazepam Hydralazine and Levodopa.
- Fluctuating drug blood levels due to first pass hepatic metabolism upon oral dosing: Nortriptyline, phenacetin, morphine, propranolol.
- Fluctuating blood levels due to enzyme induction are poor candidates for Sustained / controlled Release dosage forms: Griseofulvin, Phenytoin, Primidone, Barbiturates, Rifampicin, Meprobamate, Cyclophosphamide.
- Fluctuating blood levels due to enzyme inhibition are poor candidates for Sustained / Controlled Release dosage forms: Isoniazid, Cimetidine, Amiodarone, Erythromycin, Fluconazole, Ketoconazole, MAO – inhibitors, Para-aminosalicyclic acid, Allopurinol, Coumarins.

8.4.2.4 Dose Dependent Bio-Availability

In case of Propoxyphene, bio-availability is dose dependent. Only 18% of 65mg dose, 28% of 130 mg dose, 33% of 195 mg dose reaches the systemic circulation due to first pass effect. It makes the S.R/C.R dosage form less desirable.

8.4.2.5 Elimination Half Life

Half life is the time taken for the amount of drug in the body (or the plasma concentration) to fall by half and is determined by both clearance (Cl) and volume of distribution (Vd). $t_{1/2} = 0.693.Vd/Cl$ (assuming 1st order kinetics). Half life is increased by increasing in volume of distribution or a decrease in clearance, and vice-versa.

For drugs that follow linear kinetics, the elimination half-life is constant and does not change with dose or drug concentration. For drugs, which follow non-linear kinetics, the elimination half-life and drug clearance both change with dose or drug concentration. Drugs with short half-lives (<2hrs) and high dose impose a constraint on formulation into sustained/controlled release systems because of the necessary dose size and drugs with long half-lives (>8hr) are inherently sustained drug delivery (Birkett,1998).

8.4.2.6 Drug -Protein Binding

The drug can bind to components like blood cells, plasma, tissue proteins and macromolecules. Drug protein binding is a reversible process. As the free drug concentration in the blood decreases, the drug-protein complex dissociates to liberate the free drug and maintain equilibrium. Due to this reversible binding of a drug, the free drug levels of the drug are maintained for long time in the blood leading to a long biological half-life. A protein bound drug due to its high molecular size is unable to enter into hepatocytes, resulting in reduced metabolism. The bound drug is not available as a substrate for liver enzymes there by further reducing the rate of metabolism.

The glomerular capillaries do not permit the passage of plasma-protein and drug protein complexes. Hence only unbound drug is eliminated. The elimination half-life of drugs generally increases, when the percent of bound drug to plasma increases. Such drugs need not be formulated into sustained/controlled release formulations. Since blood proteins are mostly re-circulated, with out elimination, high drug protein binding can serve as a depot for drug producing a prolonged drug action.

8.4.2.7 Therapeutic Index

It is most widely used to measure the margin of safety of a drug.

$$TI = LD_{50} / ED_{50}$$

where, TI is the therapeutic index, LD is the lethal dose and ED is the effective dose.

The more the value of TI more the safety expected from the drug. Drugs with very small value of Therapeutic index are poor candidates for formulation into sustained release products. A drug is considered to be safe if its TI value is greater than 10.

Design and Formulation of some Oral Controlled Release Drug Delivery System

8.5 Osmotically Controlled Drug Delivery System (OCDDS)

Osmotic devices are the most reliable controlled drug delivery systems (CDDS) and can be employed as oral drug delivery systems. Osmotic pressure is used as the driving force for these systems to release the drug in a controlled manner. Osmotic pump tablet (OPT) generally consists of a core including the drug, an osmotic agent to create osmotic pressure, other excipients and semipermeable membrane as coat, which allows water to penetrate inside to dissolve the API as well as help in swelling of osmogents, which determine the release rate (Ch. Ajay Babu, *et. al.*, 2010).

Advantages

It is easy to formulate and simple in operation; improve patient compliance with reduced frequency of administration; prolonged therapeutic effect with uniform blood concentration.

Disadvantages

The disadvantage include: dose dumping; rapid development of tolerance; retrieval therapy is not possible in the case of unexpected adverse effect.

8.5.1 Types of Osmotically Controlled Drug Delivery systems

Elementary osmotic pump: consists of a core of active pharmaceutical agent (API), osmogents and a coat of Semi permeable membrane with delivery orifice. The water penetrates inside the dosage form, results in formation of saturated solution of drug within the core, which is dispensed at a controlled rate from the delivery orifice present in the membrane.

***Multi chambered osmotic pumps*:** consists of core tablet as a layer coated with semipermeable membrane, middle layer consists of osmogents to push the layer attached to API. The middle pushing layer swells and pushes the drug to release through the orifices present on two sides of the tablet. API releases from two sides of tablets.

***Push-pull osmotic pumps*:** consists of a core tablet i.e. Layer one of API, osmogents as Layer two i.e. polymeric osmotic agents and a coat of semi permeable membrane with delivery orifice. After coming in contact with the aqueous environment, polymeric osmotic layer swells and pushes the drug layer, and thus releasing drug in the form of fine dispersion through the orifice.

***Controlled-porosity osmotic pumps (CPOP)*:** consists of a core of active pharmaceutical agent (API), osmogents and a coat of Semi permeable membrane containing water soluble additives. Delivery orifice is formed by the incorporation of a leachable component. Once the tablet comes in contact with aqueous environment, water-soluble additives dissolve, lead to the formation of a microporous membrane. Water diffuses into the core through the microporous membrane, creating an osmotic gradient and thereby controlling the release of drug.

8.5. 2 Formulation and Development of Controlled-Porosity Osmotic Pumps (CPOP)

The controlled-porosity osmotic pump tablet (CPOP) is a spray-coated or coated tablet with a semipermeable membrane (SPM) containing leachable pore forming agents. The drug release is achieved through the pores, which are formed in the semipermeable wall *in situ* during the operation. In this system, the drug, after dissolution inside the core, is released from the osmotic pump tablet by hydrostatic pressure and diffuse through pores created by the dissolution of pore formers incorporated in the membrane (Fig. 8.10). The hydrostatic pressure is created either by an osmotic agent or by the drug itself or by a tablet component, after water is imbibed across the semipermeable membrane. This membrane after formation of pores becomes permeable for both water and solutes. A controlled-porosity osmotic wall can be described as having a sponge like appearance. The pores can be continuous that have micro porous lamina, interconnected through tortuous paths of regular and irregular shapes. Generally, materials (in a concentration range of 5% to 95%) producing pores with a pore size from 10 Å -100μm can be used.

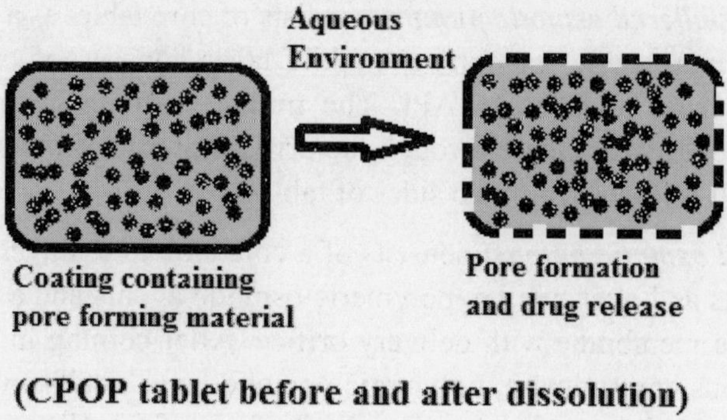

Fig. 8.10 schematic of diffuse through pores created by the dissolution of pore formers incorporated in the membrane of CPOP

This system is generally applicable for only water-soluble drugs because of poorly water soluble drugs cannot dissolve adequately in the volume of water drawn into the OPT. Recently this problem was overcome by adding agents like sulfobutyl ether-cyclodextrin $(SBE)_{7m}$- β-CD or hydroxypropyl-β-cyclodextrin (HP-β-CD) as solubilizing and osmotic agents. Several approaches have been developed to prepare the porous membrane by spray coating using polymer solutions containing dissolved or suspended water-soluble materials. The rate of drug release can also be varied by having different amounts of osmogents in the system to form different concentrations of channeling agents for delivery of the drug from the device. Incorporation of the β-cyclodextrin-drug complex has also been used as an approach for the delivery of poorly water-soluble drugs from the osmotic systems, especially controlled-porosity osmotic pump tablets.

Basic components required for controlled-porosity osmotic pump

(a) Drug; (b) Osmotic agent; (c) Semipermeable membrane; (d) Channeling agents or pore forming agents.

(a) **Criteria for selection of a drug:** which include, short biological Half-life (2- 6 hrs); high potency; required for prolonged treatment (e.g: Nifedipine, Glipizide, Verapamil and Chlorpromazine hydrochloride).

(b) **Osmotic agent:** Polymeric osmogents are mainly used in the fabrication of osmotically controlled drug delivery systems and other modified devices for controlled release of relatively insoluble drugs. Osmotic pressures for concentrated solution of soluble solutes commonly used in controlled release formulations are extremely high, ranging from 30 atm for sodium phosphate up to 500 atm for a lactose-fructose mixture. These osmotic pressures can produce high water flows across semipermeable membranes. The osmotic water flow across a membrane is given by the equation,

$$dv/dt = \frac{A\theta\Delta\pi}{l}$$

Where, dv/dt, is the rate of water flow across the membrane of area A, thickness l, permeability θ in cm^3 and Δπ is the osmotic pressure difference.

***Examples of Osmogents*:** the most commonly used osmotic agents include Sucrose, Fructose, Mannitol, Dextrose, Potassium sulphate, Potassium chloride, Sodium phosphate tribasic, Sodium phosphate dibasic, Sodium phosphate dibasic, Sodium phosphate monobasic, Anhydrous Sodium phosphate dibasic, or a combination of above agents such as: Lactose-Fructose, Dextrose-Fructose, Sucrose-Fructose, Mannitol-Fructose, Lactose-Sucrose, Lactose-Dextrose, Mannitol-Dextrose, Dextrose-Sucrose, Mannitol-Sucrose, and Mannitol-Lactose, etc.

(c) **Semipermeable Membrane:** The membrane should be stable to both outside and inside environments of the device. The membrane must be sufficiently rigid so as to retain its dimensional integrity during the operational lifetime of the device. The membrane should also be relatively impermeable to the contents of dispenser, so that there will be no loss of osmogent by diffusion across the membrane. Finally, the membrane must be biocompatible. Some examples of polymeric materials that form membranes are cellulose esters such as cellulose acetate, cellulose acetate butyrate, cellulose triacetate, ethyl cellulose and Eudragits, etc.

(d) **Channeling / Leachable pore forming agents:** These are the water-soluble components, which play an important role in the controlled drug delivery systems. When the dissolution medium comes in contact with the semipermeable membrane, it dissolves the channeling agent and forms pores on the semipermeable barrier. Then the dissolution fluid enters the osmotic system and releases the drug in a controlled manner over a long period of time by the process of osmosis. Some examples of channeling agents are polyethylene glycol (PEG) 1450, D-mannitol, bovine serum albumin (BSA), diethylphthalate, dibutylphthalate and sorbitol, etc.

8.5.3 Some Examples of Research Outputs

Roger A Rajewski *et al* (1999) studied the membrane controlling factors responsible for drug release from a controlled-porosity osmotic pump tablet (OPT) that utilizes sulfobutyl ether-cyclodextrin, [$(SBE)_{7m}$–β-CD], both as solubilizing agent and osmotic agent. The release rate of **chlorpromazine** (CLP) from OPTs containing $(SBE)_{7m}$ - β -CD increased with increasing amounts of micronized lactose and decreasing amounts of triethyl citrate.

Roger A Rajewski *et al* (2004) investigated the application of controlled-porosity osmotic pump tablet (OPT) utilizing $(SBE)_{7m}$ - β – CD, both as a solubilizer and an osmotic agent for drugs with varying physical properties. OPTs utilizing $(SBE)_{7m}$ - β -CD were prepared for five poorly soluble drugs such as **prednisolone**, **estradiol**, **naproxen**, **indomethacine** and **chlorpromazine** and for two highly water soluble drugs such as **diltiazem hydrochloride** and **salbutamol sulfate**. It was found that for the soluble drugs $(SBE)_{7m}$ -β-CD acts primarily as an osmotic and an OPT control agent. Significantly, $(SBE)_{7m}$ -β-CD not only enhances the delivery of poorly soluble drugs from OPTs, but acts as a controlling excipient for soluble drugs.

Mahalaxmi R *et al* (2009) developed the extended release controlled porosity osmotic pump formulations of model drug glipizide using a wicking agent and a solubilizing agent. The effect of different formulation variables such as wicking agent, solubilizing agent, level of pore former and membrane weight gain on *in-vitro* release were studied. Drug release was found to be affected by the level of wicking agent and solubilizing agent in the core. **Glipizide** release from controlled porosity osmotic pump was directly proportional to the level of pore former (sorbitol) and inversely proportional to membrane weight gain.

Ji-Eon Kim *et al* (2000) studied the effect of various pore formers on the controlled release of an antibacterial agent from a polymeric device. **Cefadroxil** was chosen as the model antibiotic and was incorporated into a polyurethane matrix by the solvent-casting method. Polyethylene glycol or mannitol, or bovine serum albumin (BSA) was used as a pore former. The morphological changes in the matrices before and after release studies were investigated by scanning electron microscopy (SEM). Changing the weight fraction and particle size of the pore formers/drug mixtures could control the release of cefadroxil from the matrix. The release rate of cefadroxil increased as the loading dose of the pore former increased (15<20<25%).

Gaylen. M ***et al*** **(1991)** studied on the solubility of resin-modulated method to effectively manipulate drug release kinetics from controlled porosity osmotic pumps. These solubility-modulated devices administered to dogs, release **diltiazem hydrochloride** with similar *in-vivo / in-vitro* kinetics. These approaches may be applicable to extend osmotic pump technology to drugs with intrinsic water solubility, which may be too high or low for conventional osmotic pump formulation to release the drug for twelve hrs independent of the environmental pH.

Longxiao Liu ***et al*** **(2008)** developed the bilayer-core osmotic pump tablet (OPT) for nifedipine, which does not require laser drilling to form the drug delivery orifice. The bilayer-core consisted of two layers: (a) push layer and (b) drug layer, and was made with a modified upper tablet punch, which produced an indentation at the center of the drug layer surface. The indented tablets were coated by using a conventional pan-coating process. Sodium chloride was used as osmotic agent, polyvinylpyrrolidone as suspending agent and cross-Carmellose sodium as expanding agent. The indented core tablet was coated by ethyl cellulose as semipermeable membrane containing polyethylene glycol 400 for controlling the membrane.

Sanjay Garg ***et al*** **(2003)** studied the development and evaluation of extended release formulations of **isosorbide mono nitrate** (IMN) based on osmotic technology. The release from developed formulations was independent of pH and agitational intensity, but dependent on the osmotic pressure of the release media. Results of SEM studies showed the formation of pores in the membrane from where the drug release occurred. Prediction of steady state levels, showed the plasma concentrations of IMN i.e., within the desired range permeability. The *in-vitro* drug release profiles of various formulations were evaluated by similarity factor (*f2*). It was found that the optimal OPT was able to deliver **nifedipine** by an approximately zero-order process up to 24 h, independent of both release media and agitation rates.

Examples of some marketed formulations (Table 8.1).

Table 8.1 Examples of some marketed formulations of Osmotic Pumps

Drug	Trade Name	Following System
Phenylpropanolamine	Acutrim	Elementary pump
Prazosin	Alpress LP	Push -Pull
Doxazosin	Cardura XL	Push -Pull
Verapamil	Covera HS	Push -Pull with time delay
Oxybutinin chloride	Ditropan XL	Push -Pull
Isradipine	Dynacirc CR	Push -Pull
Pseudoephiderine	Efidac 24	Elementary Pump
Chlorpheniramine maleate	Efidac 24	Elementary Pump
Glipizide	Glucotrol XL	Push - Pull

8.6 Hydrogels as controlled drug delivery systems

Hydrogels are three dimensional hydrophilic polymer networks capable of swelling in water or biological fluids with a capacity to and retain a large amount of fluids in the swollen state. Their ability to absorb water is due to the presence of hydrophilic groups such as -OH, -CONH-, -CONH $_2$, -COOH, and -SO $_3$ H. The water content in the hydrogels affect different properties such as: permeability, mechanical properties, surface properties, and biocompatibility. Hydrogels have similar physical properties as that of living tissue, and the similarity is because of the high water content, soft and rubbery consistency, and low interfacial tension with water or biological fluids. The ability of molecules of different size to diffuse into (drug loading) and out (release drug) of hydrogels, permit the use of hydrogels as drug delivery systems. Since hydrogels have high permeability for water soluble drugs and proteins, the most common mechanism of drug release in the hydrogel system, is diffusion. Factors like polymer composition, water content, crosslinking density and crystallinity can be used to control the release rate and release mechanism from hydrogels (C. S. Satish, et al., 2006).

8.6.1 Classification

Hydrogels are classified based on their nature as pH sensitive, temperature sensitive, enzyme sensitive, and electrical sensitive hydrogel.

8.6.1.1 pH Sensitive Hydrogels

pH sensitive hydrogels can be neutral or ionic in nature. The anionic hydrogels contain negatively charged moieties, cationic networks contain positively charged moieties, and neutral networks contain both positive and negatively charged moieties. In neutral hydrogels, the driving force for swelling arises from the water-polymer thermodynamic mixing contributions, and elastic polymer contributions. In ionic hydrogels, the swelling is due to the previous two contributions, as well as ionic interactions between charged polymer and free ions. The presence of ionizable functional groups like carboxylic acid, sulfonic acid or amine groups, renders the polymer more hydrophilic, and results in high water uptake.

In the case of anionic polymeric network containing carboxylic or sulphonic acid groups, ionization takes place, as the pH of the external swelling medium rises above the pKa of that ionizable moiety. The dynamic swelling change of the anionic hydrogels can be used in the design of controlled release devices for site-specific drug delivery of therapeutic proteins to large intestine, where the biological activity of the proteins is prolonged. The change in the pH of the external environment will act as a stimulus, and the response to the stimulus is the change in swelling properties of the hydrogels, causing the release of the protein.

The cationic hydrogels show swelling at pH values below pKa of the cationic group. The amine groups are protonated at pH lower than pKa, and become hydrophilic and absorb water. At pH greater than pKa, the polymer is hydrophobic, which excludes water. Hydrogels of poly (methyl methacrylate-co-dimethylaminoethyl methacrylate) were studied for diffusion coefficients of different water soluble drugs. The diffusion of water soluble drugs followed free volume theory, which suggests pore type mechanism and water insoluble drugs followed a partition or diffusion mechanism. The water transport mechanism through these hydrogels was non-Fickian in acidic pH and became Fickian in alkaline pH.

8.6.1.2 Electric Current Sensitive Hydrogels

Electric current sensitive Hydrogels are sensitive to electric current, which are made up of polyelectrolytes. It has been found that external electric field can stimulate the delivery of drugs from hydrogels, which depends on the equipment, which allows precise control with regards to the magnitude of current, duration of electric pulses, intervals between pulses, etc. There are four distinct electrochemical and electromechanical mechanisms exists for selective controlled transport of proteins and neutral solutes across hydrogel membranes such as: electrically and chemically induced swelling of a membrane to alter the effective pore size and permeability; electro-phoretic augmentation of solute flux within a membrane; electro-osmotic augmentation of solute flux within the membrane; and electrostatic partitioning of charged solutes into charged membranes.

8.6.1.3 Thermosensitive Hydrogels

Thermosensitive hydrogels are one of the widely studied responsive polymer systems. The thermosensitive polymers are characterized by the presence of hydrophobic groups, such as methyl, ethyl, and propyl groups. The most widely studied temperature sensitive polymer is poly (N-isopropylacrylamide) i.e. P(NIPAAm). P(NIPAAm) is a non-biodegradable polymer with a LCST ~32°C in water and it has been found that the cross-linked gels of this material collapse around this temperature. Pluronics® or Poloxomers® are the commercially available copolymers of poly (ethylene oxide) (PEO) and poly (propylene oxide) (PPO). These copolymers show phase change from sol-gel around body temperature and are used as injectable implants.

8.6.1.4 Enzyme Sensitive Hydrogels

Enzyme sensitive hydrogels are the hydrogels, which can be obtained by cross-linking polymerization of *N*- substituted (meth) acrylamides, *N*-*tert*-butylacrylamide and acrylic acid, with 4, [42-di (methacryloylamino)] azobenzene, 4, [42-di (*N* -methacryloyl -6-aminohexanoylamino)] azobenzene or 3, 32, 5, 52-tetrabromo-4, 4, 42, 42-tetrakis (methacryloylamino) azobenzene as the cross-linking agents. These are mainly used in the targeting of drugs to colon. The colon-specificity is achieved due to the presence of pH-sensitive monomers and azo cross-linking agents in the hydrogel structure. When, the hydrogels passed through the GI tract, the swelling capacity of the hydrogels increases as the pH increases, with the maximum swellability at around pH 7.4. Upon arrival in the colon, the hydrogels have reached a degree of swelling that makes the cross-links accessible to the enzymes (azoreductase) or mediators. Subsequently, the hydrogel network is progressively degraded via the cleavage of the cross-links and releasing the entrapped drug.

8.6.2 Preparation of Hydrogels

8.6.2.1 Solution Polymerization/Crosslinking

The co-polymerization/cross-linking reactions can be initiated by mixing of ionic or neutral monomers with the multifunctional crosslinking agent in the solution. The polymerization is initiated thermally, by UV-light, or by redox initiator system. The presence of solvent serves as heat sink, and minimizes temperature control problems. The prepared hydrogels need to be washed with distilled water to remove the unreacted monomers, crosslinking agent, and the initiator. Example of one such system is preparation of poly (2-hydroxyethyl methacrylate) hydrogels from hydroxyethyl methacrylate, using ethylene glycol dimethacrylate as crosslinking agent. The hydrogels can be made pH-sensitive or temperature-sensitive, by incorporating methacrylic acid or N-isopropylacrylamide as monomer.

8.6.2.2 Suspension Polymerization

This method is employed to prepare spherical hydrogel microparticles with size range of 1 μm to 1mm. In suspension polymerization, the monomer solution is dispersed in the non-solvent forming fine droplets, which are stabilized by the addition of stabilizer. The polymerization is initiated by thermal decomposition of free radicals. The prepared microparticles then washed to remove unreacted monomers, crosslinking agent, and initiator. Hydrogel microparticles of poly (vinyl alcohol) and poly (hydroxy ethyl methacrylate) have been prepared by this method.

8.6.2.3 Polymerization by Irradiation

High energy radiation like gamma and electron beam, have been used to prepare the hydrogels of unsaturated compounds. The radiolysis of water molecules result in the formation of hydroxyl radicals, which attack the polymer chains, resulting in the formation of macroradicals. Recombination of the macroradicals on different chain results in the formation of covalent bonds, and finally a crosslinked structure can be formed. During radiation macroradicals can cause polymerization by interacting with oxygen in an inert atmosphere using nitrogen or argon gas. Examples of polymers crosslinked by radiation method include poly (vinyl alcohol), poly (ethylene glycol) and poly (acrylic acid).

8.6.2.4 Chemical Cross-Linking

Polymers containing functional groups like -OH, -COOH, -NH_2, are soluble in water. The presence of these functional groups on the polymer chain can be used to prepare hydrogels by forming covalent linkages between the polymer chains and complement such as amine-carboxylic acid, isocyanate-OH/NH_2 or by Schiff base. Gluteraldehyde can be added as a cross-linking agent to prepare hydrogels of polymers containing -OH groups like poly-vinyl-alcohol. Also, polymers containing amine groups such as albumin, gelatin, and polysaccharides can be crosslinked using gluteraldehyde. The drugs can be loaded after the hydrogels are formed, as a result the release will be typically first order.

Crosslinking between polymers through hydrogen bond formation occur as in the case of poly(methacrylic acid) and poly(ethylene glycol). The hydrogen bond formation takes place between the oxygen of poly(ethylene glycol) and carboxylic acid group of poly(methacrylic acid). Carriers consisting of networks of poly (methacrylic acid-g-ethylene glycol) showed pH dependent swelling due to the reversible formation of inter-polymer complex, stabilized by hydrogen bonding between the ether groups of the grafted poly (ethylene glycol), and the carboxylic acid protons of the poly(methacrylic acid).

8.6.2.5 Physical Cross-Linking

In this method chitosan, a polycationic polymer can react with positively charged components, either ions or molecules, forming a network through ionic bridges between the polymeric chains. Among anionic molecules, phosphate bearing groups, particularly sodium tripolyphosphate is widely studied. Ionic crosslinking is a simple and mild procedure. Chitosan is also known to form polyelectrolyte complex with poly (acrylic acid). The polyelectrolyte complex undergoes slow erosion, which gives a more biodegradable material than covalently cross-linked hydrogels.

8.6.3 Some Examples of Research Outputs

The pH sensitivity of anionic hydrogels has been used to deliver proteins to the colon, where the activity of the proteolytic enzymes found to be comparatively lower. Calcitonin was loaded into hydrogels of poly(methacrylic acid-g-ethylene glycol), and the release mechanism was found to be relaxation controlled, and calcitonin was released in 7 h.

Hydrogels of poly(methacrylic acid-co-methacryloxyethyl glucoside) and poly- (methacrylic acid-g-ethylene glycol) were studied as a delivery system for insulin. It has been found that the insulin release from hydrogel was slow in acidic medium and rapid in alkaline pH. The hydrogels were able to provide protective effect of insulin when treated with simulated gastric fluid.

Bettini *et al,* prepared anionic copolymer of methyl methacrylate and 2-hydroxyethyl methacrylate by bulk polymerization, using ethylene glycol dimethacrylate as crosslinking agent. The prepared pH sensitive delivery system had shown an increase in swelling charactertics above the pKa (5.9) of methyl methacrylate. The drug release was relaxation controlled, since the swelling of glassy polymer was accompanied by chain relaxation process.

Hydrogels of poly (acrylic acid) (PAA), and poly (acrylic acid-co-2-hydroxyethyl methacrylate) [P(AA-co-HEMA)] hydrogels, were synthesized by **Ende and Peppas** with varying degree of hydrophilicity and crosslinking density, and were studied as potential bioadhesive controlled-release dosage forms.

Brazel and Peppas, studied hydrogels based on poly(N-isopropylacrylamide-co-methacrylic acid). Heparin and streptokinase were loaded to study the release pattern, under pulsatile conditions of varying temperature and pH. The hydrogels showed higher streptokinase release at pH 6.0 and 33°C, and collapsed at pH 5.0 and 36°C. But the same results were not observed in the case of heparin, which has smaller molecular diameter than streptokinase.

Cationic hydrogels are used in the preparation of self-regulated insulin delivery systems. The use of cationic hydrogels in the preparation of self-regulated insulin delivery systems has been reviewed by **Shivakumar and Satish**. The self-regulated insulin delivery system utilizes glucose oxidase (GOD) as the glucose sensor, and pH sensitive cationic hydrogel as the insulin release controller. In such a system, glucose is oxidized to gluconic acid, and catalyzed by GOD.

Klumb and Horbett, studied systems containing immobilized glucose oxidase (GOD) in a pH responsive cationic polymeric hydrogels such as: poly (2-hydroxyethyl methacrylate-co-N,N-dimethylaminoethyl methacrylate) poly(HEMA-co-DMAEMA) and Tetraethylene glycol dimethacrylate (TEGDMA) as cross-linking agent. Several parameters that affect the swelling and permeability of the membrane; such as concentration of amine groups (DMAEMA), cross-linking agent (TEGDMA), and glucose oxidase were studied. The result shows that swelling increased with increase in amine concentration, and decreased with decreasing in cross-linking agent concentration.

Goldraich and Kost, evaluated a matrix system, in which the drug and enzymes were uniformly distributed throughout the hydrogel poly(HEMA-co-DMAEMA). The results shows that the hydrogels with high amine content of DMAEMA (18.5 vol%) and low cross-linking agent concentration of TEGDMA (0.3 vol%), are the most sensitive to pH.

Sahawata ***et al.***, studied microparticles of sodium salt of poly(acrylic acid) as an electroresponsive delivery system, using pilocarpine as model drug. The microparticles showed 96% volume change within 50s of application of a d.c. of 0.3 mA cm^{-2}. The deswelling occurred due to diffusion of mobile cations away from the carboxylate ions, under the influence of an electric field gradient. The carboxylate anions remain undissociated under this condition, which leads to the constriction of gel. The pilocarpine release observed was, 9.8×10^{-7} mol dm^{-3} s^{-1} when d.c. was applied, and when switched off, the release decreased to 1.8×10^{-7} mol $dm^{-3}s^{-1}$.

Andersson, ***et. al.***, studied the diffusion characteristics of P(NIPAAm) gel consists of glucose and insulin with changing temperature and found that the effective diffusion coefficients for the solutes glucose and insulin results below the critical temperature: 10, 20 and 30°C. The effective diffusion coefficient for glucose increases from 2.7×10^{-10} to $4.7\times10^{-10}m^2/s$, and for insulin effective diffusion coefficient increased from $4.4\times10^{-10}$to 5.9×10^{-10} m^2/s, when the temperature was changed from 10 to 30°C.

Gumusdereliogiu and Kesgin, studied the release behaviour of bovine serum albumin from the copolymer of ethylene glycol vinyl ether, butyl vinyl ether, and acrylic acid, in the presence of crosslinking agent, diethylene glycol divinyl ether. Calculated *n* -values were ranging between 0.46 and 0.84, which indicated that the release deviates from the Fickian mode. It was concluded that the existence of some molecular relaxation process in addition to diffusion, was responsible for the observed non-Fickian behavior.

Garcva ***et al.***, studied the Timolol maleate release from pH-sensitive poly(2-hydroxyethyl methacrylate-co-methacrylic acid) hydrogels. The n value was close to 0.50, which indicated a Fickian behavior. The diffusion coefficient calculated was in the range of $3.21\times10^{-6}cm^2/sec$ to $11.52\times10^{-6}cm^2/sec$ at pH 7.4, also the diffusion coefficient depended on the methacrylic acid content in the hydrogel.

8.7 Gastro-Retentive Systems

Some drugs need to be absorbed from stomach, because of its inherent characteristics of instability in the intestine or poor absorption in the intestine; for those drugs, whose local action desired at stomach; drugs, whose absorption is maximum in the duodenal region; and for increasing bioavailability for extended period of time (Anand S., *et al.*, 2010).

To fulfill above requirements, various approaches have been proposed to increase residence time of the drugs at various regions of the stomach, which include:

- Floating drug delivery systems
- Swelling or size expansion systems
- Mucoadhesive systems
- High density systems
- Magnetic systems

8.7.1 Floating Drug Delivery Systems (FDDS)

FDDS are low density systems used to prolong the gastric residence time of the drug to improve the bioavailability, by the use of mechanism such as floating over the gastric contents and remain in the stomach for a prolonged period. Floating can be achieved by adopting different approaches such as:

8.7.1.1 Effervescent Systems

Effervescent systems in which the matrices of drugs in swellable polymers such as methocel or polysaccharides e.g., chitosan can be formed with incorporation of an effervescent components, e.g., sodium bicarbonate and citric or tartaric acid inside a floating chamber so that floating can be achieved after arrival of matrices in the stomach by the liberation of CO_2 as an effervescent reaction between the organic acids of the stomach and bicarbonate salts, which produce upward motion of the dosage forms and maintain its buoyancy. Moreover, the effervescent layer may be divided into two sub-layers to avoid direct contact between sodium bicarbonate and tartaric acid. Sodium bicarbonate may be filled inside the inner sub-layer and tartaric acid in

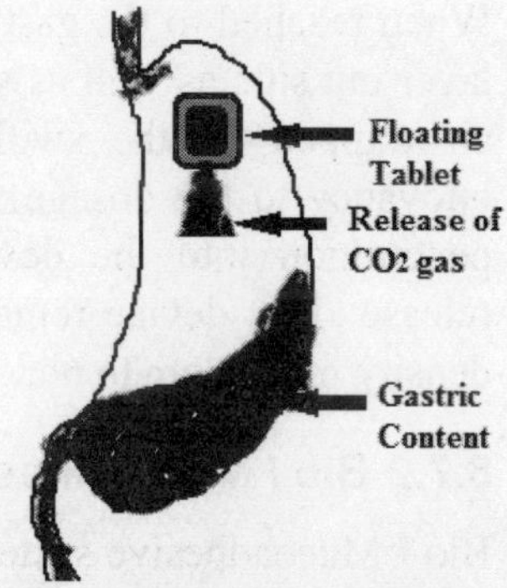

(Effervescent Floating system inside stomach)

the outer layer. When the system immersed in a buffer solution at 37°C, it sink at once in the solution and formed swollen pills, like balloons, with a density much lower than 1 g/ ml. It has been found that the system can float completely within 10 min and approximately 80% remained floating over a period of 5 hr irrespective of pH and viscosity of the test medium by utilizing above technology.

8.7.1.2 Hydrodynamically Balanced Systems (HBS)/ Non-Effervescent Systems

Hydrodynamically balanced systems (HBS)/ Non-Effervescent systems are available in the form of tablets or capsules designed to prolong the GI residence time to maximize the drug release from the GI tract for longer period, which can be prepared by incorporating a high level (20 to 80%) of one or more gel forming hydrocolloids such as hydroxylpropyl-methyl cellulose, hydroxyethylcellulose, hydroxypropylcellulose, sodium carboxymethyl cellulose to form the granules and then compressed to get the tablet or filled in the capsules. When reached to the gastric fluid, the polymer hydrates and form the gel layer out side as well as system maintain their low apparent density. The air trapped by the swollen polymer lowers the density and confers buoyancy to the dosage form. The gel layer control the rate of solvent penetration into the device in exchange controlled the rate of drug release. This device remain floating in the gastric fluid, because of low density may be up to 6hr.

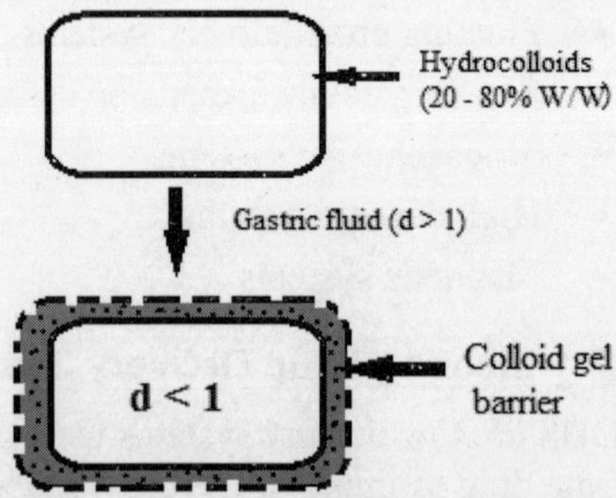

(Hydrodynamic Balanced system for intragastric drug delivery)

8.7.2 Bio / Mucoadhesive systems

Bio / Mucoadhesive systems are the systems bind to the gastric epithelial cell surface, or mucin, a glycoprotein secreted by mucus membrane and increase the gastric residence time by increasing the intimacy and duration of contact between the dosage form and the biological membrane. A bio/mucoadhesive substance is a natural or synthetic polymer capable of adhering to a biological membrane (bioadhesive polymer) or the mucus lining of the GIT (mucoadhesive polymer). The characteristics of these polymers are molecular flexibility, hydrophilic functional groups, and specific molecular weight, chain length, and conformation. Furthermore, they must be nontoxic and non-absorbable, must not form covalent bonds with the mucin on epithelial surfaces, have quick adherence to moist surfaces, easily incorporate the drug and offer no hindrance to drug release, have a specific site of attachment, and be economical. Examples of such polymers include: sodium CMC, acrylic acid derivatives (carbopol 934 and polycarbophil), hydroxylpropyl-methyl cellulose, copolymers of vinayl acetate, vinylpyrollidone, poly methyl methacrylate (Eudragid E hydrochloric acid salt), chitosan, sodium alginate, gelatin, and pectin, etc.

Mechanism of bioadhesion include: Hydration-mediated adhesion; or Bonding-mediated adhesion, which involves various bonding mechanisms such as physico mechanical bonding and chemical bonding. Physico-mechanical bonds can result from the insertion of the adhesive material into the crevices or folds of the mucosa. Chemical bonds may be either covalent (primary) or ionic (secondary) in nature. Secondary chemical bonds consist of dispersive interactions (i.e., Vander Waals interactions) and stronger specific interactions such as hydrogen bonds; or, receptor-mediated adhesion.

Approaches followed to incorporate drugs into the bioadhesive polymers include microencapsulation, granulation by matrix formulation, peletization techniques, etc.

8.7.3 Swelling/ Expanding Systems

After reaching to the stomach, these dosage forms swell to a size that prevents their passage through the pylorus, which results in retention of the system in the stomach for a longer period of time and deliver the drug accordingly. These systems are sometimes referred as *plug-type systems*, because they tend to remain lodged at the pyloric sphincter. These polymeric matrices remain in the gastric cavity for several hours even in the fed state. Sustained and controlled drug release may be achieved by selecting a polymer with the proper molecular weight and swelling properties. As and when the dosage form comes in contact with gastric fluid, the polymer imbibes water and swells. The extensive swelling of these polymers is a result of the presence of physical-chemical crosslinks in the hydrophilic polymer network. These cross-links prevent the dissolution of the polymer and thus maintain the physical integrity of the dosage form. On the other hand, a low degree of cross-linking results in extensive swelling followed by the rapid dissolution of the polymer. Therefore, an optimum amount of cross-linking is required to maintain a balance between swelling and dissolution. These systems can also be designed to erode in the presence of gastric fluid, so that after a predetermined time period, the device no longer attain or retain the expanded configuration. This device can be designed by modification of the composition of poly-acrylamide –co-acrylic acid based hydrogel to control the swell as well as controlled properties.

8.7.4 Magnetic Systems

This system is based on a simple idea that the dosage form contains a small internal magnet and a magnet placed on the abdomen over the position of the stomach. Ito et al. used this technique in rabbits with bioadhesives granules containing ultrafine ferrite (g-Fe_2O_3). They guided them to the oesophagus with an external magnet (1700 G) for the initial 2 min and almost all the granules were retained in the region after 2 h. Although to make the systems to be effective, the external magnet must be positioned with a degree of precision that might compromise patient compliance.

8.7.5 Some Examples of Research Outputs

A. K. Jain, et al., 2009, studied on the preparation and evaluation of floating microspheres using famotidine (FM) as a model drug for prolongation of the gastric retention time. The microspheres were prepared by the solvent evaporation method using different polymers, i.e. acrycoat S100 and cellulose acetate. The size or average diameter (d_{avg}) and surface morphology of the prepared microspheres were recognized and characterized by the optical and scanning electron microscopic methods, respectively. *In vitro* drug release studies were performed and the drug release kinetics was evaluated using the linear regression method. *In vitro* data obtained for floating microspheres of FM showed excellent floatability, good buoyancy and prolonged drug release.

Ozdemir et al, developed floating bilayer tablets with controlled release for furosemide. The low solubility of the drug could be enhanced by using the kneading method, preparing a solid dispersion with β cyclodextrin mixed in a 1:1 ratio. One layer contained the polymers HPMC 4000, HPMC 100, and CMC (for the control of the drug delivery) and the drug. The second layer contained the effervescent mixture of sodium bicarbonate and citric acid. Radiographic studies on 6 healthy male volunteers revealed that floating tablets were retained in stomach for 6 hours and further blood analysis studies showed that bioavailability of these tablets was 1.8 times that of the conventional tablets.

Choi et al, prepared floating alginate beads using gas-forming agents (calcium carbonate and sodium bicarbonate) and studied the effect of CO_2 generation on the physical properties, morphology, and release rates. Calcium carbonate was shown to be a less-effective gas-forming agent than sodium bicarbonate but it produced superior floating beads with enhanced control of drug release rates. In vitro floating studies revealed that the beads free of gas-forming agents sank uniformly in the media, while the beads containing gas-forming agents in proportions ranging from 5:1 to 1:1 demonstrated excellent floating (100%).

Li et al, evaluated the contribution of formulation variables on the floating properties of a gastro floating drug delivery system using a continuous floating monitoring device and statistical experimental design. The formulation was conceived using taguchi design. HPMC was used as a low-density polymer and citric acid was incorporated for gas generation. Analysis of variance (ANOVA) test on the results from these experimental designs demonstrated that the hydrophobic agent magnesium stearate could significantly improve the floating capacity of the delivery system. High-viscosity polymers had good effect on floating properties. The residual floating force values of the different grades of HPMC were in the order K4 M~ E4 M~K100 LV> E5 LV but different polymers with same viscosity, i.e., HPMC K4M, HPMC E4M did not show any significant effect on floating property. Better floating was achieved at a higher HPMC/carbopol ratio and this result demonstrated that carbopol has a negative effect on the floating behavior.

Penners et al, developed an expandable tablet containing mixture of polyvinyl lactase and polyacrylates that swell rapidly in an aqueous environment and thus reside in stomach over an extended period of time. In addition to this, gas-forming agents were incorporated. As the gas formed, the density of the system was reduced and thus the system tended to float on the gastric contents.

Fassihi and Yang, developed a zero-order controlled release multilayer tablet composed of at least 2 barrier layers and 1 drug layer. All the layers were made of swellable, erodible polymers and the tablet was found to swell on contact with aqueous medium. As the tablet dissolved, the barrier layers eroded away to expose more of the drug. Gas-evolving agent was added in either of the barrier layers, which caused the tablet to float and increased the retention of tablet in a patient's stomach.

Talwar et al, developed a once-daily formulation for oral administration of ciprofloxacin. The formulation was composed of 69.9% ciprofloxacin base, 0.34% sodium alginate, 1.03% xanthum gum, 13.7% sodium bicarbonate, and 12.1% cross-linked poly vinyl pyrrolidine. The viscolysing agent initially and the gel-forming polymer later formed a hydrated gel matrix that entrapped the gas, causing the tablet to float and be retained in the stomach or upper part of the small intestine (spatial control). The hydrated gel matrix created a tortuous diffusion path for the drug, resulting in sustained release of the drug (temporal delivery).

Michaels A S, et al., Two patents granted to Alza Corporation revealed a device having a hollow deformable unit that was convertible from a collapsed to expandable form and vice versa. The deformable unit was supported by a housing that was internally divided into two chambers separated by a pressure-sensitive movable bladder. The first chamber contained the therapeutic agent and the second contained a volatile liquid (cyclopentane, ether) that vaporized at body temperature and imparted buoyancy to the system. The system contained a bioerodible plug to aid in exit of the unit from the body.

Baumgartner et al, developed a matrix-floating tablet incorporating a high dose of freely soluble drug. The formulation containing 54.7% of drug, HPMC K4 M, Avicel PH 101, and a gas-generating agent gave the best results. It took 30 seconds to become buoyant. *In-vivo* experiments with fasted state beagle dogs revealed prolonged gastric residence time.

Moursy et al, developed sustained release floating capsules of nicardipine HCl. For floating, hydrocolloids of high viscosity grades were used and to aid in buoyancy sodium bicarbonate was added to allow evolution of CO_2. In vitro analysis of a commercially available 20-mg capsule of nicardipine HCl (MICARD) was performed for comparison. Results showed an increase in floating with increase in proportion of hydrocolloid. Inclusion of sodium bicarbonate increased buoyancy.

Atyabi and coworkers developed a floating system using ion exchange resin that was loaded with bicarbonate by mixing the beads with one molar sodium bicarbonate solution. The loaded beads were then surrounded by a semipermeable membrane to avoid sudden loss of CO_2. The in vivo behavior of the coated and uncoated beads was monitored using a single channel analyzing study in 12 healthy human volunteers by gamma radio scintigraphy. Studies showed that the gastric residence time was prolonged considerably (24 hours) compared with uncoated beads (1 to 3 hours).

Thanoo et al, developed polycarbonate microspheres by solvent evaporation technique. Polycarbonate in dichloromethane was found to give hollow microspheres that floated on water and simulated bio-fluids as evidenced by scanning electron microscopy (SEM). High drug loading was achieved and drug-loaded microspheres were able to float on gastric and intestinal fluids. It was found that increasing the drug-to-polymer ratio increased both their mean particle size and release rate of drug.

Nur and Zhang, developed floating tablets of captopril using HPMC (4000 and 15000cps) and carbopol 934P. In vitro buoyancy studies revealed that tablets of 2 kg/cm^2 hardness after immersion into the floating media floated immediately and tablets with hardness 4 kg/cm^2 sank for 3 to 4 minutes and then came to the surface. Tablets in both cases remained floating for 24 hours. The tablet with 8 kg/cm^2 hardness showed no floating capability. It was concluded that the buoyancy of the tablet is governed by both the swelling of the hydrocolloid particles on the tablet surface, when it contacts with the gastric fluids and the presence of internal voids in the center of the tablet (porosity). A prolonged release from these floating tablets was observed as compared with the conventional tablets and a 24-hour controlled release from the dosage form of captopril was achieved.

Bulgarelli et al, studied the effect of matrix composition and process conditions on casein gelatin beads prepared by emulsification extraction method. Casein by virtue of its emulsifying properties causes incorporation of air bubbles and formation of large holes in the beads that act as air reservoirs in floating systems and serve as a simple and inexpensive material used in controlled oral drug delivery systems. It was observed that the percentage of casein in matrix increases the drug loading of both low and high porous matrices, although the loading efficiency of high porous matrices is lower than that of low porous matrices.

Fell et al, prepared floating alginate beads incorporating amoxycillin. The beads were produced by drop wise addition of alginate into calcium chloride solution, followed by removal of gel beads and freeze-drying. The beads containing the dissolved drug remained buoyant for 20 hours and high drug-loading levels were achieved.

Streubel et al, prepared single-unit floating tablets based on polypropylene foam powder and matrix-forming polymer. Incorporation of highly porous foam powder in matrix tablets provided density much lower than the density of the release medium. A 17% wt/wt foam powder (based on mass of tablet) was achieved in vitro for at least 8 hours. It was concluded that varying the ratios of matrix-forming polymers and the foam powder could alter the drug release patterns effectively.

Asmussen et al, invented a device for the controlled release of active compounds in the gastrointestinal tract with delayed pyloric passage, which expanded in contact with gastric fluids and the active agent was released from a multi-particulate preparation. It was claimed that the release of the active compound was better controlled when compared with conventional dosage forms with delayed pyloric passage.

El-Kamel et al, prepared floating microparticles of ketoprofen, by emulsion solvent diffusion technique. Four different ratios of Eudragit S 100 with Eudragit RL were used. The formulation containing 1:1 ratio of the 2 above-mentioned polymers exhibited high percentage of floating particles in all the examined media as evidenced by the percentage of particles floated at different time intervals. This can be attributed to the low bulk density, high packing velocity, and high packing factor.

Illum and Ping, developed microspheres, which released the active agent in the stomach environment over a prolonged period of time. The active agent was encased in the inner core of microspheres along with the rate-controlling membrane of a water-insoluble polymer. The outer layer was composed of bioadhesive (chitosan). The microspheres were prepared by spray drying an oil/water or water/oil emulsion of the active agent, the water-insoluble polymer, and the cationic polymer.

Streubel et al, developed floating microparticles composed of polypropylene foam, Eudragit S, ethyl cellulose (EC), and polymethyl metha acrylate (PMMA) and were prepared by solvent evaporation technique. High encapsulation efficiencies were observed and were independent of the theoretical drug loading. Good floating behavior was observed as more than 83% of microparticles were floating for at least 8 hours. The in vitro drug release was dependent upon the type of polymer used. At similar drug loading the release rates increased in the following order PMMA < EC < Eudragit S. This could be attributed to the different permeabilities of the drug in these polymers and the drug distribution within the system.

Sheth and Tossounian, developed hydrodynamically balanced sustained release tablets containing drug and hydrophilic hydrocolloids, which on contact with gastric fluids at body temperature formed a soft gelatinous mass on the surface of the tablet and provided a water-impermeable colloid gel barrier on the surface of the tablets. The drug slowly released from the surface of the gelatinous mass that remained buoyant on gastric fluids.

Ushomaru et al, developed sustained release composition for a capsule containing mixture of cellulose derivative or a starch derivative that formed a gel in water and higher fatty acid glyceride and/or higher alcohol, which was solid at room temperature. The capsules were filled with the above mixture and heated to a temperature above the melting point of the fat components and then cooled and solidified.

Bolton and Desai, developed a noncompressed sustained release tablet that remained afloat on gastric fluids. The tablet formulation comprised 75% of drug and 2% to 6.5% of gelling agent and water. The noncompressed tablet had a density of less than 1 and sufficient mechanical stability for production and handling.

Kawashima et al, prepared multiple-unit hollow micro-spheres by emulsion solvent diffusion technique. Drug and acrylic polymer were dissolved in an ethanol-dichloromethane mixture, and poured into an aqueous solution of PVA with stirring to form emulsion droplets. The rate of drug release in micro balloons was controlled by changing the polymer-to-drug ratio. Microballoons were floatable in vitro for 12 hours when immersed in aqueous media. Radiographical studies proved that microballoons orally administered to humans were dispersed in the upper part of stomach and retained there for 3 hours against peristaltic movements.

Dennis et al, invented a buoyant controlled release pharmaceutical powder formulation filled into capsules. It released a drug of a basic character at a controlled rate regardless of the pH of the environment. PH-dependent polymer is a salt of a polyuronic acid such as alginic acid and a pH-independent hydrocarbon gelling agent, hydroxypropylmethyl cellulose.

Spickett et al, invented an antacid preparation having a prolonged gastric residence time. It comprised 2 phases. The internal phase consisted of a solid antacid and the external phase consisted of hydrophobic organic compounds (mono-, di-, and triglycerides) for floating and a non-ionic emulsifier.

Franz and Oth, described a sustained release dosage form adapted to release of the drug over an extended period of time. It comprised a bilayer formulation in which one layer consisted of drug misoprostal and the other had a floating layer. The uncompressed bilayer formulation was kept in a capsule and was shown to be buoyant in the stomach for 13 hours. The dosage form was designed in such a way that the entire drug was released in the stomach itself.

Wu et al, developed floating sustained release tablets of nimodipine by using HPMC and PEG 6000. Prior to formulation of floating tablets, nimodipine was incorporated into poloxamer-188 solid dispersion after which it was directly compressed into floating tablets. It was observed that by increasing the HPMC and decreasing the PEG 6000 content a decline in in-vitro release of nimodipine occurred.

Wong et al, developed a prolonged release dosage form adapted for gastric retention using swellable polymers. It consisted of a band of insoluble material that prevented the covered portion of the polymer matrix from swelling and provided a segment of a dosage form that was of sufficient rigidity to withstand the contractions of the stomach and delayed the expulsion of the dosage form from the stomach.

Mitra, developed a sustained release multilayered sheet-like medicament device. It was buoyant on the gastric contents and consisted of at least 1 dry, self-supporting carrier film of water-insoluble polymer. The drug was dispersed or dissolved in this layer and a barrier film overlaid the carrier film. The barrier film was compsosed of 1 water-insoluble layer and another water-soluble and drug-permeable polymer or copolymer layer. The 2 layers were sealed together in such a way that plurality of small air pockets was entrapped that gave buoyancy to the formulation.

Harrigan, developed an intragastric floating drug delivery system that was composed of a drug reservoir encapsulated in a microporous compartment having pores on top and bottom surfaces. However, the peripheral walls were sealed to prevent any physical contact of the drug in the reservoir with the stomach walls.

Joseph et al, developed a floating dosage form of piroxicam based on hollow polycarbonate microspheres. The microspheres were prepared by the solvent evaporation technique. In vivo studies were performed in healthy male albino rabbits. Pharmacokinetic analysis was derived from plasma concentration vs time plot and revealed that the bioavailability from the piroxicam microspheres alone was 1.4 times that of the free drug and 4.8 times that of a dosage form consisting of microspheres plus the loading dose and was capable of sustained delivery of the drug over a prolonged period.

8.8 Matrix Formulations for Controlled Release

These are the type of controlled drug delivery systems, which release the drug in a predetermined rate over the extended period of time. These release mechanism may follow the diffusion and/or dissolution controlled release profile, which depends on the characteristics of the polymers used and the solubility nature of the drug candidate.

The drug release can be controlled by dispersing the drug in swellable hydrophilic substances or in an insoluble matrix of rigid non-swellable hydrophobic materials or plastic materials.

8.8.1 Different classes of Matrix Formulations: (On the Basis of Retardant Materials Used)

8.8.1.1 Hydrophobic Matrices (Plastic Matrices)

In this method the drug is mixed with an inert or hydrophobic polymer to form the granule or a mix, which is then compressed to get a tablet. Sustained release is achieved because of drug diffusion through a network of channels that exist between compacted polymer particles and the pore size of the channel as well as hydrophobicity of the polymer determine the rate of penetration of the water into the matrix, which determine the rate of dissolution of the drug as well as rate of release of the drug from the system.

Examples of materials, which produce such hydrophobic matrices include; polyethylene, polyvinyl chloride, ethyl cellulose and acrylate polymers and their copolymers.

8.8.1.2 Lipid Matrices

These matrices can be prepared by the granulation technique using hot-melt method, in which the lipid waxes and related materials melted in the vessel and the drugs are dispersed, while in the molten state and sieved to get the granule at room temperature and air dried. The granules may compress to get the tablet. Drug release from such matrices occurs through both pore diffusion and erosion. Release characteristics are therefore more sensitive to digestive fluid composition than to totally insoluble polymer matrix. Carnauba wax, bees wax, paraffin wax in combination with stearyl alcohol or stearic acid has been utilized for retardant base for many sustained release formulation.

8.8.1.3 Hydrophilic Matrices

The formulation of the drugs in gelatinous capsules is more frequently used in tabletting technology, using hydrophilic polymers with high gelling capacities for controlled release. Infect a matrix is defined as well mixed composite of one or more drugs with a gelling agent (hydrophilic polymer). These systems are called swellable controlled release systems.

The polymers used in such preparations include: Cellulose derivatives, such as methylcellulose 400 and 4000 cPs; hydroxyethylcellulose; hydroxypropylmethylcellulose (HPMC) 25, 100, 4000 and 15000 cPs; and sodium carboxymethylcellulose; Noncellulose natural or semi-synthetic polymers, such as: agar-agar; carob gum; alginates; molasses; polysaccharides of mannose and galactose; chitosan and modified starches; polymers of acrylic acid, such as corbopol 934.

8.8.1.4 Biodegradable Matrices

These consist of the polymers, which comprised of monomers linked to one another through functional groups and have unstable linkage in the backbone. They are biologically degraded or eroded by enzymes generated by surrounding living cells or by nonenzymetic process in to olegomers and monomers that can be metabolised or excreted.

Examples are natural polymers such as proteins and polysaccharides; modified natural polymers; synthetic polymers such as aliphatic poly (esters) and poly anhydrides.

8.8.1.5 Mineral Matrices

These consist of polymers which are obtained from various species of seaweeds. Alginic acid, which is a hydrophilic carbohydrate obtained from species of brown seaweeds, which is used for sustained drug delivery.

Advantages of Matrix Tablets

- Easy to manufacture
- Versatile and cost effective
- Can be made to release high molecular weight compounds

Disadvantages of the matrix systems:

- Difficult to remove the non-degradable polymers after the drug releases inside the body.
- Release rate continuously diminishes due to an increase in diffusion resistance and/or a decrease in effective area at the diffusion front.

8.8.2 Polymers used for Matrix Tablets

Hydrogels

Polyhydroxyethyle methylacrylate (PHEMA)
Cross-linked polyvinyl alcohol (PVA)
Cross-linked polyvinyl pyrrolidone (PVP)
Polyethylene oxide (PEO)
Polyacrylamide (PA)

Soluble polymers

Polyethylene glycol (PEG)
Polyvinyl alcohol (PVA)
Polyvinyl pyrrolidone (PVP)
Hydroxypropyl methyl cellulose (HPMC)

Biodegradable polymers

Polylactic acid (PLA)
Polyglycolic acid (PGA)
Polycaprolactone (PCL)
Polyanhydrides
Polyorthoesters

Nonbiodegradable polymers

Polyethylene vinyl acetate (PVA)
Polydimethyl siloxane (PDS)
Polyether urethane (PEU)
Polyvinyl chloride (PVC)
Cellulose acetate (CA)
Ethyl cellulose (EC)

Mucoadhesive polymers

Polycarbophil
Sodium carboxymethyl cellulose
Polyacrylic acid
Tragacanth
Methyl cellulose
Pectin

Natural gums

Xanthan gum
Guar gum
Karaya gum

8.8.3 Dissolution Profile of Matrix Formulations

In-vitro dissolution has been recognized as an important element in drug development. Under certain conditions, it can be used as a surrogate for the assessment of Bio-equivalence. Several theories / kinetics model describe drug dissolution form immediate and modified release dosage forms. There are several models to represent the drug dissolution profiles where f_t is a function of t (time) related to the amount of drug dissolved from the pharmaceutical dosage system. The quantitative interpretation of the values obtained in the dissolution assay is facilitated by the usage of a generic equation that mathematically translates the dissolution curve in the function of some parameters related with the pharmaceutical dosage forms. In some cases, that equation can be deduced by a theoretical analysis of the process. In most cases with tablets, capsules, coated forms or prolonged release forms, where theoretical fundamental does not exist and some times a mode of empirical equation is used (Paulo, et al., 2001).

The kind of drug, its polymorphic form, crystallinity, particle size, solubility, an amount in the pharmaceutical dosage form can influence the release kinetic (Salmon, et al., 1980; El-Arini, et al., 1995). A water soluble drug incorporated in a matrix is mainly released by diffusion, while for a low water-soluble drug the self-erosion of the matrix will be the principal release mechanism.

8.8.4 Mathematical Models

8.8.4.1 Zero Order Kinetics

Drug dissolution from pharmaceutical dosage forms, which do not disaggregate and release the drug slowly assuming that the area does not change and no equilibrium conditions are obtained can be represented by the following equation:

$$W_0 - W_1 = Kt$$

where, W_0 is the initial amount of drug in the pharmaceutical dosage form, W_1 is the amount of drug in the pharmaceutical dosage form at time t and K is proportionality constant. Dividing this equation by W_0 and simplifying;

$$f_t = K_0 t$$

where, $f_t = 1 - (W_1/W_0)$ and f_t represents the fraction of drug dissolved in time t and K_0 the apparent dissolution rate constant or zero order release constant. In this way, a graphic of the drug-dissolved fraction versus time will be linear. This relation can be used to describe the drug dissolution of several types of modified release pharmaceutical dosage forms, as in the case of some Transdermal systems, as well as matrix tablets with low soluble drugs (Varelas, et al., 1995), coated forms, osmotic systems, etc. The pharmaceutical dosage forms following this profile release is ideal in order to achieve a prolonged action. The following equation can simply express this model as:

$$Q_t = Q_0 + K_0 t$$

where, Q_t is the amount of drug dissolved in time t, Q_0 is the initial amount of drug in the solution (most times, $Q_0 = 0$) and K_0 is the zero order release constant.

8.8.4.2 First Order Kinetics

The application of this model to drug dissolution studies was first proposed by Gibaldi and Feldman (1967) and later by Wagner (1969). This model has also been used to describe absorption and/or elimination of some drugs (Gibaldi and Perrier, 1982), although it is difficult to conceptualise this mechanism in a theoretical basis. Kitazawa et al. (1975, 1977) proposed a slightly different model, but achieved practically the same conclusions.

The dissolution phenomena of a solid particle in a liquid media imply a surface action, as can be seen by the **Noyes-Whitney** Equation as:

$$dC/dt = K\,(C_s - C)$$

where, C is the concentration of the solute in time t, C_s is the solubility in the equilibrium at experience temperature and K is a first order proportionality constant. This equation was altered by Brunner et al, (1900), to incorporate the value of the solid area accessible to dissolution, S, getting;

$$dC/dt = K_1 S(C_s - C)$$

where, K_1 is a new proportionality constant. Using the Fick's first law, it is possible to establish the following relation for the constant k_1;

$$k_1 = D/Vh$$

where, D is the solute diffusion coefficient in the dissolution media, V is the liquid dissolution volume and h is the width of the diffusion layer. Hixson and Crowell adapted the Noyes-Whitney equation in the following manner.

$$dW/dt = KS(C_s - C)$$

where, W is the amount of solute in solution at time t, dW/dt is the passage rate of the solute into solution in time t and K is a constant. This last equation is obtained from the Noyes-Whitney equation by multiplying both terms of equation by V and making K equal to k_1V. Comparing these terms, the following relation is obtained:

$$K = D/h$$

In this manner, Hixson and Crowell Equation can be rewritten as

$$dW/dt = KS/V\ (VC_s - W) = k(VC_s - W)$$

where, $k = k_1S$. If one pharmaceutical dosage form with constant area is studied in ideal conditions (sink conditions), it is possible to use this last equation that, after integration, will become;

$$W = VC_s\ (1 - e^{-kt})$$

This equation can be transformed, applying decimal logarithms in both terms, into

$$Log\ (VC_s - W) = log\ VC_s - kt / 2.303$$

The following relation can also be expressed using this model as:

$Q_1 = Q_0e^{-k_1t}$ or In $(Q_1/Q_0) = K_1t$ or In q_1 = In Q_0K_1t or in decimal logarithms:

$$Log\ Q_1 = log\ Q_0 + K_1t/2.303$$

where, Q_1 is the amount of drug released in time t, Q_0 is the initial amount of drug in the solution and K_1 is the first order release constant. In this way a graphic of the decimal logarithm of the released amount of drug versus time will be linear. The pharmaceutical dosage forms following this dissolution profile, such as those containing water-soluble drugs in porous matrices (Mulye and Turco, 1995), release the drug in a way that is proportional to the amount of drug remaining in its interior.

8.8.4.3 Higuchi Model

Higuchi (1961, 1963) developed several theoretical models to study the release of water soluble and low soluble drugs incorporated in semi-solid and/or solid matrixes. Mathematical expressions were obtained for drug particles dispersed in a uniform matrix behaving as the diffusion media. To study the dissolution from a planar system having a homogeneous matrix, the relation obtained follows the equation as:

$$f_1 - Q = \sqrt{}D(2C-C_s)C_st$$

where, Q is the amount of drug released in time t per unit area, C is the drug initial concentration, C_s is the drug solubility in the matrix media and D is the diffusivity of the drug molecules (diffusion constant) in the matrix substance.

This relation was first proposed by Higuchi to describe the dissolution of drugs in suspension from ointments bases, but is clearly in accordance with other types of dissolution from other pharmaceutical dosage forms. To these dosage forms a concentration profile, which may exist after application of the pharmaceutical system, can be represented. The solid line represents the variation of drug concentration in the pharmaceutical system, after time, t, in the matrix layer normal to the release surface, being all the drug rapidly diffused (perfect sink conditions). The total drug concentration would be expected to show a sharp discontinuity at distance h and no drug dissolution could occur until the concentration drops below the matrix drug solubility (C_s). To distances higher than h, the concentration gradient will be constant, provided $C \gg C_s$. The linearity of the gradient over the distance follows Fick's first law. At a time t the amount of drug release by the system corresponds to the shaded area in Fig. 8.11. It is evident that dQ, the amount of drug released, is related to dh, the movement of the release front:

$$dQ = Cdh - 1/2(C_s dh)$$

But, in according to the Fick first law ($dQ/dt = DC_s/h$), the following expression can be obtained:

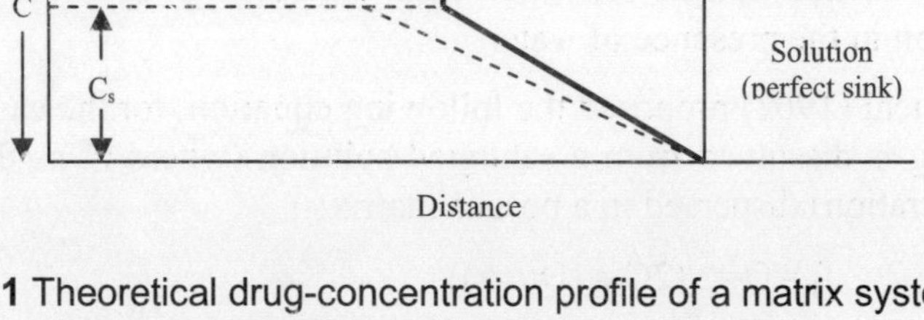

Fig. 8.11 Theoretical drug-concentration profile of a matrix system in direct contact with a perfect sink release media.

$$(C\,d/h - 1/2\,(C_s dh)) / dt = DC_s/h$$

Or, $$h(Cdh - 1/2\,(C_s dh))/DC_s = dt$$

Or, $h\,(2C\text{-}C_s)dh/2DC_s = dt$

Integrating this equation it becomes:

$$t = h^2 / 4DC_s\,(2C\text{-}C_s) + k'$$

where, k' is an integration constant and k' will be zero if time was measured from zero and then:

$$t = h^2 / 4DC_s(2C - C_s) \text{ or, } h = 2\sqrt{(tDC_s/2C - C_s)}$$

Q (amount of drug released at time t) is then:

$$Q = hC - 1/2\,(hC_s) \text{ or } Q = h(C - C_s)$$

Replacing in this equation h by the expression obtained:

$$Q = 2\sqrt{[tDC_s / 2C\text{-}C_s)]}\,(C - C_s)$$

And, finally: $Q = \sqrt{[tDC_s(2C - C_s)]}$

This relation is valid during all the time, except when the total depletion of the drug in the therapeutic system is achieved. Higuchi developed also other models, such as drug release from spherical homogenous matrix systems and planar or spherical systems having a granular (heterogeneous) matrix. To study the dissolution from a planar heterogeneous matrix system, where the drug concentration in the matrix is lower than its solubility and the release occurs through pores in the matrix, the obtained relation was the following:

$$f_1 = Q = \sqrt{[(D\varepsilon/\tau)(2C\text{-}\varepsilon C_s)C_s t]}$$

where, Q is the amount of drug released in time t by surface unity, C is the initial concentration of the drug, ε is the matrix porosity, τ is the tortuosity factor of the capillary system, C_s is the drug solubility in the matrix / excipient media and D the diffusion constant of the drug molecules in that liquid. These models assume that these systems are neither surface coated nor that their matrices undergo a significant alteration in the presence of water.

Higuchi (1962) proposed the following equation, for the case in which the drug is dissolved from a saturated solution (where C_0 is the solution concentration) dispersed in a porous matrix:

$$f_1 = Q = \sqrt{(2C_0\varepsilon\,(Dt/\tau\pi))}$$

Cobby et al. (1974) proposed the following generic, polynomial equation to the matrix tablets as :

$$f_1 = Q = G_1K_1t^{1/2} - G_2\,(K_1t^{1/2})^2 + G_3\,(K_1t^{1/2})^3$$

where, Q is the released amount of drug in time t, Ks is dissolution constant and G_1, G_2 and G_3 are shape factors.

These matrices usually have continuous channels, due to its porosity, being in this way above the first percolation threshold (in order to increase its mechanical stability) and bellow the second percolation threshold (in order to release all the drug amount), allow to apply the percolation theory (Leuenberger et al., 1989); Hastedt and Wright, 1990; Bonny and Leuenberger, 1991; Staufer and Aharony, 1994) as:

$$f_1 = Q = \sqrt{[D_B C_s t(2\phi d - (\phi + \varepsilon)C_s]}$$

where, ϕ is the volume accessible to the dissolation media throughout the network channels, D_B is the diffusion coefficient through this channel and d is the density of used drug. In a general way it is possible to resume the Higuchi model to the following expression (generally known as the simplified Higuchi model):

$$f_1 = K_u t^{1/2}$$

where, K_u is the Higuchi dissolution constant treated sometimes in a different manner by different authors and theories. Higuchi describes drug release as a diffusion process based in the Fick's law, square root time dependent. This relation canbe used to describe the drug dissolution from several types of modified release pharmaceutical dosage forms, as in the case of some transdermal systems (Costa et al., 1996) and matrix tablets with water soluble drugs (Desai et. al., 1966; Schwartz et al., 1968).

8.8.4.4 Hixson – Crowell Model

Hixson and Crowell (1931) recognized that the particle area is proportional to the cubic root of its volume derived an equation that can be described in the following manner:

$$W_0^{1/3} - W_t^{1/3} = K_s t$$

where, W_0 is the initial amount of drug in the pharmaceutical dosage form, W_1 is the remaining amount of drug in the pharmaceutical dosage form at time t and K_s is a constant incorporating the surface-volume relation. This expression applies to pharmaceutical dosage form such as tablets, where the dissolution occurs in planes that are parallel to the drug surface if the tablet dimensions diminish proportionally, in such a manner that the initial geometrical form keeps constant all the time. The above equation can be rewritten as:

$$W_0^{1/3} - W_t^{1/3} = K'N^{1/3} DCs\, t / \delta$$

where N is the no of particles, K′ is a constant (related to the surface, the shape and the density of the particle), D is the diffusion coefficient, C_s is the solubility in the equilibrium at experience temperature and δ is the thickness of the diffusion layer. The shape factors for cubic or spherical particles should be kept constant if the particles dissolve in an equal manner by all sides. This possibly will not occur to particles with different shapes and consequently this equation can no longer be applied. Dividing the above equation by $Wo^{1/3}$ and simplifying as:

$$(1-f_1)^{1/3} = 1 - K\beta t$$

where, $f_1 = 1 - (W_t / W_0)$ and f_1 represents the drug dissolved fraction at time t and K_β is a release constant. Then, a graphic of the cubic root of the unreleased fraction of drug versus time will be linear if the equilibrium conditions are not reached and if the geometrical shape of the pharmaceutical dosage form diminishes proportionally over time. When the model is used, it is assumed that the release rate is limited by the drug particles dissolution rate and not by the diffusion that might occur through the polymeric matrix. This model has been used to describe the release profile keeping in mind the diminishing surface of the drug particles leading to the dissolution (Niebergall, et al., 1963; Prista et al., 1995).

8.8.5 Evaluation of Controlled Release Tablets

Before marketing a controlled release product, it must be assure the strength, safety, stability and reliability of a product by performing *in-vitro* and *in-vivo* analysis and correlation between the two. The following approaches can be made to evaluate the above performance:

1. *In–Vitro* drug release studies: The methods include:
   a. Beaker method
   b. Rotating disc method
   c. Rotating Bottle method
   d. Rotating Basket method
   e. Stationary Basket Method
   f. Oscillating tube method
   g. Dialysis method
   h. USP dissolution method.
2. *In–Vivo* drug release studies in animal models or in human volunteers.

Once the satisfactory *in-vitro* profile is achieved, it becomes necessary to conduct *in-vivo* evaluation and establish ***in-vitro* and *in-vivo* correlation (IVIVC).**

The various in-vivo evaluation methods are:

a. Clinical response analysis
b. Blood level data analysis (plasma concentration with respect to time)
c. Urinary excretion studies
d. Nutritional studies.
e. Toxicity studies
f. Radioactive tracer techniques to determine fate of formulations *in-vivo* after administration.

8.8.6 Some Examples of Research Outputs

Ochoa L et al., 2008, prepared theophylline sustained release matrix tablets based on the combination of hydroxypropyl methylcellulose (HPMC K4M and K100M) and different meltable binders by melt granulation in a high-shear mixer. The results obtained show that the type of excipient influenced the drug release rate. The dissolution rate was found to be delayed, when lipophilic binders were used and only formulations containing Gelucire 50/13 or PEG 6000 with HPMC K4M had a profile similar to the commercial formulation. The release mechanism of theophylline from the formulations was described by Peppas's equation showing a non-Fickian release mechanism. These results suggest that melt granulation could be an easy and fast method to formulate sustained release tablets.

Lee B J et. al., 2008, formulated a drug-loaded hydroxypropyl-methylcellulose (HPMC) matrix tablet simultaneously containing drug in inner tablet core and outer coated layer using drug-containing aqueous-based polymeric Eudragit RS30D dispersions. Effects of coating levels, drug loadings in outer layers, amount and type of five plasticizers and talc concentration on the release characteristics were evaluated on the characteristics in simulated gastric fluid for 2 h followed by a study in intestinal fluids. Melatonin (MT) was selected as a model drug. The biphasic release profiles of dual drug-loaded HPMC matrix tablet were highly modified, depending on the amount and type of five plasticizers. Talc (10-30%) in coating dispersion as an anti-sticking material did not affect the release profiles. The current dual drug-loaded HPMC matrix tablet, showing biphasic release profiles may provide an alternative to deliver drugs with circadian rhythmic behaviors in the body.

Talukder M M et al., 2008, prepared a swelling matrix core containing pectin, hydroxypropyl methylcellulose (HPMC), microcrystalline cellulose and 5-aminosalicylic acid. This was subjected to a dual coating operation: an inner pH-sensitive enteric and an outer semi-permeable membrane coat with a pore former. In-vitro dissolution studies were carried out in USP apparatus-I using sequential pH media. Results indicate that this delivery system has potential for site-specific delivery of drugs to the colon irrespective of transit time and rapid changes in the proximal pH of the gastrointestinal tract.

Contoar S L et al., 2004, investigated the effectiveness of an ethylcellulose (EC) bead matrix and different film-coating polymers in delaying drug release from compacted multiparticulate systems. Formulations containing theophylline or cimetidine granulated with Eudragit(R) RS 30D were developed and beads were produced by extrusion-spheronization. Drug beads were coated using 15% wt/wt Surelease(R) or Eudragit(R) NE 30D and were evaluated for true density, particle size, and sphericity. Lipid-based placebo beads and drug beads were blended together and compacted on an instrumented Stokes B2 rotary tablet press. Although placebo beads were significantly less spherical, their true density of 1.21 g/cm(3) and size of 855 mum were quite close to Surelease(R)-coated drug beads. Although modified release profiles >8 h were achievable in tablets for both drugs using either coating polymer, Surelease(R)-coated theophylline beads released drug fastest overall.

Vueba ML et al., 2006, studied the different ketoprofen:excipient formulations, in order to determine the effect of the polymer substitution and type of diluent on the drug-release mechanism. Substituted cellulose-methylcellulose, hydroxypropylcellulose and hydroxypropyl-methylcellulose were used as polymers, while lactose monohydrate and beta-cyclodextrin were tested as diluents. Polymers such as MC25 and HPC were not found suitable for the preparation of modified release ketoprofen hydrophilic matrix tablets, while HPMC K15M and K100M shown to be advantageous.

Kapat et al., 2004, focused on the effects of different hydroxypropylmethylcellulose (HPMC) types and HPMC :direct tabletting agent (DC-agent) ratio on Verapamil Hydrochloride (VRP HCl) release from monolayered and three-layered matrix tablets.

Investigated polymers were Methocel K100LV, K15M, K100M and DC-agent was Ludipress® LCE. Eight formulations were prepared as monolayered matrix tablets while four formulations were prepared as three-layered matrix tablets by direct compression method. Drug release studies shows tablets containing low viscosity grade HPMC at inner and outer layers presented release profiles close to or within the limits of pharmacopeia. Release data of three-layered matrix tablet (F12) and the reference product (Isoptin® -KKH) which were in agreement with USP XXVII criteria, were evaluated by mathematical models (zero order, first order, Higuchi, Hixson-Crowell, Korsmeyer-Peppas), difference factor (f1) and similarity factor.

Md. Selim Reza et al., 2003, undertaken to investigate the effect of plastic, hydrophilic and hydrophobic types of polymers and their content level on the release profile of drug from matrix systems. As the physico-chemical nature of the active ingredients influence the drug retarding ability of these polymers, three different drugs were used to evaluate their comparative release characteristics in similar matrices. Matrix tablets of theophylline, diclofenac sodium and diltiazem HCl using Kollidon SR, Carnauba wax and Hydroxypropyl methylcellulose (HPMC-15cps) were prepared separately by direct compression process. Release profile showed a tendency to follow zero-order kinetics from HPMC matrix systems whereas Fickian (Case-I) transport was predominant mechanism of drug release from Kollidon SR matrix system. The mean dissolution time (MDT) was calculated for all the formulations and the highest MDT value obtained with Carnauba wax for all the drugs under investigation. The results generated in this study showed that the profile and kinetics of drug release were functions of polymer type, polymer level and physico-chemical nature of drug.

References

Abhishek Kumar Jain, CP Jain, YS Tanwar, PS Naruka. Formulation, characterization and *in vitro* evaluation of floating microspheres of famotidine as a gastro retentive dosage form. *Asian J of Pharmaceutics,* 2009, 3(3): 222-226.

AK Philip, Kamla Patha. *In situ* formed phase transited drug delivery system of ketoprofen for achieving osmotic, controlled and level a *in vitro in vivo* correlation. Ind. *J. Pharm. Sci*. 2008; 70 (6): 745-753.

Anand S. Surana and Rakhee K. Kotecha. An overview on various approaches to oral controlled drug delivery system via gastroretention. Int. J Pharma Sci Rev and Res, 2010, 2(2): 68-72.

Andersson, M, Axelsson, A., Zacchi, G., *Int. J. Pharm.,* 1997 (157): 199

Arora S, Ali J, Ahuja A, Khar RK, Baboota S. Floating Drug Delivery Systems: A Review. *AAPS PharmSciTech.* 2005, 6(03): E372-E390.

Asmussen B, Cremer K, Hoffmann HR, Ludwig K, Roreger M, inventors . Expandable gastro-retentive therapeutic system with controlled active substance release in gastrointestinal tract. US patent 6 290 989. September 18, 2001.

Atyabi F, Sharma HL, Mohammed HAH, Fell JT. In vivo evaluation of a novel gastro retentive formulation based on ion exchange resins. *J Control Release.* 1996 (42):105-113.

Baumgartner S, Kristel J, Vreer F, Vodopivec P, Zorko B. Optimisation of floating matrix tablets and evaluation of their gastric residence time. *Int J Pharm.* 2000 (195):125-135.

Bettini, R., Colombo, P. and Peppas N.A., *J. Control. Release*, 1995 (37):105

Bolton S, Desai S, inventors. Floating sustained release therapeutic compositions. US patent 4 814 179. March 21, 1989.

Bouwstra, J.A., Jungiger, H.E., In: Swarbrick, J. and Boylan, J.C., Eds., Encyclopedia of Pharmaceutical Technology. 1st Edn., Vol. 7, Marcel Dekker Inc, New York, 1993: 441.

Brazel, C.S. and Peppas, N.A., *J. Control. Release,* 1996 (39): 57.

Bulgarelli E, Forni F, Bernabei MT. Effect of matrix composition and process conditions on casein gelatin beads floating properties. *Int. J. Pharm.* 2000 (198):157-165.

C S Satish, K P Satish, H G Shivakumar. Hydrogels as controlled drug delivery systems: Synthesis, crosslinking, water and drug transport mechanism. J Pharm Sci, 2006 (68):133-40.

Ch. Ajay Babu, M. Prasada Rao and Vijay Rratna J. Controlled-Porosity Osmotic Pump Tablets - an Overview. *JPRHC*, 2010, 2 (1):114-126.

Chawla G., Gupta P., Koradia V. and Bansal A. K., Gastroretention: A Means to Address Regional Variability in Intestinal Drug Absorption, *Pharmaceutical Technology*, 2003: 50-68.

Chien Y.W., Novel Drug Delivery Systems, IInd edition, Revised and expanded, Informa healthcare USA, Inc. New York, 2009:139-196.

Choi B.Y., Park H.J., Hwang S.J., and Park J.B., Preparation of alginate beads for floating drug delivery system: effects of CO_2 gas-forming agent, *Int. J. Pharm.*, 2002(239): 81-91.

Choi BY, Park HJ, Hwang SJ, Park JB. Preparation of alginate beads for floating drug delivery: effects of CO_2 gas forming agents. *Int J Pharm*. 2002 (239):81-91.

Clarke G.M., Newton J.M., Short M.D., Comparative Gastrointestinal Transit of Pellet Systems of Varying Density, *Int. J. Pharm. 1995*, 114 (1): 1-11.

Dennis A, Timminis P, Lel K, inventors. Buoyant controlled release powder formulation. US patent 5 169 638. December 8, 1992.

Deshpande A.A. and et al., Development of a Novel Controlled-Release System for Gastric Retention, *Pharm. Res*. 1997, 14 (6): 815-819.

Deshpande A.A., Shah N.H., Rhodes C.T., Malick W., Development of a novel controlled-release system for gastric retention, *Pharm. Res.,* 1997 (14): 815-819.

El-Kamel AH, Sokar MS, Al Gamal SS, Naggar VF. Preparation and evaluation of ketoprofen floating oral drug delivery system. *Int J Pharm*. 2001 (220):13-21.

Fassihi R, Yang L, inventors. Controlled release drug delivery systems. US patent 5 783 212. July 21, 1998.

Franz MR, Oth MP, inventors. Sustained release, bilayer buoyant dosage form. US patent 5 232 704. August 3, 1993.

Garcva, D.M., Escobar, J.L., Noa, Y., Bada, N., Hernαez, E. and Katime, I. *Eur. Polym. J.* , 2004 (40): 1683.

Gaylen Zentner M., Gregory McCelland A., Steven Sutton C. Controlled porosity solubility and resin-modulated osmotic drug delivery systems for release of Diltiazem hydrochloride. *J. Control. Rel*. 1991 (16): 237-244.

Giancarlo Santus, Richard Baker W. Osmotic drug Delivery: a review of the patent literature. *J Control. Rel*. 1995(35): 1-21

Goldraich, M. and Kost, J., *Clin. Mater.,* 1993 (13): 135

GümüþderelioIu, M. and Kesgin, D., *Int. J. Pharm.,* 2005 (288): 273

Gupta P., Vermani K., and Garg S., Hydrogels: From Controlled Release to pH-Responsive. *Drug Delivery, Drug Discov Today.*, 2002, 7 (10): 569-579.

Hai Bang Lee, Longxiao Liu, Jeong Ku, Gilson Khang, Bong Lee, John M.Rhee. Nifedipine controlled delivery sandwiched osmotic tablet system. *J. Control. Rel.* 2000 (68): 145-156.

Illum L, Ping H, inventors. Gastroretentive controlled release microspheres for improved drug delivery. US patent 6 207 197. March 27, 2001.

Ito R., Machida Y., Sannan T., Nagai T., Magnetic granules: a novel system for specific drug delivery to esophageal mucosa in oral administration, *Int. J. Pharm.*, 1990, 61 (1-2): 109-117.

Ji-Eon Kim, Seung-Ryul Kim, Sun-Hee Lee, Chi-Ho Lee and Dae-Duk Kim. The effect of pore formers on the controlled release of cefadroxil from a polyurethane matrix. *Int. J. Pharm.* 2000 (201): 29-36.

Kawashima Y, Niwa T, Takeuchi H, Hino T, Ito Y. Preparation of multiple unit hollow microspheres (microballoons) with acrylic resins containing tranilast and their drug release characteristics (*in-vivo*). *J Control Release.* 1991 (16):279-290.

Kazuto Okimoto, Roger A. Rajewski, and Valentino J. Stella. Release of testosterone from an osmotic pump tablet utilizing (SBE)7m-β-cyclodextrin as both a solubilizing and an osmotic pump agent. *J. Control. Rel.* 1999 (58): 29-38.

Klumb, L.A. and Horbett, T.A., *J. Control. Release,* 1993 (27): 95.

Li S, Lin S, Chien TW, Daggy BP, Mirchandani HL. Statistical optimization of gastric floating system for oral controlled delivery of calcium. *AAPS PharmSciTech.* 2001 (2):E1.

Li S, Lin S, Daggy B P, Mirchandani H L, Chien T W. Effect of formulation variables on the floating properties of gastric floating drug delivery system. *Drug Dev Ind Pharm.*, 2002 (28):783-793.

Longxiao Liu, Xiangning Xu. Preparation of bilayer-core osmotic pump tablet by coating the indented core tablet. *Int. J. Pham.* 2008, 352 (1-2): 225-230.

M. C. Gohel, Parikh .R.K, Shah. N.Y. Osmotic Drug Delivery: An Update. Pharma info.net. 2009:7: 2.

Mahalaxmi.R, Phanidhar Sastri1, Ravikumar, Atin Kalra, Pritam Kanagale, Narkhede. Enhancement of Dissolution of Glipizide from Controlled Porosity Osmotic Pump Using a Wicking Agent and a Solubilizing Agent. *Int. J. PharmTech Res*. 2009, 1(3): 705-711.

Michaels AS, Bashwa JD, Zaffaroni A, inventors. Integrated device for administering beneficial drug at programmed rate. US patent 3 901 232. August 26, 1975.

Michaels AS, inventor. Drug delivery device with self actuated mechanism for retaining device in selected area. US patent 3 786 813. January 22, 1974.

Moursy NM, Afifi NN, Ghorab DM, El-Saharty Y. Formulation and evaluation of sustained release floating capsules of Nicardipine hydrochloride. *Pharmazie*. 2003 (58):38-43.

Nur AO, Zhang JS. Captopril floating and/or bioadhesive tablets: design and release kinetics. *Drug Dev Ind Pharm*. 2000 (26):965-969.

Ozdemir N, Ordu S, Ozkan Y. Studies of floating dosage forms of furosemide: in vitro and in vivo evaluation of bilayer tablet formulation. *Drug Dev Ind Pharm*. 2000 (26): 857-866.

P. G. Yeole, Shagupta K, and V. F. Patel. Floating Drug Delivery Systems: Need and Development. Ind j pharma sci., 2005, 67 (3): 265-272.

Penners G, Lustig K, Jorg PVG, inventors. Expandable pharmaceutical forms. US patent 5 651 985. July 29, 1997.

Peppas, N.A. and Mikos, A.G., In: Peppas, N.A., Eds,. Preparation Methods and Structure of Hydrogels, Hydrogels in Medicine and Pharmacy, Vol I, CRC Press, Boca Raton, FL, 1986.

Pradeep Vavia R., Sapna Makhija N. Controlled porosity osmotic pump-based controlled release systems of pseudoephedrine 1. Cellulose acetate as semipermeable membrane. *J. Control. Rel*. 2003(89): 5-18.

Rajan Verma K., Kivi Murali Krishna, Sanjay Garg. Formulation aspects in the development of Osmotically Controlled Oral Drug Delivery Systems (OCODDs). *J. Control. Rel.* 2002 (79): 7-27.

Roger Rajewski A., *et al.* Applicability of (SBE)7m --CD in controlled-porosity osmotic pump tablets (OPTs). *Int. J. Pharm.* 2004 (286): 81-88.

Roger Rajewski A., *et al.* Factors affecting membrane controlled drug release for an osmotic pump tablet utilizing (SBE)7m-CD as both a solubilizer and osmotic agent. *J. Control. Rel.* 1999 (60): 311-319.

Sahawata, K., Hara, M., Yasunaga, H., and Osada, Y., *J. Control. Release,* 1990 (14): 253.

Sanjay Garg, Rajan Verma k., Aditya Kaushal M. Development and evaluation of extended release formulations of Isosorbide mononitrate based on osmotic technology. *Int. J. Pharm.*, 2003 (263): 9-24.

Shailesh Sharma. Osmotic controlled drug delivery. Pharmainfo.net. 2008, 6(3).

Sheth PR, Tossounian JL, inventors. Novel sustained release tablet formulations -4167558. September 11, 1979.

Shivakumar, H.G. and Satish C.S., *Indian J. Pharm. Sci.*, 2004 (66): 137.

Singh B.M., Kim K.H., Floating drug delivery systems: an approach to oral controlled drug delivery via gastric retention, *J. Control. Rel.*, 2000 (63): 235-259.

Spickett RGW, Vidal JLF, Escoi JC, inventors. Antacid preparation having prolonged gastric residence. US patent 5, 288, 506. February 22, 1993.

Streubel A, Siepmann J, Bodmeier R. Floating matrix tablets based on low density foam powder: effect of formulation and processing parameters on drug release. *Eur J Pharm Sci*. 2003 (18):37-45.

Streubel A, Siepmann J, Bodmeier R. Floating microparticles based on low density foam powder. *Int J Pharm*. 2002 (241): 279-292.

Suresh Vyas P., Prabakaran D., Paramjit Singh, Parijat Kanaujia, Jaganathan K.S., Amith Rawat. Modified push-pull osmotic system for simultaneous delivery of Theophylline and Salbutamol: development and *in vitro* characterization. *Int. J. Pharm*. 2004 (284): 95-108.

Talwar N, Sen H, Staniforth JN, inventors. Orally administered controlled drug delivery system providing temporal and spatial control. US patent 6 261 601. July 17, 2001.

Thanoo BC, Sunny MC, Jayakrishnan A. Oral sustained release drug delivery systems using polycarbonate microspheres capable of floating on the gastric fluids. *J Pharm Pharmacol*. 1993 (45): 21-24.

Ushomaru K, Nakachimi K, Saito H, inventors. Pharmaceutical preparations and a method of manufacturing them. US patent 4 702 918. October 27, 1987.

Whitehead L, Collett JH, Fell JT. Amoxycillin release from a floating dosage form based on alginates. *Int J Pharm*. 2000 (210): 45-49.

Wong PSL, Dong LC, Edgren DE, Theeuwes F, inventors. Prolonged release active agent dosage form adapted for gastric retention. US patent 6 120 803. September 19, 2000.

Wu W, Zhou Q, Zhang HB, Ma GD, Fu CD. Studies on nimodipine sustained release tablet capable of floating on gastric fluids with prolonged gastric resident time. *Yao Xue Xue Bao*. 1997 (32): 786-790.

Yang L, Esharghi J, Fassihi R. A new intra gastric delivery system for the treatment of helicobacter pylori associated gastric ulcers: in vitro evaluation. *J Control Release,* 1999 (57):215-222.

Nishihata T, Tahara K, Yamamoto K. Overall mechanisms behind matrix sustained release (SR) tablets prepared with hydroxypropyl cellulose 2910, *J Controlled Release,* 1995 (35): 59-66.

Vyas SP, Khar RK. Controlled Drug Delivery: Concepts and Advances. I[st] ed. vallabh prakashan, 2002:156-189

Pina ME, Salsa T, Veiga F. Oral controlled-release dosage forms. I. cellulose ether polymers in hydrophilic matrices, *Drug Dev Ind Pharm,* 1997, 23(9): 929- 938.

Martini L, Close M, Gravell K. Use of a hydrophobic matrix for the sustained release of a highly water soluble drug, *Drug Dev Ind Pharm,* 2000, 26(1): 79- 83.

Borguist P, Korner A, Larsson A. A model for the drug release from a polymeric matrix tablets-effect of swelling and dissolution, *J Controlled Release,* 2006 (113): 216-225.

Reja M, Quadir MA, Haider SS. Comparative evaluation of plastic, hydrophobic and hydrophilic polymers as matrices for controlled-release drug delivery, *J Pharm Sci,* 2003 (692): 274-291.

Siepmann J, Peppas NA, HPMC matrices for controlled drug delivery: new model combining diffusion, swelling and dissolution mechanisms and predicting the release kinetics, *Pharm Research,* 2000 (16):1748-1756.

Horvath S, Julien JS, Lapeyre F. Influence of drug solubility in the formulation of hydrophilic matrices, *Drug Dev Ind Pharm,* 1989, 15(14-16):2197-2212

Renolds TD, Tajeer J. Polymer erosion and drug release characterization of HPMC matrices, *Pharm Research,* 1991 (11):1115-1119.

Aoki S, Uesugi K, Tatsuishi K, Ozawa H. Evaluation of the correlation between in vivo and in vivo release of phenylpropanolamine hydrochloride from controlled-release tablets, *Int J Pharm,* 1992 (85): 65-73.

Rao KVR, Devi KP. Swelling controlled-release systems: recent development and applications, *Int J Pharm,* 1988 (48):1-13

Hogan, JE. Hydroxypropyl methylcellulose sustained release technology, *Drug Dev Ind Pharm,* 1989, 15(6-7):975-999

Nakhat PD, Yeole PG, Galgatte UC, Babla IB. Design and evaluation of xanthan gum-based sustained release matrix tablets of diclofenac sodium, *Int J Pharm Sci,* 2006, 68(2): 185-189.

Prakash SS, Niranjan PC, Kumar PH, Santanu C, Devi V. Design and evaluation of verapamil hydrochloride controlled release tablets using hydrogel polymers, *J Pharm Research, 2007*, 6(2):122-125.

Amaral MH, Lobo JM, Ferreira DC. Effect of HPMC and hydrogenated castor oil on naproxen release from sustained-release tablets, *AAPS Pharm Sci Tech,* 2001, 2(2):1-8.

Levina M, Palmer F, Rajabi-Siahboomi A. Investigation of directly compressible metformine HCl 500 mg extended release formulation based on hypromellose, Controlled Release Society Annual Meeting 2005:1-3.

George M., Grass IV, Robinson J. "Sustained and controlled release drug delivery systems" chapter 6 in "Modern Pharmaceutics" edited by Banker G.S., Rhodes C.T., 2nd edition, Marcel Dekker, 1990: 639-658.

Lee BJ, Ryu SG. Formulation and release characterstic of hydroxyl propyl cellulose matrix tablet containing metformin. *Drug ind. Pharma.*, 1999, 25(4):493-501.

Ochoa L, Igartua M. Preparation of sustained release hydrophilic matrices by melt granulation in a high-shear mixer. *AAPS PharmSciTech.*, 2008, 9(3):1016-1024

Talukder RM, Fassihi R. Development and in-vitro evaluation of a colon-specific controlled release drug delivery system. *Drug Dev Ind Pharm.*, 2008 (31):1-15.

Cantor SL, Hoag SW. Formulation and Characterization of a Compacted Multiparticulate System for Modified Release of Water-Soluble Drugs-Part II Theophylline and Cimetidine. *Eur J Pharm Biopharm.*, 2004, 58(1):51-59.

Paulo, C. and Jose, M.S.G. Modeling and comparison of dissolution profiles. 2001 (13): 123-133.

Salmon, J. L. and Doelker, E. Formulation des comprimes a libera-tion prolongee. *Pharma. Acta Helv.,* 1980 (55): 174-182.

El- Arini, S. K. and Leuenberger, H. Modeling of drug release from polymer matrices : effect of drug loading. *Int. J. Pharm.* 1995 (121): 141-148.

Verelas, C. G., Dixon, D.G. and Steiner, C. Zero-order release from biphasic polymer hydrogels. *J. Control. Release,* 1995 (34): 185-192.

Gibaldi, M. and Perrier, D., 2nd Edition, Pharmacokinetics. Drugs and the Pharmaceutical Sciences. Marcel Dekker, Inc., New York and Basel. Vol. 15, 1982.

Kitazawa, S., Johno, L., Ito, Y., Teramura, S. and Okada, J., 1975. Effects of hardness on the disintegration time and the dissolution rate of uncoated caffeine tablets. *J. Pharm. Pharmacol.* 1975 (27): 765-770.

Mulye, N.V. and Turco, S. J. A simple model based on first order kinetics to explain release of highly water soluble drugs from porous dicalcium phosphate dehydrate matrices. *Drug Dev. Ind. Pharma.*, 1995 (21): 943-953.

Higuchi, T. Rate of release of medicaments from ointment bases containing drugs in suspension, *J. Pharm. Sci*., 1961(50): 874-875.

Higuchi, T. Mechanism of sustained-action medication. Theoretical analysis of rate of release of solid drugs dispersed in solid matrices. *J. Pharm. Sci*., 1963 (52):1145-1149.

Higuchi, W.I. Analysis of data on the medicament release from ointments. *J Pharm. Sci,* 1962 (52): 1145-1149.

Hixson, A. W. and Crowell, J. H. Dependance of reaction velocity upon surface and agitation. *Ind. Eng. Chem*., 1931 (23): 923-931.

Niebergall, P. J., Milosovich, G. and Goyan, J.E. Dissolution rate studies. II. Dissolution of particles under conditions of rapid agitation. *J. Pharm. Sci*., 1963 (52): 236-241.

9

Parenteral Controlled Release Systems

Introduction

Controlled release drug delivery systems are so designed to control the pharmacokinetics profile of the drug to improve the patient compliance by maintaining the steady state plasma level drug concentration over the longer period of time. They reduce the side effect, because of controlling the release rate of the drug as well as targeting the release at the desired site for local and/or systemic action as per the patient's requirement. Controlled release (CR) parenteral drug delivery systems may be formulated in the form of suspensions (micro/nano), emulsions (micro/nano), liposomal formulations, microspheres, resealed erythrocytes, neosomes, gels or can be placed as subcutaneous implants to modify the drug release over periods of days to months or years. All drugs may not serve the purpose as a candidate for controlled delivery through parenteral route. The candidate drug should be potent with known toxicity and pharmacokinetic profiles. Critical formulation and process variables for individual products must be identified to develop the necessary characterization studies that under gird the substance, excipient, and product specifications that allow batch release.

A CR parenteral dosage form may be selected for administration, when there are problems associated with oral delivery (e.g., gastric irritation, first-pass effects or poor absorption) and a need for extended release or targeted delivery (e.g., rapid clearance) and both systemic and localized delivery is required.

Reason for Development of PDS (Parenteral Depot System)

Parenteral administration of a drug in depot formulation in an aqueous suspension or in an oleaginous solution into subcutaneous or muscular tissue results in the formation of a depot at the site of injection, which acts as reservoir to release the drug molecules continuously at a controlled rate leading to the prolongation of absorption from the formulation and sustain the therapeutic drug level with a reduction in the frequency of injection.

9.1 Approaches used in Depot formulation

9.1.1 Dissolution-controlled Depot Formulations

In this depot formulation the rate limiting step of drug absorption is the dissolution of drug particles in the formulation or in the tissue fluid surrounding the drug formulation. So drug absorption can control by slow dissolution of drug particle. The rate of drug dissolution $(Q/t)_d$ under sink conditions is defined by the equation as follows:

$$\left(\frac{Q}{t}\right)_d = \frac{Sa\,Ds\,Cs}{h_d} \qquad \text{.....(1)}$$

where, Sa is the surface area of the drug particles in contact with the medium; Ds is the diffusion coefficient of drug molecules in the medium; Cs is the saturation solubility of drug in the medium; and h_d is the thickness of the hydrodynamic diffusion layer surrounding each of the drug particle.

Basically, two approaches can be utilized to control the dissolution of drug particle to prolong the absorption and hence the therapeutic activity of the drug.

(i) Formation of salt or complexes with low aqueous solubility. Typical examples are preparations of penicillin G procaine (Cs = 4 mg/ml) and penicillin G benzathine
(Cs = 0.2 mg/ml) from the highly water-soluble alkali salts of penicillin G and preparations of naloxone pamoate and naltrexone-Zn-tannate from the water-soluble hydrochloride salts of naloxone and naltrexone, respectively.

(ii) Suspension of macrocrystals. Macrocrystals (large crystals) are known to dissolve more slowly than microrystals (small crystals). Typical example is the aqueous suspension of testosterone isobutyrate for intramuscular administration and diethyl stilbesterol monocrystals for subcutaneous injection.

9.1.2 Adsorption-type Depot Preparations

This type of depot preparation is formed by the binding of drug molecules to adsorbents. In this case only the unbound, free species of the drug is available for absorption. The rate of release of the drug depends upon the binding affinity between the drug and the binding excipients used in the formulations. The rate of availability of unbound form of the drug at equilibrium state $(C)_f$ is determined by the Langmuir equation as:

$$\frac{(C)f}{(C)b} = \frac{1}{a(C)b.m} + \frac{(C)_f}{(C)_{b.m}} \quad .....(2)$$

where, $(C)_b$ is the amount of drug absorbed (mg) adsorbed by 1 g adsorbent; $(C)_{b.m}$ is the maximum amount of drug (mg) adsorbed by 1 g adsorbent, which can be estimated from the slope obtained from the linear plot of $(C)_f/(C)_b$ versus $(C)_f$ and a is the constant, can be determined from the intercept and $(C)_{b.m}$.

Examples include depot preparation of vaccine in which the antigens are bound to highly dispersed aluminum hydroxide gel to sustain their release.

9.1.3 Encapsulation-type Depot Preparations

This depot preparation is prepared by encapsulating drug solids within a permeation barrier or dispersing drug particles in a diffusion matrix. The release of drug molecule is controlled by the rate of permeation across the permeation barrier and the rate of biodegradation of the barrier macromolecules. Both permeation barrier and diffusion matrix are fabricated from biodegradable or bioabsorbable macromolecules, such as gelatin, dextran, polylacticacid, lactide-glycolide copolymers, phospholipids, and long-chain fatty acids and glycerides. Typical examples are notrexone pamoate-releasing biodegradable microcapsules, liposomal formulation of doxorubicin, and norethindrone-releasing biodegradable lactide-glycolide copolymer beads.

9.1.4 Esterification-type Depot Preparations

This depot preparation is produced by esterifying a drug to form a bioconvertible Prodrug-type ester and then formulating it in an Injectable formulation. This chemical approach depends upon number of enzyme (esterase) present at the injection site, which forms a drug reservoir at the site of Injection. The rate of drug absorption is controlled by the interfacial partitioning of drug esters from the reservoir to the tissue fluid and the rate of bioconversion of drug esters to regenerate active drug molecules. Examples of such systems include: fluphenazine enanthate, nandrolone decanoate in oleaginous solution.

9.1.5 Polymeric Drug Delivery Systems

Many classes of cross-linked polymer gels display phase transition characteristics i.e. abrupt change in swollen volume in response to small environmental changes like pH, light, temperature, electric field intensity, ionic strength, and even specific stimuli like glucose concentration. Drugs containing charged hydrogel networks have been recognized as useful matrices for sustained or controlled delivery of drugs, because their response to external pH variation, such hydrogel systems include: glucose sensitive insulin releasing devices, and osmotic insulin pump, etc.

9.1.5.1 Uses of Biodegradable Polymers in Parenteral Depot System

Bio-degradable polymer may be defined as synthetic or natural polymer which is degradable *in-vivo,* either enzymatically or non-enzymatically to produce bio-compatible or non-toxic by products, which can be further metabolized or excreted through normal physiological pathway. The advantage of the biodegradable polymers in implants or drug delivery devices is that it releases drug by diffusion controlled mechanism hence predetermined drug delivery rate can be achieved easily.

Biodegradable polymers investigated for controlled drug delivery include:

Poly lactide / poly glycolide polymers; Poly anhydrides; Poly caprolactones; Poly orthoesters; Psuedo polyamino acid; Poly phosphazenes, etc.

9.1.5.2 Factors Affecting Drug Eelease from the Polymeric Sustained Release Dosage Forms

The chemical and physical characteristics of polymers can be determined by its average molecular weight and chemical structure (*i.e.,* the functional groups attached to its backbone). Smaller sized polymers are suitable for formulation of coating and as co-solvents. Higher molecular weight polymers are covalently attached to pharmaceutical actives (*e.g.,* polymers for pegylation). In addition, polymers can have various architectures, shapes, and linkers. The architecture can be one-dimensional (threads), two-dimensional (sheets), or three-dimensional (networks). The shape of a three-dimensional configuration can vary as well. For example, polyethylene glycols (PEGs) used for biopharmaceuticals can be straight chains, branched, or forked. Some types of drugs work better with certain classes of polymers. For example: fragmented antibodies work well with forked PEGs. Polymers used as excipients or in biodegradable implants may contain ionic groups, in which the polymer may be pH dependent. Biodegradable polymers (*e.g.*, polylactides, polyglycolides, and poly-lactide-co-glycolides (PLGAs) are commonly used for parenteral controlled release microspheres, for polymeric micelles, and implantable drug-device systems.

9.1.6 Implants

Implants are used as depot formulations either to limit high drug concentrations to the immediate area surrounding the pathology or to provide sustained drug release for systemic therapy. Clinically, implant systems have been used in situations, where chronic therapy is indicated, such as hormone replacement therapy and chemical castration in the treatment of prostate cancer. Parenteral implants may take the form of highly viscous liquids or semi-solid formulations, both of which may be injected with a needle. Alternatively, implants may be in the form of tiny rods impregnated with drug substances or a liquid, which form gels *in situ*. *In situ* forming gels forms gel, when the polymer solubilising solvent diffuses away from the injection site, leaving the polymer in contact with an aqueous environment *in-vivo*, or gel on cooling after being injected at an elevated temperature. In situ forming gels may be used to prepare sustained release formulations of oligonucleotides and non-steroidal anti-inflammatory agents. Implants are prepared from a variety of polymeric materials such as: polysaccharides, polylactic acid co-glycolic acid, and the non-biodegradable methacrylates. Biodegradable materials, such as polylactic acid co-glycolic acids are often preferred as they don't need surgical removal of the implant after treatment. Polylactic acid co-glycolic acid implants of gancyclovir for the treatment of cytomegalo virus infection maintained sufficient therapeutic levels of the drug in the vitreous humor and retina/choroids for three to five months. However, non-biodegradable materials do provide therapeutic levels of drug for up to one year *in-vivo*.

Drug release may also be controlled by various stimuli, such as electrical stimuli in polyelectrolyte system. This allows the fabrication of pulsatile delivery systems such as the electrically triggered release of insulin from poly-di-methylaminopropyl acryl amide gels.

9.2 Development of Controlled Release Injectable Formulations

9.2.1 *In Situ* Forming Drug Delivery Systems (ISFD)

In situ forming polymeric formulations are drug delivery systems that are in sol form before administration to the body, but once administered, undergo gelation *in situ*, to form a gel. The formation of gels depends on factors like temperature modulation, pH change, presence of ions and ultra violet irradiation, from which the drug gets released in a sustained and controlled manner.

Injectable *in situ* forming implants are classified into five categories, according to their mechanism of depot formation: (i) Thermoplastic pastes (ii) *In situ* cross linked systems (iii) *In situ* polymer precipitation (iv) Thermally induced gelling system (v) *In situ* solidifying organogels.

(i) ***Thermoplastic pastes*** **(TP):** Thermoplastic pastes are semisolid polymers, which after injection as a melt forms a depot upon cooling at body temperature. They characterized as a low melting point or Tg (glass transition temperature) in the range of 25-65°C and an intrinsic viscosity in the range of 0.05-0.8 dl/g. It has been found that no release observed below the viscosity of 0.05 dl/g, whereas above 0.8 dl/g, the ISFD was no longer injectable using a needle. At injection temperature above 37°C but below 65°C these polymers behave like viscous fluids which after solidification forms high viscous depots. Drugs are incorporated into the molten polymer by mixing without the application of solvents. Bioerodible thermoplastic pastes can be prepare from monomers such as D,L-lactide, glycolide,

E-caprolactone, dioxanone, and orthoesters. Polymers and copolymers of this monomer have been extensively used in surgical sutures, ocular implants, soft tissue repair, etc.

(ii) ***In situ cross linked polymer systems*****:** The cross-linked polymer controls the drug release by following diffusion of the hydrophilic macromolecules. Cross-linked polymer network can be found in situ by free radical reactions initiated by heat (thermosets) or absorption of photon or ionic interactions between small cation and polymer anions.

Dunn et al, used biodegradable copolymers of D, L-lactide or L-lactide with E-caprolactone to prepare a thermosetting system for prosthetic implants for slower release drug delivery systems.

Hibbell et al., described a photopolymerizable biodegradable hydrogel as a tissue contacting material and controlled release carrier. This system consists of a macromer i.e. PEG(polyethylene glycol)-oligo-glycol-acrylate with a photo initiator, such as eosin and visible light. The controlled release of protein was observed over a period of several days by using this technique.

(iii) **In situ polymer precipitation:** In this system a water-insoluble and biodegradable polymer is dissolved in a biocompatible organic solvent to which a drug is added forming a solution or suspension after mixing. When this formulation is injected into the body, the water miscible organic solvent dissipates and water penetrates into the organic phase, which leads to phase separation and precipitation of the polymer forming the depot at the site of injection (Dunn et al.). One of the examples of such system is Eligard©, which contain the leuteinizing hormone releasing hormone(LHRH) agonist leuprolide acetate and poly-lactide-co-glycolic acid dissolved in N-methyl-2-pyrrolidone (NMP) in a 45:55 (polymer:NMP) ratio. This system led to suppression of testosterone levels in dogs for approximately 91days.

One of the problems with this system is the possibility of a burst in drug release especially during the first few hours after injection into the body. In order to control the burst effect, four factors have been examined, which are: the concentration of polymer in the solvent, the molecular weight of the polymer, the solvent used and the addition of surfactant.

(iv) ***Thermally induced gelling system*:** Many polymers undergo abrupt changes in solubility as a function of environmental temperature. The thermosensitive polymers such as poly-N-isopropylacrylamide (NIPAAM) exhibit sharp lower critical solution temperature (LCST) at about 32˚C, which can be shifted to body temperature by formulating poly NIPAAM based gels with salt and surfactant. But, poly NIPAAM is not suitable for biomedical applications due to its well known cytotoxicity (activation of platelets) and non-biodegradability Triblock poly(ethylene oxide)-poly(propylene oxide)- poly(ethylene oxide) copolymer i.e. PEO-PPO-PEO(pluronics or poloxamers), have shown gelation at body temperature, when highly concentrated polymer solution>15% w/w were injected. These polymer concentration shown disadvantage of changing the osmolarity of the formulation, kinetics of the gelation, and causes discomfort in ophthalmic applications due to vision blurring and crusting.

Thermosensitive Chitosan-β-glycerophosphate (C-GP) formulation containing liposomes demonstrated the *in-vitro* controlled delivery of carboxyfluorescein over 2 weeks. The release rate strongly depended on the liposome size and composition (i.e. addition of cholesterol), and on the presence of phospholipase in the release medium.

(v) ***In situ Solidifying Organogel*****:** Organogels are composed of water insoluble amphiphilic lipids, which swell in water and forms various types of lyotropic liquid crystals. The amphiphilic lipids examined for drug delivery are glycerol monooleate, glycerol monopalmitostearate, glycerol monolinoleate, sorbitan monostearate (SMS) and different gelation modifiers (polysorbates 20 and 80) in various organic solvents and oils. These compound forms a cubic liquid crystal phase upon injection into an aqueous medium which is gel like and highly viscous.

SMS organogels containing either w/o or vesicular in water in oil (v/w/o) emulsion were investigated *in-vivo* as delivery vesicles for vaccines using albumin and haemagglutin as model anigens. Intramuscular administration of the v/w/o gel yields the long lasting depot effect up to 48hr. Controlled releases of contraceptive steroids levonorgestrel and ethinyl estradiol were achieved by Gao et al. In these work biodegradable organogel formulations prepared from glycerol palmitostearate (precirol) in derivatized vegetable oil, produced *in-vitro* drug release of levonorgestrel up to 14 days.

9.2.2 Microspheres

Several biodegradable polymers have been investigated for preparation of microspheres as depot formulation. The application of biodegradable microspheres to deliver small molecules, proteins, and macromolecules using multiple routes of administration has been widely investigated and several products have been brought to market.

For peptide or protein containing microspheres mainly three processes were studied more intensively, namely the emulsion solvent evaporation technique, caocervation and phase separation methods and to some extent spray drying.

Lupron Depot®, microsphere containing the LHRH (leuprolide) acetate with PLGA (75/25)-14000 and PLA-15000, prepared by w/o/w emulsion-solvent evaporation method. The microsphere release drug in a zero order fashion over 1 to 3 months after intramuscular or subcutaneous injection in animals. PLGA microsphere had been also used for delivery of glycoprotein (GP) IIb/IIIa antagonist, plasmid DNA, Interleukin-1α and prolidase enzyme, etc.

9.2.3 Liposomes

Liposomes are lyotropic liquid crystals composed of relatively biocompatible and biodegradable materials and consist of an aqueous core entrapped in one or more bilayered natural and/or synthetic lipids. Drugs with widely varying lipophilicities can be encapsulated in liposomes either in the phospholipid bilayer and/or entrapped in the aqueous core. The most common phospholipid used is phosphatidylcholine (PC) molecule. PC is an amphipathic molecule in which a glycerol bridge links a pair of hydrophobic acyl-hydrocarbon chains with a hydrophilic polar head groups i.e., phospho choline. Formulation of drugs in liposomes has provided an opportunity to enhance the therapeutic indices of various agents mainly through the alteration of bio-distribution. They are versatile drug carriers, which can be used to control retention of entrapped drugs in the presence of biological fluids and enhanced vesicle uptake by target cells.

Lipid complex such as: Abelcet®, Amphotec® and liposomal formulations such as: Ambisome®, Daunosome®, and a stealth liposome Doxil® had got approval for human use. These products have been developed for intravascular administration, for enhancement of circulation times and reducing toxicity by lipid encapsulation. The encapsulation of drug into multivesicular liposomes i.e., Depo Foam® offers a novel approach to sustained release drug delivery. Drug into unilamellar and multilamellar liposomes and complexation of drug with lipids, resulted in products with better performance over period of several hours to a few days after intravascular administration, where as Depo Foam® encapsulation has been result in sustained release over several days to weeks.

9.2.4 Suspensions

Suspensions are widely used pharmaceutical dosage forms, which offer a potential use as a parenteral sustained release system. Subcutaneous administration of a drug as an aqueous or oil suspension results in the formation of a depot at the injection site (Davis et al). The depot act as a drug reservoir, slowly releasing the drug continuously at a rate dependent upon both the intrinsic aqueous solubility of the drug and the dissolution of the drug particles into tissue fluid surrounding the drug particle in the subcutaneous tissue. Oleaginous suspension of micronized crystal of penicillin procaine in vegetable oil, such as peanut or sesame oil, gelled with 2% aluminum monostearate was reported to produce therapeutic blood level of penicillin in both animal and human for 162hr.

Aqueous thixotropic suspension of penicillin procaine (40-70%w/w), produced by Abbot Lab. such as Duracillin® (Lilly), Crystacillin® (Squibb), which on intramuscular injection tends to form compact and cohesive depot, leading to the slow release of penicillin procaine. Insulin has long been formulated with zinc as a suspension for subcutaneous delivery (for example, Humulin®, Iletin®, Lente® and Novolin®, developed and manufactured by Lilly) to produce action up to 36hr.

9.2.5 Solid Lipid Nanoparticle (SLN)

SLN are colloidal particles composed of a biocompatible/biodegradable lipid matrix, which is solid at body temperature and exhibit size range in between 100 and 400 nm. Upon parenteral administration SLN shows excellent physical stability, protection of incorporated labile drugs from degradation, controlled drug release (fast or sustained) depending on the incorporation of model, good tolerability and site specific targeting. Techniques utilized for preparation of SLN are high pressure homogenization (HPH), nano-emulsion, emulsification-solvent evaporation or diffusion, w/o/w multiple emulsion method and high speed stirring and/or ultrasonication methods.

SLN loaded with prednisolone by HPH, released the drug *in-vitro* (i.e. in absence of enzyme) over a period of more than 5 weeks. For stearic acid SLN containing cyclosporine-A, Cavalla et al., determined an *in-vitro* release of <4% after 2hr compared to >60% from solution.

Yang et al., was performed an in-vivo studies of encapsulating anticancer agent camptothecin with stearic acid SLN in the year 1995. They prepared SLN loaded camptothecin by HPH (average particle size

197 nm) and administered intravenously to mice. These SLN demonstrated higher AUC and maximum retention time (18 fold enhancement) compared to camptothecin solution.

9.3 Evaluation of Parenteral Depot System

9.3.1 *In-Vitro* Release Methods

Current USP apparatus for *in vitro* release testing are designed for oral and transdermal products and may not be optimal for controlled release parenteral products. USP apparatus 1 (basket) and 2 (paddle) were designed for immediate- and modified-release oral formulations. USP apparatus 3 (reciprocating cylinder) and 4 (flow through cell) were designed for extended-release oral formulations. These latter two methods may be the most relevant to CR parenterals and may be suitable after appropriate modification. Alternative apparatus, such as small sample vials and vessels, with and without agitation, are currently used for CR parenterals.

Problems that may be associated with these alternative apparatus include lack of sink conditions and sample aggregation. Research is required to determine the scientific basis for the tests, procedures, including apparatus (e.g., geometry and hydrodynamics), and acceptance criteria for CR parenterals.

General agreement is that sink conditions should be used for *in-vitro* testing for quality control purposes provided that the study design allowed for discrimination between formulation variants with different *in-vivo* release profiles. However, non-sink conditions may be necessary if the purpose of the *in-vitro* test is to establish IVIVC.

The apparatus and media used should take into account the release mechanism and the physical properties of the product (e.g., size and stability). In addition, *in-vitro* release tests must also discriminate between the performances of different formulation variants and ideally should have bio-relevance.

The rationale for this understanding is that the ultimate purpose of quality control testing is to ensure the clinical performance, i.e., efficacy and safety of the product. To achieve *in-vivo* relevance, physiologic variables at the site need to be considered including body temperature and metabolism (both can significantly affect blood flow), muscle pH, buffer capacity, vascularity, level of exercise, as well as volume and osmolarity of the products.

Any tissue response, such as inflammation and/or fibrous encapsulation of the product, may need to be considered. *In-vitro* release methods should be designed on the basis of *in Vivo* Release Mechanisms

The following general approaches to be taken into consideration or *in-vitro* test method design: (i) identification of release media and conditions result in reproducible release rates; (ii) preparation formulation variants that are expected to have different biologic profiles; (iii) testing of formulation variants *in-vitro* as well as *in-vivo;* and (iv) modification of *in-vitro* release methods to allow discrimination between formulation variants that have different *in-vivo* release profiles.

9.3.2 Development of IVIVC for CR Parenterals

The principles used in IVIVC of oral extended-release products may be applied to parenterals with appropriate modification, justified on a scientific basis. IVIVC modeling and measurements may be different for different types of products (e.g., targeted-release Vs. extended release products). Similarly, *in-vitro* release methods and media are likely to vary, depending on the product and should be developed based on *in-vivo* relevance. For example, *in-vitro* cellular tests may be acceptable as long as they are reproducible and can be validated. Similarly, *in-vivo* measurements may vary and may include plasma concentrations, efficacy/safety data, surrogate end-point data, as well as tissue concentrations.

In the case of some products, such as liposomes, it may be necessary to measure *in-vivo* concentrations of both free and encapsulated drug. Models that represent multiple processes (e.g., physical and biologic) should be considered, as appropriate.

The use of animals was considered to be acceptable to prove that an *in-vitro* release system is discriminating. However, the use of animal models was considered inappropriate to prove an IVIVC for regulatory purposes. Instead, bio-relevance should be developed by using clinical data.

Use of Animal Models in Release Testing

In the development of *in-vitro* release methods, animal data may be used to obtain tissue distribution and pharmacokinetic information. Plasma levels may not be the best measure of *in-vivo* behavior for CR parenteral products intended for local delivery or targeted release. Animal models were considered to be invaluable, and serial tissue samples might be used to compare product performance before and after manufacturing changes for CR parenterals with tissue-specific delivery. Although data will be useful in initial development, but ultimately human data must be used to establish an IVIVC. Selection of an appropriate animal model with comparative studies be performed between injection sites in humans and animals to establish interspecies differences in drug release. Larger animals such as sheep and dogs may be more representative of humans with regard to interspecies differences than would small laboratory animals. Because inter-subject variability significantly impacts *in vivo* data, inbred animals may be useful in identifying variables that affect the drug release and absorption processes. Extensive inter- and intra-subject variability may mask critical formulation, and manufacturing variables unless very large human populations are used. The identification of an appropriate animal model for CR parenteral products was recommended as a research project, possibly for investigation through Product Quality Research Initiative (PQRI). Research study should include different animals as well as different sites and should attempt to establish correlations between human and animal data relating the findings to physiologic parameters.

Accelerated *in-vitro* Release Testing

Accelerated release testing is desirable for routine quality control purposes.

These tests should have relevance to real-time *in-vitro* release tests conducted under conditions, which simulate the *in-vivo* situation as closely as possible. Real-time *in-vitro* tests for the full product duration should be conducted during product development and are essential for validating accelerated release rate tests.

Accelerated tests should be bio-relevant and the mechanism of drug release should not be altered in accelerated tests; rather, it should only be speeded up. For example, in the case of PLGA micro-spheres that release drug primarily through polymer erosion, the accelerated test should speed up the polymer erosion process.

In the design of accelerated *in-vitro* test, factors such as polymer transition and degradation temperature should be considered to avoid any change in the mechanisms of drug release.

9.3.3 Stability

Shelf Life Stability

The initial shelf life stability of entrapped drug is essential because of manufacturing conditions for some CR parenterals (e.g., some microsphere products) may vary. Some time in protein drugs chemical breakdown may result conformational changes, which can affect activity. Therefore, activity must be demonstrated as part of the stability testing using a pharmacodynamic method. Stability testing of many of these products requires extraction of the drug. The method of extraction (e.g., solvent system) should be selected to avoid any potential alteration in drug stability. Shelf life stability should be conducted at room temperature as occurs for other products.

In-Vivo Drug Stability

In-vivo drug stability is an issue for controlled release parenterals, especially those intended for long-term extended release. In large implants, drug stability could be determined by analyzing the drug remaining in explanted systems. This method could only be feasible for dispersed system CR parenterals by using an appropriate animal model, where tissue samples could be excised.

An alternative approach that might be acceptable would be an *in vitro* test that simulated *in-vivo* conditions (e.g., 37°C at 100% humidity).

In-Vivo Product Stability

In-vivo evaluation of CR parenterals include: evaluation of product stability and tissue response to the product as well as drug stability. This is another area, where appropriate animal models may be useful.

Particle Size

Particle size may affect release rates of extended release products such as microspheres. It may also affect targeting ability and reticuloendothelial system (RES) uptake of liposome products. Acceptable particle size ranges may vary for different CR parenteral systems. The particle size range as well as average particle size plays major role in syringability of the product as well as product performance and safety. Particl size can be measured by instrumentation techniques such as particle size analyzer, coulter counter, etc.

9.3.4 Sterilization, Sterility Assurance, and Foreign Particulate Matter

Sterilization and Sterility Assurance

CR parenterals are complex products usually containing polymers and/or lipids with glass transition temperatures below the temperature required for heat sterilization. However some products, which are going to break down if subjected to heat sterilization, require aseptic technique. To reduce drug crystal growth (which could affect performance and bioavailability) during heat sterilization, a cloud point modifier can be introduced.

Because CR parenterals are not liquids, terminal filtration is not an option; in such cases gamma irradiation can be used as an alternative method of terminal sterilization. Because product breakdown (e.g., via polymer degradation) is a potential problem with gamma irradiation, continuing product integrity would have to be demonstrated after gamma radiation. This could be demonstrated by maintenance of polymer weight before and after irradiation. Low energy gamma radiation may be effective in sterilizing without damaging the product and may be an acceptable method of terminal sterilization for some CR parenterals, provided that the long-term quality of the product is not compromised.

To confirm the sterility, the products of CR parenterals must be tested for sterility.

9.3.5 Syringeability or Resuspendability

The syringeability or injectability depends on the viscosity (Newtonian vs. non-Newtonian) of the product, syringe size, needle size, particle size, and morphology of the suspension. The methods of analysis include determination of viscosity by using viscometer appropriate for Newtonian or Non-Newtonian fluids. The viscosity and rheological properties of *in situ* forming drug delivery systems may be assesses using Brookfield rheometer or some other type of viscometers such as Ostwald's viscometer. The particle size can be determined by coulter counter. The resuspendability can be measured by measuring rate of sedimentation, using sedimentation tank following Stoke's equation.

9.3.6 Texture Analysis

The firmness, consistency and cohesiveness of hydrogels are assessed using texture analyzer which mainly indicates the syringeability of sol, so that the formulation can be easily administered *in-vivo*. Higher values of adhesiveness of gels are needed to maintain an intimate contact with surfaces like tissues.

9.4 Applications of Parenteral Controlled Release Systems

Applications of CR parenteral delivery against the disease include fertility, hormone therapy, protein therapy, infections (antibiotics and antifungal), cancer therapy, orthopedic surgery and postoperative pain, chronic pain, vaccination/immunization, CNS disorders, and immunosupression, etc. In the present era the biodegradable injectables offer attractive opportunities for protein delivery. Modified release (MR) parenteral drug products are available in several dosage forms, including microspheres, liposomes, gels, suspensions, in situ forming implants, lipophilic solutions, solid-lipid nanoparticles (SLN), etc. Some of the commercial formulations are cited in the Table 9.1.

Table 9.1 Examples of CR Parenteral Products

Trade Name	Drug/Active ingredient
Suspension products	
Depo-Medrol	Methylprednisolone
Depo-Provera	Medoxyprogestrone
Celestone Soluspan	Betamethasone
Lente NPH	Insulin
Microsphere Product	
Lupron Depot	Luprolide
Sandostatin LAR	Octreotide
Nutropin Depot	Somatropin
Trelstar Depot	Triptorelin
Risperdal	Risperidone
Vivitrol	Naltrexone
Trenatone	Luprolide
Sandostatin LAR	Octreotide
Suprecur MP	Buserelin
Parlodel LAR	Bromocriptine
Atridox	Doxycycline
Gliadel	Carmustin

Trade Name	Drug/Active ingredient
Arestin	Minocycline
Liposomal Product	
Doxil	Daunorubisin
Daunoxome	Daunorubisin
Ambisome	Amphotericin B
Depocyt	Cytarabine
Lipid Complex Product	
Ambilect	Amphotericin B
Amphotec	Amphotericin B
Visudyne	Verteporfin
Depo Foam	Insulin
Implant Product	
Norplant	Levonorgestrel
Gliadel	Carmustine
Zoladex	Goserelin
Viadur	Luprolide
Gel formulations	
Timoptic-XE	Timolol maleate
Oncogel	paclitaxel
Cytoryn	Interleukin-2

9.5 Some Examples of Research Outputs

Packhaeuser C. B. et al., 2007, were investigated the feasibility to generate in situ forming parenteral depot systems from insulin loaded dialkylaminoalkyl-amine poly(vinyl alcohol)-g-poly(lactide-co-glycolide) nanoparticles. Biodegradable nanoparticles formed polymeric semi-solid depots upon injection into isotonic phosphate buffered saline (PBS) with no additional initiators. The release showed a triphasic profile with an initial burst, pore diffusion and diffusion from the swollen matrix over more than two weeks.

Chourasia M. K. et. al., 2004, developed an implantable delivery system, which when comes in contact with biological fluid, solidifies immediately and release the drug in a controlled manner for prolonged period of time. Polylactic acid, polyglycolic acid and copolymers of polylactic acid and polyglycolic acid were synthesized and characterized for various physicochemical attributes. In vivo studies revealed sustained release of the entrapped drug from formulations, which is reflected from the persistence of drug in blood for longer period of time.

Pechenov S. et. al., 2004, worked for development of ready-to-inject in situ formable controlled release gel systems for proteins. He developed and characterized injectable controlled release systems composed of crystals of amylase, a model protein, suspended in solutions of polymeric and non-polymeric matrix materials in organic solvents. In his study, alpha-amylase derived from Aspergillus oryzae was crystallized and crystals were suspended in a poly(DL-lactide-co-glycolide) (PLGA) solution in acetonitrile (PLGA/acetonitrile), or in sucrose acetate isobutyrate (SAIB) plasticized with ethanol (SAIB/ethanol) systems. The results indicate that the protein crystals could be incorporated in these *in situ* formable gels without the need for initial drying. The crystals withstand organic solvents and water/organic solvent interfaces, and provide high protein loading (>30%). He concluded that protein crystals might offer greater feasibility in developing sustained release injectable in situ formable protein depot systems.

Shenoy D. B. et. al., 2003, developed an injectable, depot-forming drug delivery system for insulin based on micro particle technology to maintain constant plasma drug concentrations over prolonged period of time for the effective controlled blood sugar levels. Formulations were optimized with two well-characterized biodegradable polymers namely, Poly (DL-Lactide-co-glycolide) and poly-epsilon-caprolactone and evaluated in vitro for physicochemical characteristics, drug release in phosphate buffered saline (pH 7.4), and evaluated in vivo in sterptozotocin- induced hypoglycemic rats.

Shenoy, D. B. et. al., 2002, studied a formulation of Poly (DL-co-glycolide) (PLG) – based, microspheric depot system for Bleomycin (BLM) and evaluated in-vivo in mice bearing transplantable melanoma murine solid tumor. The micro particulate delivery systems were formulated employing water in oil in water emulsion solvent evaporation technique and characterized in vitro.

Periti , P. et. al., 2002, Developed a sustained release parentreral depot formulation of leuprorelin acetate a synthetic agonist analogue of gonadotropin-releasing hormone in which the hydrophilic leuprorelin is entrapped in biodegradable highly lipophillic synthetic polymer microspheres, and found to be releasing for several days.

Kumar, V. et. al., 2002, formulated prolonged release biodegradable microspheres for treatment of inflammation. Natural biodegradable polymers, namely, bovine serum albumin and chitosan were used to encapsulate curcumin to form a depot forming drug delivery system. Microspheres were prepared by emulsion-solvent evaporation method coupled with chemical cross-linking of the natural polymers.

Ravivarapu, H. B., 2000, studied about sustained suppression of pituitary gonadal axis with an injectable, in situ forming implant of leuprolide acetate.

Jain, R. A. et. al., 2000, compared various injectable protein loaded biodegradable poly (lactide co-glycolide) (PLGA) devices: In-situ-formed implant versus in-situ formed micro spheres versus isolated microspheres.

Chenite, A. et. al., 2000, developed a novel approach to provide, thermally sensitive neutral solutions based on chitosan/ polyol salt combinations. These formulations possess a physiological pH and can be held liquid below room temperature for encapsulating living cells and therapeutic proteins; they form monolithic gels at body temperature. When injected in vivo the liquid formulations turn into gel implants in situ for sustaining the drug delivery.

Miyazaki, et. al. attempted to formulate *in situ* gels for ocular delivery using xyloglucan (1.5% w/w) as the natural polymer. These *in situ* forming polymeric systems were observed to show a significant mitotic response for a period of 4h when instilled into lower cul-de-sac of rabbit eye.

Ito, et. al. designed and synthesized injectable hydrogels that are formed *in situ* by cross-linking of hydrazide modified hyaluronic acid with aldehyde modified versions of cellulose derivatives such as carboxymethylcellulose, hydroxypropylmethylcellulose and methylcellulose. These *in situ* forming gels were used for preventing postoperative peritoneal adhesions thus avoiding pelvic pain, bowel obstructions and infertility. *In vivo* experiments using rabbit side wall defect-bowel abrasion model, a significant reduction of peritoneal adhesions was observed as compared to that resulting from the administration of saline solution.

Wu, et. Al., designed a new thermosensitive hydrogel by simply mixing N-[(2-hydroxy-3-methyltrimethylammonium)propyl]chitosan chloride and poly (ethylene glycol) with a small amount of α-β-glycerophosphate; for nasal delivery of insulin. The formulation was in solution form at room temperature that transformed to a gel form when kept at 37 °C. Animal experiments demonstrated hydrogel formulation to decrease the blood-glucose concentration by 40-50% of the initial values for 4-5 h after administration with no apparent cytotoxicity.

References

A. J. McHugh. *Journal of Controlled Release.* 2005, 109(1-3): 211-221

Anand Babu Dhanikula and Ramesh Panchagnula. Development and characterization of biodegradable chitosan films for local delivery of paclitaxel. *AAPS J.* 2004 September; 6(3): 88–99.

Bharadwaj Sudhir. Bio-degradable Parenteral Depot System: A Best Approach of Controlled Release. www.pharmainfo.net/latestreviews, 2008, 6(1).

Cao S, Ren X, Zhang Q, Chen E, Xu F, Chen J, *et al. In situ* gel based on gellan gum as new carrier for nasal administration of mometasone furoate. *Int J Pharm,* 2009(365):109-15

Chandrashekhar G, Udupa N. Biodegradable injectable implant system for long term drug delivery using poly (lactic-co-glycolic) acid copolymers. *J Pharm Pharmacol*, 1998 (48):669-74

Chen S, Jagdish S. Controlled delivery of testosterone from smart polymer solution based systems: *In vitro* evaluation. *Int J Pharm,* 2005 (295):183-90

Chenite A, Chaput C, Wang D, Combes C, Buschmann MD, Hoemann CD *et al*. Novel injectable solution of chitosan form biodegradable gels *in situ*. *Biomaterials,* 2000 (21): 2155-61

Chitkara Deepak, Shikanov A., Kumar N. and Domb A. J. Biodegradable Injectable In Situ Depot-Forming Drug Delivery Systems. *Macromolecular Bioscience*, 2006 (6): 977–990.

Chourasia M K, Ashawat M S, et. al., *Ind j of pharm sci*, 2004, 66(3): 322-328

Diane J. Burgess, Ajaz S. Hussain, Thomas S. Ingallinera, and Mei-Ling Chen. Assuring Quality and Performance of Sustained and Controlled Release Parenterals: AAPS Workshop Report, Co-Sponsored by FDA and USP. *Pharmaceutical Research,* 2002, 19 (11): 1761-1768.

Diane J. Burgess, Ajaz S. Hussain, Thomas S. Ingallinera and Mei-Ling Chen. Assuring Quality and Performance of Sustained and Controlled Release Parenterals: Workshop Report. *AAPS PharmSci,* 2002, 4 (2), Art-7: 1-11

Diane J. Burgess, Daan J.A. Crommelin, Ajaz S. Hussain, and Mei-Ling Chen. Assuring Quality and Performance of Sustained and Controlled Release Parenterals: EUFEPS Workshop Report. *AAPS Pharm Sci,* 2004, 6 (1), Art-11: 1-12

Dunn R L, English J P, Cowsar D R, Vanderbelt D D. Biodegradable *in situ* forming implants and methods for producing the same. US Patent 5340849, 1994.

Hatefi A, Amsden B. Biodegradable injectable *in situ* forming drug delivery systems. *J Control Release,* 2002 (80):9-28.

He S, Yaszemski MJ, Yasko AW, Engel PS, Mikos AG. Injectable Biodegradable polymer composites based on poly-(propylene fumarate) crosslinked with poly (ethylene glycol)-dimethacrylate. *Biomaterials*, 2000 (21):2389-94

Hitesh Bari. A prolonged release parenteral drug delivery system– An overview. *Int JPharm Sci Rev and Res.,* 2010, 3(1): 1-11.

Ismail FA, Napaporn J, Hughes JA, Brazean GA. *In situ* gel formulation for gene delivery: release and myotoxicity studies. *Pharm Dev Technol*, 2000 (5):391-7

Ito T, Yeo Y, Highley CB, Bellas E, Benitez CA, Kohane DS. The prevention of peritoneal adhesions by *in situ* cross-linking hydrogels of hyaluronic acid and cellulose derivatives. *Biomaterials*, 2007(28):975-83

Jain R.A. et al., *Eur. J.Pharm. Biopharm.*, 2000, 50 (2): 257-263.

Jeong B, Bae YH, Kin SW. *In situ* gelation of PEG-PLGA-PEG triblock copolymer aqueous solutions and degradation thereof. *J Biomed Mater Res*, 2000(50):171-7

Kumar MT, Bharathi D, Balasubramaniam J, Kant S, Pandit JK. pH induced *in situ* gelling systems of indomethacin for sustained ocular delivery. *Indian J Pharm Sci,* 2005(67):327-33.

Kumar, V. Eld. *Indian J. Physiol. Pharmacol,* 2002, 46 (2): 209-217.

Lambert WJ, Peck KD. Development of an *in situ* forming biodegradable poly lactide-co-glycolide system for the controlled release of proteins. *J Control Release*, 1995(33):189-95

Madan M, Bajaj A, Lewis S, Udupa N, Baig JA. *In situ* forming polymeric drug delivery systems. *Indian J Pharm Sci*, 2009 (71): 242-251

Mahesh V. Chaubal. Role of Excipients in Parenteral Sustained-Release. *Drug Delivery Technology*, 2003, 7(3).

Miyazaki S, Hirotatsu A, Kawasaki N, Wataru K, Attwood D. *In situ* gelling gellan formulations as vehicles for oral drug delivery. *J Control Rel*, 1999 (60): 287-95.

Miyazaki S, Kawasaki N. Comparison of *in situ* gelling formulations for the oral delivery of cimetidine. *Int J Pharm*, 2001 (220): 161-8.

N Swapna, AV Jithan. Preparation, characterization and *in vivo* evaluation of parenteral sustained release microsphere formulation of zopiclone. *J Young Pharmacists*, 2010, 2(3): 223-228

Pechenov S et al. *J Control Release*. 2004, Apr 16, 96(1):149-58.

Periti, P., Mazzei,T., Mini E., *Clin. Pharmaceokinet.*, 2002, 41 (7): 485-504.

Ravivarapu HB, Moyer KL, Dunn RL. Sustained Activity and Release of Leuprolide Acetate from an In Situ Forming Polymeric Implant. *AAPS Pharm Sci Tech*, 2000, 89 (6): 732-741.

Raghavnaveen. Parenteral Controlled Drug Delivery System, www.pharmainfo.net/ reghavnaveen, 04/28/2009.

Ramesh CR, Zentner GM, Jeong B. Macro med, Inc. Biodegradable low molecular weight triblock poly (lactide-co- glycolide) polyethylene glycol copolymers having reverse thermal gelation properties. US patent 6201072. 2001.

Rathi R, Zentner C, Gaylen M, Jeong B. Macro med, Inc. Biodegradable low molecular weight triblock poly (lactide-co-glycolide) polyethylene glycol copolymers having reverse thermal gelation properties. US patent 6117949. 2000.

S. Apte and S.O. Ugwu, "A Review and Classification of Emerging Excipients in Parenteral Medications," *Pharm. Technol.* 2003, 27 (3):46–60.

Sau-Hung S L, Joseph R. Robin and Vincent H. L. Lee. Parenteral products. Robinson's Controlled drug delivery. 2nd Edn., Informa Healthcare, USA, Inc. 2009(20):442-480.

Shah NH, Railkar AS, Chen FC, Tarantino R, Murjani M, Palmer D, *et al*. A biodegradable injectable implant for delivering micro and macro molecules using poly (lactide-co-glycolide) acid copolymers. *J Control Release*, 1993(27):139-47.

Shenoy, D.B., D'souza, R.J. Udupa N., Poly (DL-lactide-co-glycolide) microporous microsphere-based depot formulation of a peptide-like antineoplastic agent. *J. Microencapsulation*, 2002, 19 (4): 525-535.

Siegel RA, Firestone BA. pH dependent equilibrium swelling properties of hydrophobic poly electrolyte copolymer gels. *Macromolecules*, 1988 (21): 3254-3259.

Tuncer Degim and Nevin Çelebi. Controlled Delivery of Peptides and Proteins. *Current Pharmaceutical Design*, 2007(13): 99-117

Wang P, Johnston T P. Kinetics of sol to gel transition for poloxamer polyols. *J Appl Polymer Sci,* 1991(43): 283-92

Sachiro K, Taguchi T, Hirofumi S, Tanaka J, Tateishi T. Injectable *in situ* forming drug delivery systems for cancer chemotherapy using a novel tissue adhesive: Characterization and *in-vitro* evaluation. *Eur J Pharm Biopharm*, 2007 (295): 183-90

Wu J, Wei W, Wang LY, Su ZG, Ma G. A thermosensitive hydrogel based on quaternized chitosan and poly (ethylene glycol) for nasal delivery system. *Biomaterials,* 2007(28): 2220-32

Yew W. Chien. Parenteral Drug Delivery and Delivery Systems. Novel Drug Delivery Systems, 2nd Edn., Informa Healthcare, USA, Inc. 2009: 381-528.

Zhidong L, Jaiwei L, Shufang N, Hui L, Pingtian D, Weisan P. Study of an alginate/HPMC based *in situ* gelling ophthalmic delivery system for gatifloxacin. *Int J Pharm*, 2006 (315):12-17.

10

Transdermal Drug Delivery Systems

Introduction

Transdermal drug delivery systems are topically administered drug formulations in the form of patches, which deliver drugs for systemic effects at a predetermined and controlled rate. Transdermal drug delivery is the non-invasive delivery of medications from the surface of skin to the circulatory system. It reduces the drawbacks of poor bioavailability due to hepatic metabolism (first pass effect) and degradation in oral route or causing irritation to the stomach due to inherent characteristics of drugs like NSAIDS. The development of TDDS is multidisciplinary activity that encompasses fundamental feasibility studies, starting from the selection of drug molecule to the demonstration of sufficient drug flux in an *ex-vivo* and *in-vivo* model followed by fabrication of a drug delivery system that meets all the stringent needs, which are specific to the drug molecule (physicochemical and stability factors), the patient (comfort and cosmetic appeal), the manufacturer (scale up and manufacturability) and most important the economy.

There are two important layers in skin: the dermis and the epidermis. The outermost layer, the epidermis, is approximately 100 to 150 micrometers thick, has no blood flow and includes a layer within it known as the stratum corneum. This is the layer most important to transdermal delivery as its composition allows it to keep water within the body and foreign substances out. Beneath the epidermis, the dermis contains the system of capillaries that transport blood throughout the body. If the drug is able to penetrate the stratum corneum, it can enter the blood stream. A process known as passive diffusion, which occurs too slowly for practical use, is the only means to transfer normal drugs across this layer. The stratum corneum develops a thin, tough, relatively impermeable membrane which usually provides the rate limiting step in transdermal drug delivery system. Sweat ducts and hair follicles are also paths of entry, but they are considered rather insignificant (Stanley Scheindlin)

A transdermal drug delivery device, which may be of an active or a passive design, is a device which provides an alternative route for administering medication. These devices allow pharmaceuticals to be delivered across the skin barrier. In theory, transdermal patches work very simply. A drug is applied in a relatively high dosage to the inside of a patch, which is worn on the skin for an extended period of time. The drug enters the blood stream directly through the skin following diffusion process. Since there is high concentration on the patch and low concentration in the blood, the drug will keep diffusing into the blood for a long period of time, maintaining the constant concentration of drug in the blood to elicit therapeutic activity (Fig. 10.1).

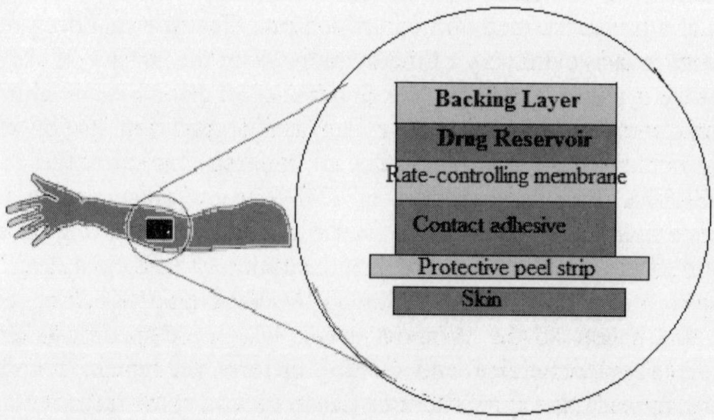

Fig. 10.1 Transdermal Patches containing the reservoir type device. The protective peel has to be removed prior to use.

Transdermal drug delivery enables to avoid gastrointestinal absorption, with its associated pitfalls of enzymatic and pH associated deactivation. This method also allows for reduced multi-day therapy with a single application, rapid notification of medication in the event of emergency, as well as the capacity to terminate drug effects rapidly via patch removal

However this system has its own limitations in which the drug that require high blood levels cannot be administered and may even cause irritation or sensitization reaction to the skin. The adhesives may not adhere well to all types of skin and may be dis-comfortable to wear.

Along with these limitations, the high cost of the product is also a major drawback for the wide acceptance of this product.

10.1 Transdermal Permeation

The various steps involved in transport of drug from patch to systemic circulation are as follows:

- Diffusion of drug from drug reservoir to the rate controlling membrane.
- Diffusion of drug through rate limiting membrane of stratum corneum.
- Sorption by stratum corneum and penetration through viable epidermis.
- Uptake of drug by capillary network in the dermal papillary layer.
- Effect on target organ.

10.1.1 Drug Delivery Routes across Human Skin

Drug molecules in contact with the skin surface can penetrate by three potential pathways: through the sweat ducts, *via* the hair follicles and sebaceous glands (collectively called the shunt or appendage route), or directly across the stratum corneum (Fig. 10.2).

The relative importance of the shunt or appendageal route versus transport across the stratum corneum has been debated by scientists over the years and is further complicated by the lack of a suitable experimental model to permit separation of the three pathways.

In-vitro experiments tend to involve the use of hydrated skin or epidermal membranes so that appendages are closed by the swelling associated with hydration. Scheuplein, et al., proposed that a follicular shunt route was responsible for the pre-steady-state permeation of polar molecules and flux of large polar molecules or ions that have difficulties in diffusing across the intact stratum corneum. However it is generally accepted that as the appendages comprise a fractional area for permeation of approximately 0.1%, their contribution to steady state flux of most drugs is minimal. This assumption has resulted in the majority of skin penetration enhancement techniques being focused on increasing transport across the stratum corneum rather than via the appendages. Exceptions are iontophoretic drug delivery which uses an electrical charge to drive molecules into the skin primarily via the shunt routes as they provide less electrical resistance, and vesicular delivery (Heather A. and E. Benson).

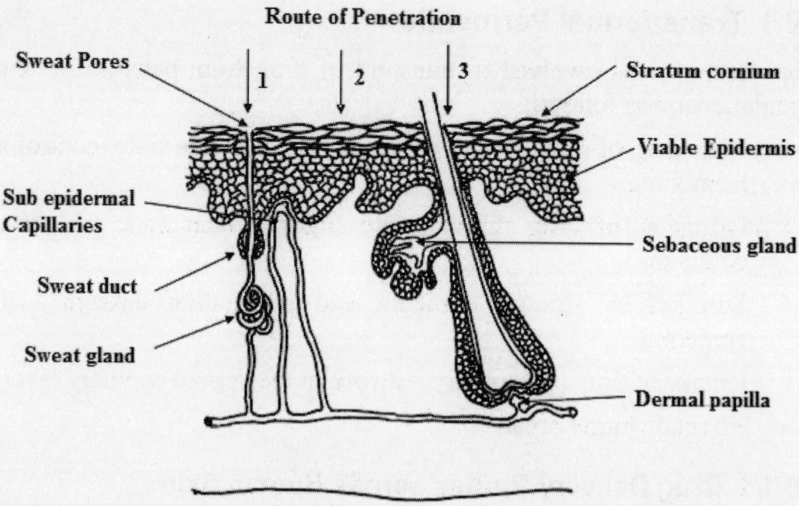

Fig. 10.2 Structure of skin showing route of penetration; 1: through the sweat ducts; 2: directly across the stratum corneum; 3: via hair follicles.

Considerable research effort has been directed towards gaining a better understanding of the structure and barrier properties of the stratum corneum. The stratum corneum consists of 10-15 layers of corneocytes and varies in thickness from approximately 10-15 μm in the dry state to 40 μm, when hydrated. It comprises a multi-layered "brick and mortar" like structure of keratin-rich corneocytes (bricks) in an intercellular matrix (mortar) composed primarily of long chain ceramides, free fatty acids, triglycerides, cholesterol, cholesterol sulfate and sterol/wax ester. However it is important to view this model in the context that the corneocytes are not brick shaped but are polygonal, elongated and flat (0.2-1.5 μm thick, 34-46 μm in diameter). The intercellular lipid matrix is generated by keratinocytes in the mid to upper part of the stratum granulosum discharging their lamellar contents into the intercellular space. In the initial layers of the stratum corneum this extruded material rearranges to form broad intercellular lipid lamellae, which then associate into lipid bilayers, with the hydrocarbon chains aligned and polar head groups dissolved in an aqueous layer (Fig. 10.3). The lipid composition of the stratum corneum behaves in different from than that of other biological membranes. The hydrocarbon chains are arranged into regions of crystalline, lamellar gel and there by the lamellar liquid crystal phases create various domains within the lipid bilayer. The presence of intrinsic and extrinsic proteins, such as enzymes, may also affect the lamellar structure of the stratum corneum. Water is an essential component of the stratum corneum, which acts as a plasticizer to prevent cracking of the stratum corneum and is also involved in the generation of natural moisturizing factor (NMF), which helps to maintain suppleness. The hydrophilic chemicals diffuse within the aqueous regions near the outer surface of intracellular keratin filaments (intracellular or trans-cellular route) whilst lipophilic chemicals diffuse through the lipid matrix between the filaments (intercellular route; Fig. 10.3).

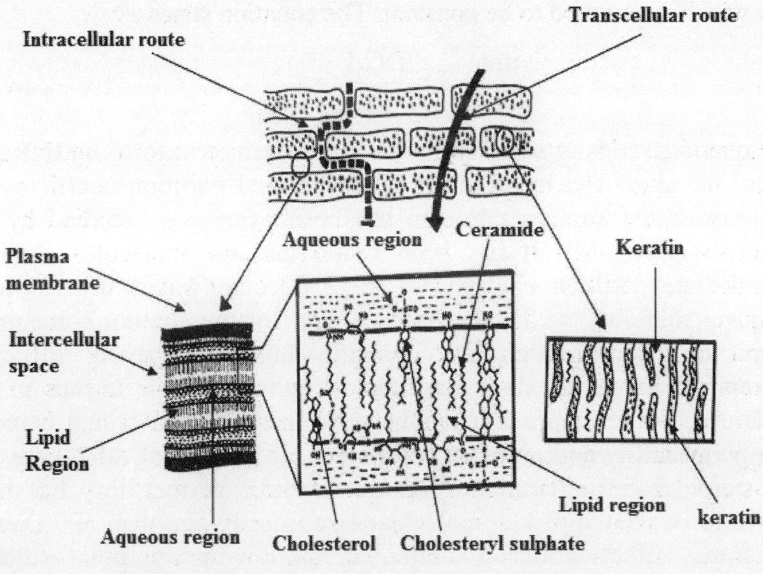

Fig. 10.3 Diagrammatic representation of the stratum corneum and the intercellular and trans-cellular routes of penetration (adapted from: Barry, B.W. Eur. J. Pharm Sci., 2001, 14, 101-14).

However, this is an oversimplification of the situation as each route cannot be viewed in isolation. A molecule traversing via the trans-cellular route must partition into and diffuse through the keratinocyte, but in order to move to the next keratinocyte, the molecule must partition into and diffuse through the estimated 4-20 lipid lamellae between each keratinocyte. This series of partitioning into and diffusing across multiple hydrophilic and hydrophobic domains is unfavorable for most drugs. The intercellular route is now considered to be the major pathway for permeation of most drugs across the stratum corneum, for which the majority of techniques to optimize permeation of drugs across the skin are directed towards manipulation of solubility in the lipid domain or alteration of the ordered structure of this region (Fig. 10.3).

10.1.2 Penetration Enhancement through Optimization of Drug and Vehicle Properties

Drug permeation across the stratum corneum obeys Fick's first law (equation 1) where steady-state flux (J) is related to the diffusion coefficient (D) of the drug in the stratum corneum over a diffusion path length or membrane thickness (h), the partition coefficient (P) between the stratum corneum and the vehicle, and the applied drug concentration (C_0) which is assumed to be constant. The equation states as:

$$\frac{dm}{dt} = J = \frac{DC_0P}{h} \qquad .....(1)$$

Equation 1 aids in identifying the ideal parameters for drug diffusion across the skin. The influence of solubility and partition coefficient of drug across the stratum corneum has been extensively studied by the scientists (Katz, M). It has been found that the molecules showing intermediate partition coefficients [logP (Octanol/water) of 1/3] have adequate solubility within the lipid domains of the stratum corneum to permit diffusion through this domain whilst still having sufficient hydrophilic nature to allow partitioning into the viable tissues of the epidermis. For example a parabolic relationship was obtained between skin permeability and partition coefficient for a series of salicylates and non-steroidal anti-inflammatory drugs. Optimal permeability has been shown to be related to low molecular size (ideally less than 500 Dalton, which may affects diffusion coefficient, and low melting point which is related to solubility. When a drug possesses these ideal characteristics (as in the case of nicotine and nitroglycerin), transdermal delivery is feasible. However, where a drug does not possess ideal physicochemical properties, manipulation of the drug or vehicle to enhance diffusion, becomes necessary.

10.2 Kinetics of Transdermal Permeation

It is very essential to know the skin permeation kinetics for successful development of transdermal therapeutic systems (Misra A N). This permeation can be possible only if the drug possesses certain physiochemical properties. The rate of permeation across the skin is given by:

$$\frac{dQ}{dt} = P_s\left(C_d - C_r\right) \qquad .....(2)$$

where, C_d and C_r are the concentration of the skin penetrant in the donor compartment i.e. on the surface of stratum corneum and in the receptor compartment i.e. body respectively. Ps is the overall permeability coefficient of the skin tissue to the penetrant. This permeability coefficient is given by the relationship:

$$P_s = \frac{K_s D_{ss}}{h_s} \qquad .....(3)$$

where, K_s is the partition coefficient for the interfacial partitioning of the penetrant molecule from a solution medium or a transdermal therapeutic system to the stratum corneum, D_{ss} is the apparent diffusivity for the steady state diffusion of the penetrant molecule through a thickness of skin tissues and h_s is the overall thickness of skin tissues. As $K_{s,}$ D_{ss} and h_s are constant under given conditions the permeability coefficient Ps for a skin penetrant can be considered to be constant. From equation (2) it is clear that a constant rate of drug permeation can be obtained only when $C_d >> C_r$ i.e. the drug concentration at the surface of the stratum corneum C_d is consistently and substantially greater than the drug concentration in the body C_r. The equation becomes:

$$\frac{dQ}{dt} = P_s C_d \qquad .....(4)$$

And the rate of skin permeation is constant provided the magnitude of C_d remains fairly constant throughout the course of skin permeation. For keeping C_d constant the drug should be released from the device at a rate R_r i.e. either constant or greater than the rate of skin uptake R_a, i.e. $R_r >> R_a$. Since $R_r >> R_a$, the drug concentration on the skin surface C_d is maintained at a level equal to or greater than the equilibrium solubility of the drug in the stratum corneum C_s .i.e. $C_d >> C_s$. Therefore a maximum rate of skin permeation is obtained and is given by the equation:

$$\frac{dQ}{dt} = P_s C_s \qquad .....(5)$$

From the above equation it can be seen that the maximum rate of skin permeation depends upon the skin permeability coefficient P_s and is equilibrium solubility in the stratum corneum C_S. Thus skin permeation appears to be stratum corneum limited.

10.3 Factors Affecting Transdermal Bioavailability

- Physiological Factors, which include: Stratum corneum layer of the skin; anatomic site of application on the body; skin condition and disease state; age of the patient; skin metabolism; desquamation (peeling or flaking of the surface of the skin); skin irritation and sensitization; and racial difference.
- Formulation Factors, which Include: Physico-chemical nature of route; structure and nature of vehicles and membrane; characteristics of penetration enhancers; and method of application.

10.4 Mechanical Methods of Permeation Enhancement

10.4.1 Iontophoresis

Iontophoresis involves the application of electromotive force to drive or repel oppositely charged ions through the dermal layers into the area to be treated, either into the surrounding tissues for localized treatment or into the circulatory system for systemic treatment (Samir Mitragotri). Positively charged ions are driven into skin at the anode while negatively charged ions are driven into skin at the cathode. Studies have shown increased skin permeation of drugs at anodic/cathodic electrodes regardless of predominant molecular ionic charge. Iontophoresis facilitates drug delivery across the barrier. For example: pilocarpine delivery to cystic fibrosis diagnostic test, lidocaine appears to be a promising approach for rapid onset of anesthesia. Keller *et al*. (2000) administered dyes /markers/ anaesthetics encapsulated within a lipid vesicle, by iontophoresis. Conventional liposome anaesthetic products normally have lag time 30-60 min. However, the lag time of charged liposomal anaesthetic product was reduced to 5-10 min.

10.4.2 Electroporation

Electroporation is a method of application of short, high-voltage electrical pulses to the skin. After electroporation, the permeability of the skin for diffusion of drugs increases by four folds of magnitude. The electrical pulses are believed to form transient aqueous pores in the stratum corneum, through which drug transport occurs. It is safe and the electrical pulses can be administered painlessly using closely spaced electrodes to constrain the electric field within the nerve-free stratum corneum. Phipps *et al.* 2005, improved electrotransport drug delivery system for fentanyl and sufentanil. The transdermal electrotransport flux of fentanyl and sufentanil was found to depend on their respective concentration in aqueous solution.

10.4.3 Use of Ultrasound

Application of ultrasound, particularly low frequency ultrasound, has been shown to enhance transdermal transport of various drugs including macromolecules. It is also known as sonophoresis. Katz et al. reported on the use of low-frequency sonophoresis for topical delivery of EMLA cream.

10.4.4 Use of Microscopic Projection

Transdermal patches with microscopic projections called micro-needles were used to facilitate transdermal drug transport. Needles ranging from approximately 10-100 μm in length are arranged in arrays. When pressed into the skin, the arrays make microscopic punctures that are large enough to deliver macromolecules, but small enough that the patient does not feel the penetration or pain. The drug is surface coated on the micro-needles to aid in rapid absorption. They are used in development of cutaneous vaccines for tetanus and influenza.

10.4.5 Skin Abrasion

The abrasion technique involves the direct removal or disruption of the upper layers of the skin to facilitate the permeation of topically applied medicaments. Some of these devices are based on techniques employed by dermatologists for superficial skin resurfacing (e.g. microdermabrasion) which are used in the treatment of acne, scars, hyperpigmentaion and other skin blemishes.

10.4.6 Laser Radiation

This method involves direct and controlled exposure of a laser to the skin that results in the ablation of the stratum corneum, without significant damage to the underlying epidermis. Removal of the stratum corneum using this method has been shown to enhance the delivery of lipophilic and hydrophilic drugs.

10.5 Formulation and Development of Transdermal Drug Delivery Systems

10.5.1 Basic Components of Transdermal Drug Delivery Systems

- Polymer matrix / Drug reservoir
- The drug
- Permeation enhancers
- Pressure sensitive adhesive (PSA)
- Backing laminates
- Release liner
- Other excipients like plasticizers and solvent

10.5.1.1 Polymer Matrix / Drug Reservoir

Polymers are the backbone of TDDS, which control the release of the drug from the device. Polymer matrix can be prepared by dispersion of drug in liquid or solid state synthetic polymer base. Polymers used in TDDS should have biocompatibility and chemical compatibility with the drug and other components of the system such as penetration enhancers and PSAs. Additionally they should provide consistent and effective delivery of a drug throughout the product's intended shelf life and should be of safe status.

The polymers utilized for TDDS can be classified as:

- **Natural Polymers**: e.g. cellulose derivatives, zein, gelatin, shellac, waxes, gums, natural rubber and chitosan *etc*.
- **Synthetic Elastomers**: e.g. polybutadiene, hydrin rubber, polyisobutylene, silicon rubber, nitrile, acrylonitrile, neoprene, butylrubber *etc*.
- **Synthetic Polymers**: e.g. polyvinyl alcohol, polyvinylchloride, polyethylene, polypropylene, polyacrylate, polyamide, polyurea, polyvinylpyrrolidone, polymethylmethacrylate *etc*.

The polymers such as: cross linked polyethylene glycol, eudragits, ethyl cellulose, polyvinylpyrrolidone and hydroxypropylmethylcellulose are used as matrix formers for TDDS. Other polymers such as: EVA, silicon rubber and polyurethane are used as rate controlling membrane.

10.5.1.2 Drug

It is generally accepted that the best drug candidates for passive adhesive transdermal patches must be non ionic, of low molecular weight (less than 500 Daltons), have adequate solubility in oil and water (log P in the range of 1-3), a low melting point (less than 200°C) and are potent (dose in mg per day). Examples of such categories of drugs include: fentanyl, nitroglycerine, nicotine, testosterone, clonidine, lidocaine, estradiol, scopolamine, etc. In addition drugs like rivastigmine for alzheimer's and parkinson dementia, rotigotine for parkinson, methylphenidate for attention deficit hyperactive disorder and selegiline for depression are recently approved as TDDS

10.5.1.3 Permeation Enhancers

These are the chemical compounds that increase permeability of stratum corneum so as to attain higher therapeutic levels of the drug candidate. Penetration enhancers interact with structural components of stratum corneum *i.e.,* proteins or lipids. They alter the protein and lipid packaging of stratum corneum, thus chemically modifying the barrier functions leading to increased permeability. Examples of permeation enhancers are:

Solvents

These compounds increase penetration possibly by swelling the polar pathway and/or by fluidizing lipids. Examples include: water alcohols e.g. methanol and ethanol; alkyl methyl sulfoxides e.g. dimethyl sulfoxide, alkyl homologs of methyl sulfoxide dimethyl acetamide, dimethyl formamide; pyrrolidones e.g. 2 pyrrolidone, N-methyl, 2-purrolidone; laurocapram (Azone); and miscellaneous solvents e.g. propylene glycol, glycerol, silicone fluids, isopropyl palmitate, etc.

Surfactants

These compounds are proposed to enhance polar pathway transport, especially of hydrophilic drugs. The ability of a surfactant to alter penetration is a function of the polar head group and the hydrocarbon chain length. Examples include: Anionic Surfactants: e.g. Dioctyl sulphosuccinate, Sodium lauryl sulphate, Decodecylmethyl sulphoxide, etc.; Nonionic Surfactants: e.g. Pluronic F127, Pluronic F68, etc.; Bile Salts: e.g. Sodium taurocholate, Sodium deoxycholate, Sodium tauroglycocholate; Binary system: These systems apparently open up the heterogeneous multilaminate pathway as well as the continuous pathways e.g. Propylene glycol-oleic acid and 1, 4-butane diol-linoleic acid.

Miscellaneous Chemicals

These include urea, a hydrating and keratolytic agent; N, N-dimethyl-m-toluamide; calcium thioglycolate; anticholinergic agents; eucalyptol; di-o-methyl-ß-cyclodextrin and soyabean casein, etc.

10.5.1.4 Pressure Sensitive Adhesives

A PSA is a material that helps in maintaining an intimate contact between transdermal system and the skin surface. It should adhere with not more than applied finger pressure, be aggressively and permanently tachy, and exert a strong holding force. Additionally, it should be removable from the smooth surface without leaving a residue. Polyacrylates, polyisobutylene and silicon based adhesives are widely used in TDDSs.

Penetration enhancer should not cause instability of the drug. In case of reservoir systems that include a face adhesive, the diffusing drug must not affect the adhesive. In case of drug-in-adhesive matrix systems, the selection will be based on the rate at which the drug and the penetration enhancer will diffuse through the adhesive. Ideally, PSA should be physicochemically and biologically compatible and should not alter drug release.

10.5.1.5 Backing Laminate

While designing a backing layer, the consideration of chemical resistance of the membrane is most important. Excipient compatibility should also be considered because of the prolonged contact between the backing layer and the excipients. The most comfortable backing will be the one, which exhibits lowest modulus or high flexibility, good oxygen transmission and a high moisture vapor transmission rate. Examples of some backing materials include: vinyl, polyethylene and polyester films.

10.5.1.6 Release Liner

During storage the patch is covered by a protective liner that is removed and discharged immediately before the application of the patch to skin. It is therefore regarded as a part of the primary packaging material rather than a part of dosage form for delivering the drug. However, as the liner is in intimate contact with the delivery system, it should comply with specific requirements regarding chemical inertness and permeation to the drug, penetration enhancer and water. Typically, release liner is composed of a base layer which may be non-occlusive (*e.g.* paper fabric) or occlusive (*e.g.* polyethylene, polyvinylchloride) and a release coating layer made up of silicon or teflon. Other materials used for TDDS release liner include polyester foil and metallized laminates.

10.1.5.7 Other Excipients

Various solvents such as chloroform, methanol, acetone, isopropanol and dichloromethane are used to prepare drug reservoir. In addition plasticizers such as dibutylpthalate, triethylcitrate, polyethylene glycol and propylene glycol are added to provide plasticity to the transdermal patch.

10.6 Types of Trans Dermal Patches

10.6.1 Drug in Adhesive Type Transdermal Patch(s)

The drug and other selected excipients, if any, are directly incorporated into the organic solvent based pressure sensitive adhesive solution, mixed, cast as a thin film and dried to evaporate the solvents, leaving a dried adhesive matrix film containing the drug and excipients. This drug in adhesive matrix is sandwiched between release liner and backing layer. Drug -in -adhesive patch may be single layer or multi layer. The multi layer system is different from single layer in that it adds another layer of drug-in-adhesive, usually separated by a membrane.

Some examples of suitable pressure sensitive adhesives are polysiloxanes, polyacrylates and polyisobutylene. These pressure sensitive adhesives are hydrophobic in nature and are prepared as solutions of polymer dissolved in organic solvents. Hence, this type of system is preferred for hydrophobic drugs as it is to be incorporated into organic solvent based hydrophobic adhesive.

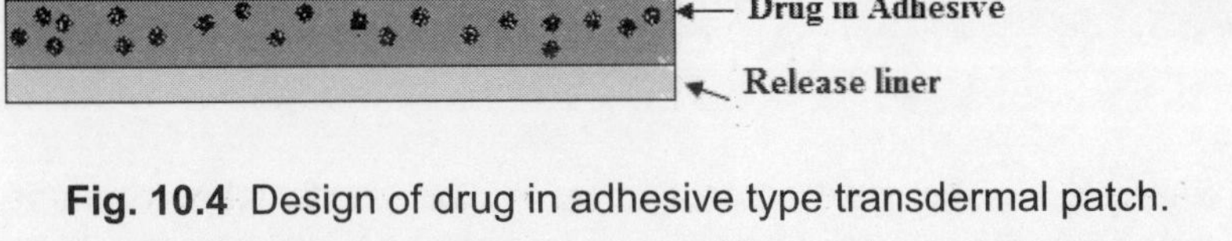

Fig. 10.4 Design of drug in adhesive type transdermal patch.

Single-Layer Drug-in-Adhesive

The single-layer Drug-in-Adhesive system is characterized by the inclusion of the drug directly within the skin-contacting adhesive. In this transdermal system design, the adhesive not only serves to affix the system to the skin, but also serves as the formulation foundation, containing the drug and all the excipients under a single backing film.

The rate of release of drug from this type of system is dependent on the diffusion across the skin.

The intrinsic rate of drug release from this type of drug delivery system is defined by the equation:

$$dQ/dT = \frac{Cr}{1/Pm + 1/Pa}$$

where, Cr is the drug concentration in the reservoir compartment and Pa and Pm are the permeability coefficients of the adhesive layer and the rate controlling membrane, Pm is the sum of permeability coefficients and simultaneous penetrations across the pores and the polymeric material. Pm and Pa, respectively, and are defined as follows:

$$P_m = \frac{K_{m/r}.D_m}{h_m}; P_a = \frac{K_{a/m}.D_a}{h_a}$$

where, $K_{m/r}$ and $K_{a/m}$ are the partition coefficients for the interfacial partitioning of drug from the reservoir to the membrane and from the membrane to adhesive respectively; Dm and Da are the diffusion coefficients in the rate controlling membrane and adhesive layer, respectively; and hm and ha are the thicknesses of the rate controlling membrane and adhesive layer, respectively.

Multi-layer Drug-in-Adhesive

The Multi-layer Drug-in-Adhesive is similar to the Single-layer Drug-in-Adhesive in, which the drug is incorporated directly into the adhesive. However, the multi-layer encompasses either the addition of a membrane between two distinct drug-in-adhesive layers or the addition of multiple drug-in-adhesive layers under a single backing film. The rate of drug release in this system is defined by:

$$dQ/dt = \frac{K_{a/r}.D_a}{h_a} C_r$$

where, $K_{a/r}$ is the partition coefficient for the interfacial partitioning of the drug from the reservoir layer to adhesive layer.

Rachel *et al.,* (2004) prepared drug in adhesive patches of green tea extract and it was observed that major catechins and caffeine extracted from green tea were successfully delivered through transdermal from drug-in-adhesive patches. Kannikkanan *et al.,* (2004) prepared and evaluated monolithic drug in adhesive type transdermal patches of melatonin and used eudragit E100 as adhesive polymer. Lake and

Pinnock (2000) proved that once a week drug in adhesive patch of estrogen is more patient compliant as compared to twice a week reservoir patch. Characteristics of drug in adhesive patch may account for improved patient çompliance due to ease of remembering once weekly patch application, improved cosmetic acceptance and better adhesion. Examples of marketed preparations of drug-in-adhesives patches are Climara®, Nicotrol® and Deponit®. Design of this system is shown in Figure 10.4.

10.6.2 Matrix Type Transdermal Patch(s)

Drug reservoir is prepared by dissolving the drug and polymer in a common solvent. The insoluble drug should be homogenously dispersed in hydrophilic or lipophillic polymer. The required quantity of plasticizer like dibutylphthalate, triethylcitrate, polyethylene glycol or propylene glycol and permeation enhancer is then added and mixed properly. The medicated polymer formed is then molded into rings with defined surface area and controlled thickness over the mercury on horizontal surface followed by solvent evaporation at an elevated temperature. The film formed is then separated from the rings, which is then mounted onto an occlusive base plate in a compartment fabricated from a drug impermeable backing. Adhesive polymer is then spread along the circumference of the film. The rate drug release from polymer matrix follows the equation:

$$\frac{dQ}{dt} = \left(\frac{L_d C_p D_p}{2t} \right)^{1/2}$$

where, L_d is the drug loading dose initially dispersed in the polymer matrix; C_p and D_p are the solubility and diffusivity of the drug in the polymer matrix, respectively.

Commonly used polymers for matrix are cross linked polyethylene glycol, eudragits, ethyl cellulose, polyvinylpyrrolidone and hydroxypropylmethylcellulose.

The dispersion of drug particles in the polymer matrix can be accomplished by either homogenously mixing the finely ground drug particles with a liquid polymer or a highly viscous base polymer followed by cross linking of polymer chains or homogenously blending drug solids with a rubbery polymer at an elevated temperature.

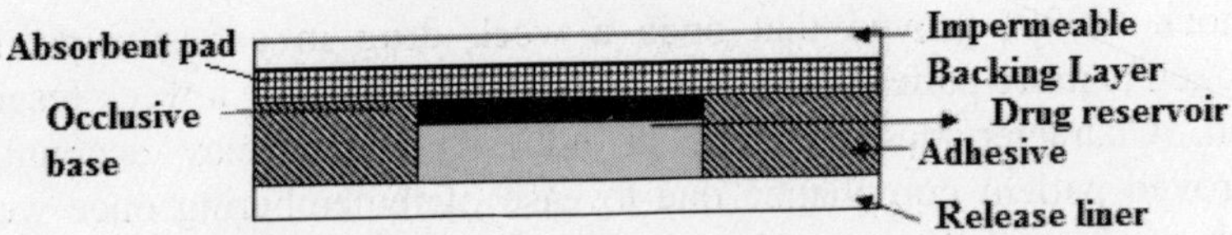

Fig. 10.5 Design of matrix type transdermal patch.

The matrix system is exemplified by the development of Nitro-Dur®. Advantages of matrix patches include absence of dose dumping, direct exposure of polymeric matrix to the skin and no interference of adhesive. Design of matrix type patch is shown in Fig. 10.5.

10.6.3 Reservoir Type Transdermal Patch(s)

The drug reservoir is made of a homogenous dispersion of drug particles suspended in an unleachable viscous liquid medium (e.g. silicon fluids) to form a paste like suspension or gel or a clear solution of drug in a releasable solvent (e.g. ethanol). The drug reservoir formed is sandwiched between a rate controlling membrane and backing laminate.

The rate controlling membrane can be nonporous so that the drug is released by diffusing directly through the material, or the material may contain fluid filled micro-pores in which the drug may additionally diffuse through the fluid, thus filling the pores. In the case of nonporous membrane, the rate of passage of drug molecules depends on the solubility of the drug in the membrane and the thickness of membrane. Hence, the choice of membrane material is dependent on the type of drug being used. By varying the composition and thickness of the membrane, the dosage rate per unit area of the device can be controlled. The drug release from reservoir type device follows the equation:

$$dQ/dt = \frac{K_{a/r}.D_a}{ha(l)} L_d(h_a)$$

where, Ld is the loading dose; $K_{a/r}$ is the partition coefficient; ha(l) is the thickness of the diffusion path length; $L_d(h_a)$ is the drug loading level in the multi-laminate adhesive layer.

Mostly EVA, ethyl cellulose, silicon rubber and polyurethanes are used to prepare rate controlling membranes. Polyurethane membranes are suitable especially for hydrophobic polar compounds having low permeability through hydrophobic polymers such as silicon rubber or EVA membrane.

Liang *et al.*, 1990, studied controlled release of scopolamine through EVA membrane in transdermal patch formulations and release rates were compared with uncontrolled reservoirs. It was found that an EVA membrane patch released scopolamine at a constant rate for more than 72 hours.

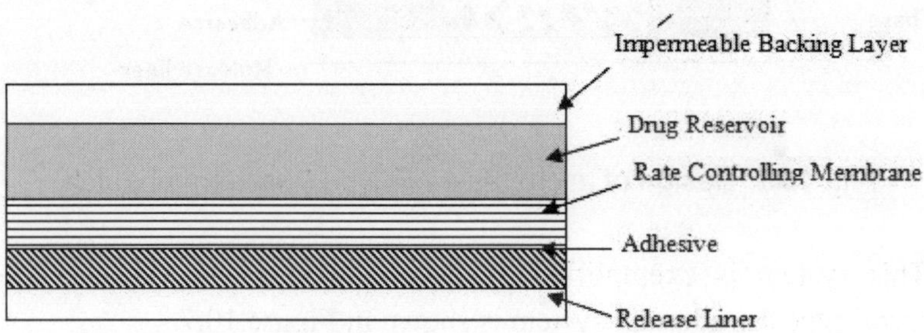

Fig. 10.6 Design of reservoir type transdermal patch.

The main advantage of reservoir type patches is that the patch design can provide a true zero order release pattern to achieve a constant serum drug level. Examples of marketed preparations are Duragesic®, Estradem® and Androderm®. Fig. 10.6, illustrates the design of reservoir type of device of transdermal patches.

10.6.4 Membrane Matrix Hybrid Type Patch(s)

These are the modification of reservoir type transdermal patch. The liquid formulation of the drug reservoir is replaced with a solid polymer matrix (e.g. polyisobutylene) which is sandwiched between rate controlling membrane and backing laminate. Examples of marketed preparations are Catapress® and TransdermScop®.

10.6.5 Micro-reservoir Type Transdermal Patch(s)

The drug reservoir is formed by suspending the drug solids in an aqueous solution of water miscible drug solubilizer e.g. polyethylene glycol. The drug suspension is homogenously dispersed by a high shear mechanical force in lipophillic polymer, forming thousands of unleachable microscopic drug reservoirs (micro reservoirs). The dispersion is quickly stabilized by immediately cross linking the polymer chains in-situ, which produces a medicated polymer disc of a specific area and fixed thickness. Occlusive base plate mounted between the medicated disc and adhesive form backing prevents the loss of drug through the backing membrane.

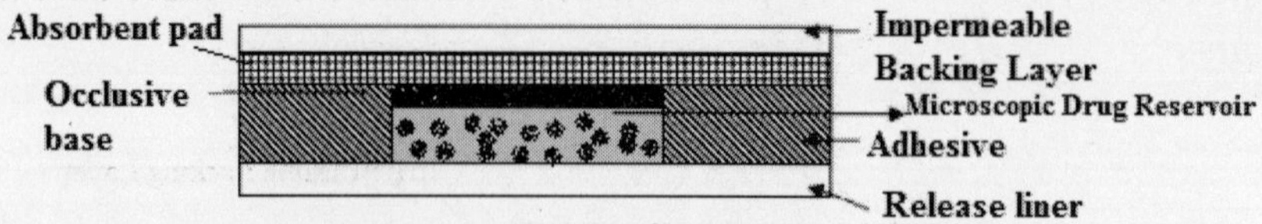

Fig. 10.7 Design of micro reservoir type transdermal patch.

This system is exemplified by development of Nitrodisc®. Micro reservoir type transdermal system is shown in Figure 10.7.

10.6.6 Gel-Matrix Adhesive

Aveva and Nitto Denko, produced the first and only marketed transdermal patch using a revolutionary gel matrix adhesive system for an unequaled balance of adhesion. Because the gel matrix adhesive doesn't cause a disruption of the stratum corneum (skin) during removal, these patches can be removed and reapplied with minimal skin irritation with patient acceptability.

10.6.7 Crystal Reservoir Technology

One of the efficient way of releasing a drug is based on the over saturation of an adhesive polymer with medication, thus forcing a partial crystallization of the drug. The presence of both molecular solute and solid crystal form allows a considerable higher concentration and consistent supply of drug in each patch. As the skin absorbs the molecular solute, crystals re-dissolve to maintain maximum thermodynamic activity at the site of contact.

10.7 Applications

Table 10.1 Currently available medications for transdermal delivery

Drug	Trade name	Type of transdermal patch	Manufacturer	Indication
Fentanyl	Duragesic	Reservoir	Alza / Janssen Pharmaceutica	Moderate/ Severe pain
Nitroglycerine	Deponit Minitran Nitrodisc Nitrodur TransdermNitro	Drug in adhesive Drug in adhesive Micro reservoir Matrix Reservoir	Schwarz Pharma 3M Pharmaceuticals Searle, USA Key Pharmaceuticals Alza/Novartis	Angina Pectoris
Nicotine	Prostep Nicotrol Habitraol	Reservoir Drug in adhesive Drug in adhesive	ElanCorp/Lederie Labs Cygnus Inc./McNeil Consumer Products Ltd. Novartis	Smoking Cessation
Testosterone	Androderm Testoderm TTS	Reservoir Reservoir	Thera Tech/ GlaxoSmithKline Alza	Hypogonadism in males
Clonidine	Catapres-TTS	Membrane matrix hybrid type	Alza/Boehinger Ingelheim	Hypertension
Lidocaine	Lidoderm	Drug in adhesive	Cerner Multum, Inc.	Anesthetic
Scopolamine	Transderm Scop	Membrane matrix hybrid type	Alza/Novartis	Motion sickness
Estradiol	Climara Vivelle Estraderm Esclim	Drug in adhesive Drug in adhesive Reservoir Drug in adhesive	3M Pharmaceuticals/ Berlex Labs Noven Pharma/Novartis Alza/Novartis Women First Healthcare, Inc.	Postmenstrual Syndrome
Ethinyl Estradiol	Ortho Evra	Drug in adhesive	Johnson & Johnson	

Product name	Drug	Manufacturer	Indication
Alora	Estradiol	TheraTech/Proctol and Gamble	Postmenstrual syndrome
Climaderm	Estradiol	Ethical Holdings/Wyeth-Ayerest	Postmenstrual syndrome
Combipatch	Estradiol/ Norethindrone	Noven, Inc./Aventis	Hormone replacement therapy
Fematrix	Estrogen	Ethical Holdings/Solvay Healthcare Ltd.	Postmenstrual syndrome

Table *Contd...*

Product name	Drug	Manufacturer	Indication
FemPatch	Estradiol	Parke-Davis	Postmenstrual syndrome
Nicoderm	Nicotine	Alza/GlaxoSmithKline	Smoking cessation
Nuvelle TS	Estrogen/ Progesterone	Ethical Holdings/Schering	Hormone replacement therapy
Ortho-Evra	Norelgestromin/ estradiol	Ortho-McNeil Pharmaceuticals	Birth control
Transderm Scop	Scopolamine	Alza/Norvatis	Motion sickness

10.8 Evaluation of Transdermal Patches

In order to evaluate desired performance and reproducibility under the specified environmental conditions following studies are predictive for transdermal dosage forms (Yie W. Chien) and can be classified into following types:

- Physicochemical evaluation
- *In vitro* evaluation
- *In vivo* evaluation

10.8.1 Physicochemical Evaluation

Thickness: The thickness of transdermal film can be determined by traveling microscope, dial gauge, screw gauge or micrometer at different points of the film.

Uniformity of weight: Weight variation can be studied individually, weighing 10 randomly selected patches and calculating the average weight. The individual weight should not deviate significantly from the average weight.

Drug content determination: An accurately weighed portion of film (about 100 mg) is to be dissolved in 100 mL of suitable solvent in which drug is soluble and then the solution is to be shaken continuously for 24 h in shaker incubator. Then the whole solution is to be sonicated. After sonication and subsequent filtration, drug in solution can be estimated spectrophotometrically by appropriate dilution.

Content uniformity test: To determine the content uniformity, 10 patches are to be selected and content can be determined for individual patches. If 9 out of 10 patches have content within 85% to 115% of the specified value and one has content not less than 75% to 125% of the specified value, then transdermal patches pass the test of content uniformity. But if 3 patches have content in the range of 75% to 125%, then additional 20 patches are to be tested for drug content. If these 20 patches have ranges from 85% to 115%, then the transdermal patches passes the test.

Moisture content: The prepared films are to be weighed individually and kept in a desiccator containing calcium chloride at room temperature for 24 h. The films are to be weighed again after a specified interval, until they show a constant weight. The percent moisture content can be calculated using following formula:

$$\%\,\text{Moisture content} = \frac{\text{Initial weight} - \text{Final weight}}{\text{Final weight}} \times 100$$

Moisture Uptake: Weighed films are to be kept in a desiccator at room temperature for 24 h. These are then to be taken out and exposed to 84% relative humidity using saturated solution of Potassium chloride in a desiccator, until a constant weight is achieved. % moisture uptake can be calculated by using following formula:

$$\%\,\text{Moisture uptake} = \frac{\text{Final weight} - \text{Initial weight}}{\text{Initial weight}} \times 100$$

Flatness: A transdermal patch should possess a smooth surface and should not constrict with time. This can be demonstrated with flatness study. For flatness determination, one strip is to be cut from the centre and two from each side of patches. The length of each strip is measured and variation in length is measured by determining percent constriction. Zero percent constriction is equivalent to 100 percent flatness.

$$\%\,\text{constriction} = \frac{I_1 - I_2}{I_2} \times 100$$

I_2 = Final length of each strip

I_1 = Initial length of each strip

Folding Endurance: Evaluation of folding endurance involves determining the folding capacity of the films subjected to frequent extreme conditions of folding. Folding endurance is determined by repeatedly folding the film at the same place until it break. The number of times the films could be folded at the same place without breaking is folding endurance value.

Tensile Strength: To determine tensile strength, polymeric films are sandwiched separately by corked linear iron plates. One end of the films is kept fixed with the help of an iron screen and other end is connected to a freely movable thread over a pulley. The weights are added gradually to the pan attached with the hanging end of the thread. A pointer on the thread is used to measure the elongation of the film. The weight just sufficient to break the film is noted. The tensile strength can be calculated using the following equation.

$$\text{Tensile strength} = F/a.b\ (1+L/l)$$

F is the force required to break; 'a' is width of film; 'b' is thickness of film; L is length of film; l is elongation of film at break point

Water Vapor Transmission Studies (WVT)

The water vapor can be measured by putting the patch inside the vial and placing the vial inside the desiccator, in which 200mL of saturated sodium bromide and saturated potassium chloride solution is to be placed. The desiccator is to be tightly closed and humidity inside the desiccator can be measured by using hygrometer. The water vapor can be measures as:

$$WVT = W/\ ST$$

W, is the increase in weight in 24 h; S is area of film exposed (cm^2); T is exposure time.

Microscopic studies: Distribution of drug and polymer in the film can be studied using scanning electron microscope. For this study, the sections of each sample are to be cut and then mounted onto stubs using double sided adhesive tape. The sections are then to be coated with gold palladium alloy using fine coat ion sputter to render them electrically conductive. Then the sections are to be examined under scanning electron microscope.

10.8.2 Adhesive Studies

The therapeutic performance of TDDS can be affected by the quality of contact between the patch and the skin. The adhesive properties of a TDDS can be characterized by considering the following factors.

- **Peel Adhesion properties**: It is the force required to remove adhesive coating from test substrate. It is tested by measuring the force required to pull a single coated tape, applied to substrate at 180° angle. The test is passed if there is no residue on the substrate.

- **Tack properties**: It is the ability of the polymer to adhere to substrate with little contact pressure. Tack is dependent on molecular weight and composition of polymer as well as on the use of tackifying resins in polymer.
- **Thumb tack test**: The force required to remove thumb from adhesive is a measure of tack.
- **Rolling ball test:** This test involves measurement of the distance traveled by a stainless steel ball travels along an upward facing adhesive. The less tacky the adhesive, the more distance the ball will travel.
- **Quick stick (Peel tack) test**: The peel force required breaking the bond between an adhesive and substrate, which can be measured by pulling the tape away from the substrate at 90^0 at the speed of 12 inch/min.
- **Probe tack test**: Force required pulling a probe away from an adhesive at a fixed rate, which can be recorded as tack.
- **Shear strength properties or creep resistance**: Shear strength is the measurement of the cohesive strength of an adhesive polymer, can be determined by measuring the time, it takes to pull an adhesive coated tape off a stainless plate. Minghetti *et al.*, (2003) performed the test with an apparatus which was fabricated according to PSTC-7 (pressure sensitive tape council) specification.

10.8.3 *In-vitro* Release Studies

Drug release mechanisms and kinetics are two characteristics of the dosage forms which play an important role in describing the drug dissolution profile of a controlled release dosage forms and hence their *in-vivo* performance. A number of mathematical model have been developed to describe the drug **dissolution kinetics** from controlled release drug delivery system *e.g.*, **Higuchi, First order, Zero order and Peppas and Korsenmeyer's model.** The dissolution data is fitted to these models and the best fit is obtained to describe the release mechanism of the drug.

Diffusion Cells (Franz Diffusion Cell or its modification i.e., Keshary-Chien Cell)

In this method transdermal system is placed in between receptor and donor compartment of the diffusion cell. The transdermal system faces the receptor compartment in which receptor fluid *i.e.,* buffer is placed. The agitation speed and temperature are kept constant. The whole assembly is kept on magnetic stirrer and solution in the receiver compartment is constantly and continuously stirred throughout the experiment using magnetic beads. At predetermined time intervals, the receptor fluid is removed for analysis and is replaced with an equal volume of fresh receptor fluid. The concentration of drug is determined spectrophotometrically. The pH of the dissolution medium ideally should be adjusted to pH 5 to 6, reflecting physiological skin conditions. For the same reason, the test temperature is typically set at 32°C (even though the temperature may be higher when skin is covered). PhEur considerd 100 rpm a typical agitation rate and also allows for testing an aliquot patch section.

In-vitro Permeation Studies

The amount of drug available for absorption to the systemic pool is greatly dependent on the drug released from the polymeric transdermal films. The drug reached at skin surface is then passed to the dermal microcirculation by penetration through cells of epidermis, between the cells of epidermis through skin appendages.

Usually permeation studies are performed by placing the fabricated transdermal patch with rat skin or synthetic membrane in between receptor and donor compartment in a vertical diffusion cell such as Franz diffusion cell or Keshary-Chien diffusion cell. The transdermal system is applied to the hydrophilic side of the membrane and then mounted in the diffusion cell with lipophilic side in contact with receptor fluid. The receiver compartment is maintained at specific temperature (usually 32±5°C for skin) and is continuously stirred at a constant rate. The samples are to be withdrawn at different time intervals and equal amount of buffer is to be replaced each time. The samples are to be diluted appropriately and absorbance can be determined spectrophotometrically for drug release determination.

Preparation of skin for permeation studies: Hairless animal skin and human cadaver skin can be used for permeation studies. Human cadaver skin may be a logical choice as the skin model, because the final product will be used in humans. But it is not easily available. So, hairless animal skin may generally favor for experimental work, because of its easy availability.

Separation of epidermis from full thickness skin: The epidermis is to be separated by using a cotton swab moistened with distilled water. Then epidermis sheet is to be cleaned by washing with distilled water and dried under vacuum. Dried sheets are to be stored in desiccators until further use.

10.8.4 *In-vivo* Studies

In-vivo evaluations are the true depiction of the drug performance. The variables which cannot be taken into account during *in-vitro* studies can be fully explored during *in-vivo* studies. *In-vivo* evaluation of TDDS can be carried out using:

- Animal models
- Human volunteers

Animal models

Animal studies are preferred at small scale. The most common animal species used for evaluating transdermal drug delivery system are mouse, hairless rat, hairless dog, hairless rhesus monkey, rabbit, guinea pig etc. Various experiments conducted lead us to a conclusion that hairless animals are preferred over hairy animals in both *in-vitro* and *in-vivo* experiments. Rhesus monkey is one of the most reliable models for *in vivo* evaluation of transdermal drug delivery in man.

Human models

The final stage of the development of a transdermal device involves collection of pharmacokinetic and pharmacodynamic data following application of the patch to human volunteers. Clinical trials to be conducted to assess the efficacy, where risk is involved; side effects; and patient compliance are to be taken into consideration.

Skin irritation studies: White albino rats, mice or white rabbits can be used to study any hypersensitivity reaction on the skin. Mutalik and Udupa N. 2005, carried out skin irritation test using mice. After visual evaluation of skin irritation, the animals may be sacrificed for histological examination.

10.8.5 Stability Studies

The stability studies are to be conducted to investigate the influence of temperature and relative humidity on the drug content in different formulations. The transdermal formulations are subjected to stability studies as per ICH guidelines following the protocols (i.e., for accelerated stability study) as stated in the table 10.2.

Table 10.2 Stability study condition for finished pharmaceutical Products (FPP) as per ICH guide lines

Study	Storage condition	Minimum time period
Long-term[a]	25 °C ± 2°C/60% RH ± 5% RH or 30 °C ± 2°C/65% RH ± 5% RH or 30 °C ± 2°C/75% RH ± 5% RH	12 months
Intermediate[b]	30 °C ± 2°C/65% RH ± 5% RH	6 months
Accelerated	40 °C ± 2°C/75% RH ± 5% RH	6 months

(a) Whether long-term stability studies are performed at 25 °C ± 2 °C/60% RH ± 5% RH or 30 °C ± 2 °C/65% RH ± 5% RH or 30 °C ± 2 °C/75% RH ± 5% RH is determined by the climatic zone in which the finished pharmaceutical products is intended to be marketed . Testing at a more severe long-term condition can be an alternative to storage at 25 °C/60% RH or 30 °C/65% RH.

(b) If 30 °C ± 2 °C/65% RH ± 5% RH or 30 °C ± 2 °C/75% RH ± 5% RH is the long-term condition, there is no intermediate condition.

References

Kandavilli S, Nair V, Panchagnula R. Polymers in transdermal drug delivery systems, *Pharmaceutical Technology*, 2002: 62-78.

Gannu R, Vamshi Vishnu Y, Kishan V, Madhusudan Rao Y. Development of nitrendipine transdermal patches: In vitro and ex vivo characterization, *Current Drug Delivery*, 2007 (4): 69-76.

Guy RH. Current status and future prospects of transdermal drug delivery, *Pharm Res*, 1996(13): 1765-1769.

Guy RH, Hadgraft J, Bucks DA. Transdermal drug delivery and cutaneous metabolism, *Xenobiotica*, 1987(7): 325-343.

Chein YW. Transdermal Drug Delivery and Delivery systems. Novel Drug Delivery Systems. Informa Healthcare, USA, Inc., Special Editon, 2009 (50): 301-380.

Keith AD. Polymer matrix considerations for transdermal devices, Drug *Dev. Ind. Pharm,* 1983(9): 605.

Girish chopda. Transdermal Drug delivery Systems: A review. www.pharmainfo. net/2006/Vol-4-Issue-1.

Shreeraj Shah. Transdermal drug Delivery Technology Revised & recent advances. www.pharmainfo. net/2008/Vol-6-Issue-5.

Geeta Aggarwal. Development, Fabrication and Evaluation of Transdermal Drug Delivery systems. www.pharmainfo. net/2009/Vol-7-Issue-5.

Guyot M, Fawaz F. Design and in vitro evaluation of adhesive matrix for transdermal delivery of propranolol, *Int J Pharm,* 2000(204): 171-182.

Verma PRP, Iyer SS. Transdermal delivery of propranolol using mixed grades of eudragit: Design and in vitro and in vivo evaluation, *Drug Dev Ind Pharm*, 2000 (26): 471-476.

Oquiso T, Iwaki M, Paku T. Effect of various enhancers on transdermal penetration of indomethacin and urea and relationship between penetration parameters and enhancement factors, *J Pharm Sci*, 1995 (84):482-488.

Parikh DK, Tapash KG. Feasibility of transdermal delivery of fluoxetine, *AAPS Pharm Sci Tech*., 2005 (6): E144-149.

Mukherjee B, Kanupriya, Mahapatra S, Das S, Patra B. Sorbitan monolaurate 20 as a potential skin permeation enhancer in transdermal patches, *J Applied Research,* 2005 (5): 96-107.

El-Kattan AF, Asbill CS, Kim N, Mickniak BB. Effect of formulation variables on the percutaneous permeation of ketoprofen from gel formulations, *Drug Delivery*, 2000 (7): 147-153.

Shin SC, Shin EY, Cho CY. Enhancing effects of fatty acids on piroxicam permeation through rat skins, *Drug Dev Ind Pharm,* 2000 (26):563-566.

Tan HS, Pfister WR. Pressure sensitive adhesives for transdermal drug delivery, *Pharm Sci Technol Today,* 1999 (2): 60-69.

S. Mitragotri, D. Ray, J. Farrell, H. Tang, B. Yu, J. Kost, D. Blankschtein, and R. Langer. Synergistic effect of ultrasound and sodium lauryl sulfate on transdermal drug delivery. *J. Pharm. Sci.,* 2000 (89): 892–900.

Gondaliya D, Pundarikakshudu K. Studies in formulation and pharmacotechnical evaluation of controlled release transdermal delivery system of bupropion, *AAPS PharmSciTech,* 2003, 4: Article3.

Lewis S, Pandey S, Udupa N. Design and evaluation of matrix type and membrane controlled transdermal delivery systems of nicotine suitable for use in smoking cessation, *Ind J Pharm Sci,* 2006 (68): 179-184.

Walter M. Transdermal therapeutic system (TTS) with fentanyl as active ingredient. European Patent EP 1418951, 2004.

Venkateshwaran S, Fikstad D, Ebert CD. Pressure sensitive adhesive matrix patches for transdermal delivery of salts of pharmaceutical agents. US Patent 5985317, 1999.

Arabi H, Hashemi SA, Ajdari N. Preparation of a transdermal delivery system and effect of membrane type for scopolamine drug. *Iranian Polymer J,* 2002, 11(4): 245-249.

Misra AN. Transdermal Drug Delivery. In: Jain NK, ed. Controlled and novel Drug Delivery. CBS Publications, New Delhi, 2008: 100-146.

Yie W. Chien. Transdermal Therapeutic Systems. Robinson's Controlled Drug Delivery, 2nd Edn., 2009 (20): 523-552.

Panchagnula R, Bokalial R, Sharma P, Khandavilli S. Transdermal delivery of naloxone: skin permeation, pharmacokinetic, irritancy and stability studies, *Int J Pharm,* 2005 (293): 213-223.

Murthy SN, Hiremath SRR. Physical and chemical permeation enhancers in transdermal delivery of terbutaline sulfate, *AAPS Pharm SciTech*, 2001 (2): Article -1.

ASTM. Standard test method for pressure-sensitive adhesive using an inverted probe machine. American Society for testing and materials, Philadelphia, USA. (1971) 2979-3071.

PSTC. Test methods for pressure sensitive adhesive tapes, 7th ed., Pressure sensitive adhesive tape council, Glenview, IL. 1976.

Ho KY, Dodou K. Rheological studies on pressure sensitive adhesives and drug in adhesive layers as a means to characterize adhesive performance, *Int J Pharm,* 2007 (333): 24-33.

Dimas DA, Dalles PP, Rekkas DD, Choulis NH. Effect of several factors on the mechanical properties of pressure sensitive adhesives used in transdermal therapeutic systems, *AAPS PharmSciTech*, 2000 (1):E16.

Sood A, Panchagnula R. Role of dissolution studies in controlled release drug delivery system, *STP Pharma Sci,* 1999 (9):157-168.

P. K. Gaur, S. Mishra3, S. Purohit, K. Dave. Transdermal drug delivery system: a review. *Asian Journal of Pharmaceutical and Clinical Research,* 2009, 2(1): 14-20

Heather A.E. Benson. Transdermal Drug Delivery: Penetration Enhancement Techniques. *Current Drug Delivery,* 2005 (2): 23-33

Stanley Scheindli. Transdermal Drug Delivery: Past Present and Future. Reflections, 2004, 4(6): 308-312.

Samir Mitragotri1. Synergistic Effect of Enhancers for Transdermal Drug Delivery. *Pharmaceutical Research*, 2000, 1(17), 11:1354-1359.

Ashok K. Tiwary, Bharti Sapra and Subheet Jain. Innovations in Transdermal Drug Delivery: Formulations and Techniques. *Recent Patents on Drug Delivery & Formulation,* 2007 (1): 23-36.

J. Ashok kumar1, Nikhila pullakandam1, S.Lakshmana prabu, V.Gopal. Transdermal drug delivery system: An overview. *International Journal of Pharmaceutical Sciences Review and Research*, 2010, 3(2):49-54.

Katz, M.; Poulsen, B.J., In *Handbook of Experimental Pharmacology*, Brodie, B.B.; Gilette, J., Eds. Springer Verlag: Berlin., 1971:103-174.

Index

D

E

F

G

H

I

L

M

N

O

P

R

S

T

V

Y

Z